What is Land For?

What is Land For?

The Food, Fuel and Climate Change Debate

Edited by

Michael Winter and Matt Lobley

earthscan
from Routledge

First published by Earthscan in the UK and USA in 2009

For a full list of publications please contact:
Earthscan
2 Park Square, Milton Park, Abingdon, Oxfordshire OX14 4RN
711 Third Avenue, New York, NY 10017

First issued in paperback 2015

Earthscan is an imprint of the Taylor & Francis Group, an informa business

Notices

Practitioners and researchers must always rely on their own experience and knowledge in evaluating and using any information, methods, compounds, or experiments described herein. In using such information or methods they should be mindful of their own safety and the safety of others, including parties for whom they have a professional responsibility.

ISBN 13: 978-1-138-88122-8 (pbk)
ISBN 13: 978-1-8440-7720-5 (hbk)

Typeset by 4word Ltd, Bristol
Cover design by Susanne Harris

Cover images: main photo, Freshwater tidal wetland © Andrea Gingerich/istockphoto.com; Corn field © Bill Grove/istockphoto.com; Tasmanian farm landscape © Keiichi Hiki/istockphoto.com; Bog in autumn © Norbert Bieberstein/istockphoto.com; Digester © Lianem/Fotolia.com

A catalogue record for this book is available from the British Library

Library of Congress Cataloging-in-Publication Data

What is land for? : the food, fuel and climate change debate / edited by Michael Winter and Matt Lobley.
p. cm.
Includes bibliographical references and index.
ISBN 978-1-84407-720-5
1. Land use. 2. Climatic changes. I. Winter, Michael. II. Lobley, Matt.
HD111.W43 2009
333.73'13—dc22

2009022939

Contents

Part 2 Emerging Issues and New Perspectives

List of Figures and Tables

Figures

Tables

Acknowledgements

We are grateful to the Rural Economy and Land Use Programme (Relu) and the Commission for Rural Communities (CRC) for funding a workshop to discuss the ideas that form the basis of many of the chapters of this book. In addition, we are grateful for contributions and feedback from Geoffrey Hammond, Marcelle McManus, Hazel Evans, Gordon Stokes, Vicki Swales, Alastair Johnson and John Creedy at the workshop.

List of Contributors

Michael Winter, Centre for Rural Policy Research, Department of Politics, University of Exeter, Exeter, EX14 4RY, UK

Matt Lobley, Centre for Rural Policy Research, Department of Politics, University of Exeter, Exeter, EX14 4RY, UK

Katy J. Appleton, School of Environmental Sciences, University of East Anglia, Norwich, NR4 7TJ, UK

Charles Banks, School of Civil Engineering & the Environment, University of Southampton, Southampton, UK

Nesha Beharry, Sustainability Research Institute, University of Leeds, Leeds, UK

David A. Bohan, Centre for Bioenergy and Climate Change, Rothamsted Research, Harpenden, West Common Harpenden, Hertfordshire, AL5 2JQ, UK

Alan J. Bond, School of Environmental Sciences, University of East Anglia, Norwich, NR4 7TJ, UK

Aletta Bonn, Moors for the Future Partnership, Peak District National Park, UK

Tim Burt, Department of Geography, Durham University, Durham, UK

Peter Carruthers, University of Worcester and Commission for Rural Communities, UK

Suzanne J. Clark, Centre for Mathematical and Computational Biology, Rothamsted Research, West Common Harpenden, Hertfordshire, AL5 2JQ, UK

Mark D. Cunningham, Game and Wildlife Conservation Trust, Fordingbridge, Hampshire, SP5 1EF, UK

Trudie Dockerty, School of Environmental Sciences, University of East Anglia, Norwich, NR4 7TJ, UK

Claire A. Dunlop, Department of Politics, University of Exeter, Exeter, UK

Jon W. Finch, Centre for Ecology and Hydrology, Maclean Building, Benson Lane, Crowmarsh Gifford, Wallingford, Oxfordshire, OX10 8BB, UK

Les Firbank, North Wyke Research, Okehampton, Devon, EX20 2XB, UK

David Gowing, Department of Life Sciences, The Open University, UK

Bruce Griffith, North Wyke Research, Okehampton, Devon, EX20 2XB, UK

David Haley, MIRIAD, Manchester Metropolitan University, Righton Building, Cavendish Street, Manchester, M15 6BG, UK

Helen Mayer Harrison, The Harrison Studio, 417 Linden Street, Santa Cruz, California 95062, USA

Newton Harrison, The Harrison Studio, 417 Linden Street, Santa Cruz, California 95062, USA

Alison J. Haughton, Centre for Bioenergy and Climate Change, Rothamsted Research, Harpenden, West Common Harpenden, Hertfordshire, AL5 2JQ, UK

Tim Hess, Department of Natural Resources, Cranfield University, UK

Joseph Holden, School of Geography, University of Leeds, Leeds, UK

John J. Hopkins, Natural England, Peterborough, UK

Klaus Hubacek, Sustainability Research Institute, University of Leeds, Leeds, UK

Angela Karp, Centre for Bioenergy and Climate Change and Centre for Mathematical and Computational Biology, Rothamsted Research, West Common Harpenden, Hertfordshire, AL5 2JQ, UK

Anne Liddon, Rural Economy and Land Use Programme (Relu), Centre for Rural Economy, University of Newcastle, Newcastle upon Tyne, UK

Andrew A. Lovett, School of Environmental Sciences, University of East Anglia, Norwich, NR4 7TJ, UK

Philip Lowe, Rural Economy and Land Use Programme (Relu), Centre for Rural Economy, University of Newcastle, Newcastle upon Tyne, UK

Mark D. Mallott, Centre for Bioenergy and Climate Change, Rothamsted Research, Harpenden, West Common Harpenden, Hertfordshire, AL5 2JQ, UK

Victoria E. Mallott, Centre for Bioenergy and Climate Change, Rothamsted Research, Harpenden, UK

Joe Morris, Department of Natural Resources, Cranfield University, UK

Jeremy Phillipson, Rural Economy and Land Use Programme (Relu), Centre for Rural Economy, University of Newcastle, Newcastle upon Tyne, UK

Guy Poppy, School of Biological Sciences, University of Southampton, Southampton, UK

Helena Posthumus, Department of Natural Resources, Cranfield University, UK

Clive Potter, Centre for Environmental Policy, Imperial College, London, UK

Neil Ravenscroft, School of Environment & Technology, University of Brighton, Brighton, UK

Federica Ravera, Institute for Environmental Sciences and Technologies (ICTA), Autonomous University of Barcelona, Spain

Mark Reed, Centre for Environmental Sustainability and Centre for Planning and Environmental Management, University of Aberdeen, Aberdeen, UK

Andrew B. Riche, Centre for Bioenergy and Climate Change, Rothamsted Research, Harpenden, West Common Harpenden, Hertfordshire, AL5 2JQ, UK

Jim Rouquette, Department of Life Sciences, The Open University, UK

Rufus B. Sage, Game and Wildlife Conservation Trust, Fordingbridge, Hampshire, SP5 1EF, UK

Pete Smith, Institute of Biological & Environmental Sciences, School of Biological Sciences, Cruickshank Building, St Machar Drive, University of Aberdeen, Aberdeen, AB24 3UU, UK

Lindsay Stringer, Sustainability Research Institute, University of Leeds, Leeds, UK

Gilla Sünnenberg, School of Environmental Sciences, University of East Anglia, Norwich, NR4 7TJ, UK

Alan Swinbank, Centre for Agricultural Strategy, University of Reading, Reading, UK

David Tarrasón, Centre for Ecological Research and Forestry Applications (CREAF), Autonomous University of Barcelona, Spain

Becky Taylor, Plumpton College, Sussex, UK

Martin M. Turner, Department of Geography, University of Exeter, Laver Building, New North Road, Exeter, Devon, EX4 4QE, UK

Alan Woods, Consultant Land-Use Policy Analyst, UK

List of Acronyms and Abbreviations

AAFC	Agriculture and Agri-food Canada
ABI	Association of British Insurers
ABPR	Animal By-products Regulations
AD	anaerobic digestion
ARBRE	Arable Biomass Renewable Energy
AWS	automatic weather station
BAP	Biodiversity Action Plan
BBSRC	Biotechnology and Biological Sciences Research Council
BERR	Department for Business, Enterprise and Regulatory Reforms
BSE	bovine spongiform encephalitis
CALM	Carbon Aware Land Management
CAP	Common Agricultural Policy
CCA	Climate Change Agreements
CCL	Climate Change Levy
CCS	carbon capture and storage
CCX	Chicago Climate Exchange
CDM	Clean Design Mechanism
CEH	Centre for Ecology & Hydrology
CERs	Certified Emission Reductions
CET	Central England Temperature
CHP	combined heat and power systems
CLT	Community Land Trust
COACH	Co-operation Action within CCS China–EU programme
COGECA	General Committee for Agricultural Cooperation in the European Union
COP	cereal, oilseed and protein
COPA	Committee of Professional Agricultural Organizations
CPRE	Campaign to Protect Rural England
CRC	Carbon Reduction Commitment
CSA	community-supported agriculture
CSS	Countryside Stewardship Scheme
DECC	Department of Energy and Climate Change
Defra	Department for Environment, Food and Rural Affairs

DG Agriculture	Directorate-General for Agriculture
DOC	dissolved organic carbon
EA	Environment Agency
ECCP	European Climate Change Programme
ECS	Energy Crops Scheme
ECTSC	European Commission's Technical Standards Committee
ECX	European Climate Exchange
EIA	Environmental Impact Assessment
EIT	Economies in Transition
ELS	Entry Level Stewardship [scheme]
ERDP	European Rural Development Programme
ESA	Environmentally Sensitive Area
ESRC	Economic and Social Research Council
ETS	EU Emissions Trading Scheme
EU	European Union
FAO	Food and Agriculture Organization
FBS	Farm Business Survey
FiT	feed-in tariff
FSE	farm scale evaluation
FWAG	Farming and Wildlife Advisory Group
GDP	gross domestic product
GHG	greenhouse gas
GIS	Geographical Information Systems
GVA	gross value added
HGCA	Home Grown Cereals Authority
HLS	Higher Level Stewardship [scheme]
IDB	Internal Drainage Board
IEA	International Energy Agency
IEEP	Institute for European Environmental Policy
IFPRI	International Food Policy Research Institute
IFS	integrated farming systems
IIASA	International Institute for Applied Systems Analysis
IPCC	Intergovernmental Panel on Climate Change
IPS	Industrial & Provident Society
IAASTD	International Assessment of Agricultural Knowledge, Science and Technology for Development
IFAD	International Fund for Agricultural Development
IRGA	infra red gas analyser
JCA	Joint Character Area
JNCC	Joint Nature Conservation Committee
LAI	leaf area index
LCM2000	Land Cover Map 2000

LCMGB	Land Cover Map of Great Britain
LEAF	Linking Environment and Farming
LEC	levy exemption certification
LPG	liquefied petroleum gas
LUPG	Land Use Policy Group
MA	Millennium Assessment
MAFF	Ministry of Agriculture, Fisheries and Food
MMU	Manchester Metropolitan University
MONARCH	Modelling Natural Resource Responses to Climate Change
MOU	Memorandum of Understanding
NDEP	Nevada Division of Environmental Protection
NEGTAP	National Expert Group on Transboundary Air Pollution
NERC	National Environment Research Council
NFFO	Non Fossil Fuel Obligation
NFU	National Farmers Union
NPP	net primary productivity
NZEC	Near Zero Emissions Coal
ODM	organic dry matter
OECD	Organization for Economic Co-operation and Development
OS	Ordnance Survey
PSA	public service agreement
PERRL	Pilot Emission Removals, Reductions and Learnings
R&D	research and development
RCUK	Research Councils UK
Relu	Rural Economy and Land Use (Programme)
RGGI	Regional Greenhouse Gas Initiative
RO	renewables obligation [for licensed electricity suppliers]
ROC	Renewable Obligation Certificate
RSPB	Royal Society for the Protection of Birds
RTFO	Renewable Transport Fuels Obligation
SA	sustainability appraisal
SAC	Special Area of Conservation
SAIN	Sustainable Agriculture Innovation Network
SEA	Strategic Environmental Assessment
SEEDA	South East England Development Agency
SOC	soil organic carbon
SPA	Special Protection Area
SPS	Single Payment Scheme
SRC	short-rotation coppice
SSCA	Saskatchewan Soil Conservation Association
SSSI	Site of Special Scientific Interest
TSEC	towards a sustainable energy economy

UKCIP	United Kingdom Climate Impact Programme
UNECE	United Nations Economic Commission for Europe
UNEP	United Nations Environment Programme
UNFCCC	United Nations Framework Convention on Climate Change
USD	US dollars
USDA	US Department of Agriculture
WFD	Water Framework Directive
WRE	wide reflective equilibrium
WTO	World Trade Organization

1

Introduction: Knowing the Land

Matt Lobley and Michael Winter

Introduction

Land and the use of land provide a key link between human activity and the natural environment. Our use of land is one of the principal drivers of global environmental change, and, in turn, environmental change, particularly climate change, will increasingly influence the use made of land as communities strive to adapt to, and mitigate, the effects of a changing climate. For instance, as farmers and land managers are increasingly positioned as 'carbon stewards' and new environmental bastions in the struggle against climate change, there is growing pressure to adapt land use and land management practices in order to minimize carbon losses, maximize carbon storage (see Smith in this volume) and provide substitutes for fossil fuels. At the same time, a series of long-term trends (such as changing global dietary patterns) and shorter-term 'events' (such as recent poor harvests and the ongoing drought in Australia) have led to constrained global food supply and stimulated pronounced changes in global agricultural commodity prices, putting further pressure on agriculturally productive land.

Consequently, land and food are at the forefront of the domestic policy agenda in the UK to an extent unprecedented since the 1950s. Climate change lies at the heart of the new debate and it was the climate change agenda that prompted the UK environment minister David Miliband to launch a national debate on land use in 2006. 'Food security', until very recently seen as the last refuge of a backward-looking agricultural fundamentalism, has reappeared in the political vocabulary. With scarcely a backward glance at the 'old environmentalism' of multifunctional agri-environments and its emphasis on biodiversity and landscapes, agricultural supply-chain interests have embraced the 'new environmentalism' of climate change with enthusiasm. They proudly proclaim the readiness of the industry to produce both food and bio-crops, and to do so with a neo-liberal confidence in markets to determine the balance between food and non-food crops in land use. For instance, in his speech to the National Farmers Union (NFU) Centenary Conference in February 2008,

Gordon Brown stressed the 'core responsibility' of British farmers to 'grow and produce the majority of food consumed by the British people', alongside a 'front line' role adapting and reacting to the challenges and opportunities of climate change and exploiting the potential of farmers to become 'energy exporters'. Farmers and their advisors have been quick to embrace the 'new productivism', with the agricultural consultants Andersons stating that the 'PR battle is being won, and farmers, as producers of food and fuel in a dangerous world, are being valued once again.' (Andersons, 2007).

A recent collection of essays entitled *Feeding Britain*, with a foreword by the government minister Hilary Benn (Bridge and Johnson, 2009), contains papers by representatives of the key sector development bodies, such as the Home Grown Cereals Authority (HGCA) and the Horticultural Development Company, and presents a bullish outlook. For example, Jonathan Cowens, Chief Executive of the HGCA, is emboldened to suggest that environmental cross-compliance measures (modest though these may be in the eyes of most environmentalists) could lead market-orientated cereal farmers to forgo the Single Farm Payment so as to avoid the restrictions. In a SWOT analysis, he identifies 'environmental use of land' as one of the threats to the cereal sector, alongside 'loss of pesticides due to legislation or resistance'.

But policy (and politics), characterized by incrementalism, has not necessarily caught up with these market- and industry-led changes, nor the changing risks associated with new circumstances (see Dunlop in this volume). Agri-environment schemes, organic farming and sensitive river-catchment planning all continue to figure highly within European rural policy. Non-governmental organizations such as the Royal Society for the Protection of Birds (RSPB) initiate schemes to take land out of production to recreate wildlife-rich reserves. Local and slow food movements challenge the logic and ethics of global markets. Moreover, the far-sightedness of the old environmentalists is beginning to challenge some of the assumptions of the new proponents of food security, particularly their inherent 'productivism'. Is it axiomatic, they ask, that agriculture's best contribution to tackling climate change is to grow bio-crops, or invest in anaerobic digesters, or make land over for wind farms? Might not there be an equally important role in maximizing the carbon sequestration or water-holding properties of biodiverse land? Some have even suggested that biodiverse-rich ecosystems allow for maximum carbon sequestration.

This book does not set out to provide definitive answers to these questions. It is too soon to do that and much of the science is too immature. Rather we seek to establish and to explore the contours of the new debate. In no small measure the book emerges from a strong commitment from both of us to interdisciplinarity which has been strengthened and nurtured by the Rural Economy and Land Use (Relu) programme of the UK research councils. Each of us is involved in Relu projects and several of the contributors to this volume are Relu project leaders too. Relu helped to fund a workshop exploring the themes of the book in which most of the contributors participated. We are also committed to policy relevance and application. The Commission for Rural Communities, an advisor to the UK government, co-funded the workshop as part of its climate change work, in which it is seeking to establish both the implications of climate change for rural communities and the 'rural offer' in dealing with climate change.

This chapter, indeed the whole of this book, has three premises. The first premise is that food and energy security issues now occupy centre stage in policy thinking about land use and this is likely to remain the case for some time to come. The second is that this new emphasis on food and energy security will not mean an abandonment of a continued public policy emphasis on multifunctionality and ecosystem services. Indeed this emphasis is likely to continue to grow. The third premise is that there will be 'local' trends that may on occasions seem counterintuitive in a global context.

These three premises need to inform decisions that society makes on how to pose the right questions, determine the right research priorities, collect the right data and conduct the right analysis. These will require normative judgements and will be subject to contestation. We hope that the chapters in this book will collectively help to make the case for putting food and energy security, ecosystem services and localism centre stage not only in the land debate but in the climate change debate too. But first what is our justification for attributing such importance to these three issues?

Food and energy security

For three decades agricultural commodity surpluses in Europe and the developed world contributed to a dominant discourse of 'land surplus' in which set-aside, extensification, alternative land uses, even managed land-abandonment and 'wilding' were totemic terms in debates over land. Quite suddenly all this changed as a consequence of rapidly shifting commodity markets. The era of land abundance and commodity surpluses that dominated policy thinking, at least in terms of the European Common Agricultural Policy, for most of the 1980s and 1990s, is well and truly over. Some would argue that the land surplus debate was, in any case, an artificial construct, emerging out of the peculiarities of European agricultural politics. Indeed, there is a curious mismatch between the Euro-centric policy concerns of the 1980s and 1990s and the concerns of various international agencies and pressure groups over poverty and development. Much of the academic discussion on *global* food and land issues in the 1990s, although cautiously optimistic, was certainly not so sanguine as to assume that land abundance was in any way a global problem. Leading writers such as Gordon Conway (1997) and Tim Dyson (1996) were critical of neo-Malthusianism on the basis that it underestimated the capacity of the human species to adapt and innovate in response to new challenges. But both Conway and Dyson were acutely aware of the challenges of, for example, seasonal weather fluctuations, so that a poor harvest in one part of the world affects markets many thousands of miles away. For example, poor harvests in the Soviet Union and elsewhere in 1972 led to a massive undercover operation to purchase cereals on the international grain markets, an action which, during the Cold War, had major geo-political consequences. The 1974 World Food Conference in Rome was held in an atmosphere of Malthusian gloom about future prospects for world food supplies. Yet just three years later burgeoning production led to world wheat prices lower, in real terms, than at any time since 1945 (Goodman and Watts, 1997). This was not so much an outcome of better weather conditions across the world but a direct result of farmers and nation states responding to market conditions resulting from the cereal shortages in the context of an increasingly international economy. Dyson pins his optimism on this demand and supply response being a recurring pattern. He acknowledges that research, development and investment will be needed and that these cannot necessarily be guaranteed, especially, perhaps, in those parts of the world where they are most urgently required. However, his analysis underplays two trends – first, the impact of climate change itself, both in terms of direct impacts on food production and the potential implications of adaptation and mitigation; and secondly the dependence of agriculture on a finite energy source, oil. It is these concerns that have led to such a powerful re-emergence of food security in the policy arena.

In June 2007, US wheat prices were at their highest for a decade, and in the UK the price of milling wheat doubled during that year. The Food and Agriculture Organization (FAO) Food Price Index for 2007 averaged 157, a 34 per cent increase from 2005, and by May 2008 the index stood at 209, the highest recorded monthly average since the current index started in 1990. Four main drivers of the rapid escalation in food prices have been identified (Nellemann, 2009):

- cyclical factors such as poor harvests due to extreme weather conditions leading to very low global commodity stocks;
- a rapid increase in the share of non-food crops, particularly biofuels;
- high oil prices affecting agricultural input costs, food distribution costs and, ultimately, food prices;
- speculation in food commodity markets.

These drivers have added to the impact of more deep-seated, structural change such as the increased demand for food crops and livestock products from developing and emerging economies. Commentators have debated the relative contribution of the different drivers and, although it is hard to disentangle the impact of new crops compared with other causes of market price increases, what is clear is the emergence of new pressures on land from Amazonian Brazil, where there are reports of a rapid escalation of deforestation, to the European Union (EU), where set-aside was reduced to 0 per cent in 2007/2008, hence its elimination for the first time since its introduction as a voluntary scheme in 1988. Even though agricultural incomes will remain subject to volatility, in their 2008 *Agricultural Outlook* the Organization for Economic Co-operation and Development (OECD) and FAO predicted that the conjoined temporary and structural factors identified above may keep prices above historic equilibrium levels over the next ten years and that this will kindle continuing debate on the 'food versus fuel' issue.

Of course, the global recession has to some extent slowed or even reversed this trend. By the beginning of 2009 the FAO Food Price Index stood at a level similar to that in 2006–07, but this was still above the 2004 index (FAO, 2009). As the director general of the International Food Policy Research Institute, Joachim von Braun, has written in *Nature*:

> the worldwide credit crunch has let some air out of the commodity price bubble, providing a little relief ... But recession also threatens to cut the income and employment of the most vulnerable and undermine investment in agricultural production. The economic bailout and suggested market regulations now being discussed will not protect food prices from future spikes. The world's food worries are by no means over. (von Braun, 2009; see also FAO, 2008a).

In particular, von Braun argues that the economic downturn could have adverse consequences for investment in agricultural research and development (R&D), thereby eventually increasing global food prices beyond the level they would have been without the recession. Moreover, there are some parts of the world, notably China and India, where the recession may have limited impact on long-term structural changes and the rapid pace of economic transformation with its impact on diet. The longer-term population trends are challenging, with world population projected to grow from six to nine billion by 2050. As John Bridge (Bridge and Johnson, 2009) has recently suggested, this growth and, critically, the expected associated changing patterns of demand will require world food production to double. In the context of such predictions, the renewed scholarly and policy focus on food security issues is hardly surprising (see also Ambler-Edwards et al, 2009).

A multifunctional countryside

If we were facing only shortages of food and energy, then a modern-day equivalent of the war-time 'dig for victory' would be the order of the day, and in some quarters, as we have seen already, there is a palpable sense of 'back to business' within the agro-food lobby. However, there are reasons why that is not, nor should be, the case. Politically and culturally, as the chapters in this volume by Dunlop, Lowe et al, Potter, and Ravenscroft and Taylor all demonstrate in different ways, the arguments for seeing the countryside as much more than a site for food production remain powerful. They are deeply embedded in decades of public interest and intervention. A multifunctional countryside in this context encompasses, in particular, recreational, nature conservation and landscape interests. In a society such as Britain – characterized by a high population, a large middle class, a low relative contribution of agriculture to gross value added (GVA), and a deep and well-established tradition of counterurbanization – these interests will not just disappear with increased food and energy demands. They are embedded in public policies and in various expressions of public interest, including pressure group membership. Thus, when speaking to the 2009 Oxford Farming Conference, Secretary of State for Environment, Food and Rural Affairs Hilary Benn stated that 'I want British agriculture to produce as much food as possible' (Benn, 2009) but went on to say that this must be consistent with systems of production that both sustain the environment and safeguard the landscape, as well as producing the type of food that consumers want. If this generalized public interest were not enough, the importance of multifunctionality is massively reinforced by the emerging policy and scientific consensus in the debate on the importance of land management practices for the matter of mitigating and adapting to climate change (even if the precise cause-and-effect relationships have yet to be fully understood).

Although the focus on climate change and land use has so far attracted most popular attention in terms of the potential competition between food and energy cropping (see Karp et al in this volume), there are a number of other potentially significant land use implications of moves to tackle climate change and also to cope with declining availability of oil for fuel and other products once peak oil production is reached. The use of land for flood alleviation is tackled by Morris and colleagues, and Hubacek and colleagues consider the range of ecosystem services provided by upland areas in this context.

Localism

There is a danger that the emphasis on global markets and global environmental change, hugely important though these trends are, can sometimes lead to the neglect of local responses. Local land use trends may run counter to what might be expected from a simplistic downwards extrapolation of macro trends. This is a fundamental point about both the reach of globalization, which although great may not be universal, and the spatially differentiated responses to global trends. In other words the specificities of national, regional and local social, economic, political and cultural contexts will impact on land management practices. These specificities include variations in consumer taste and demand and contrasting regulatory requirements in different places.

Much of this local difference is captured in efforts to make regions and localities competitive even in a globalized context. In other words localism can be seen as the reverse side of the globalization coin. Agriculture's contributory role to landscape and biodiversity and the re-territorialization of food has contributed to the rapidly emerging agenda of regional and local competitiveness. A growing sense of place pervades agricultural and food policy discourses.

What is land?

We have been talking about land as though its defining characteristics are self-evident. But what is land, what are the right questions to ask about it and what are the appropriate data that we need to understand land? This section reviews approaches to understanding the meaning of land. Definitions are important and here we outline the differences between land, land use and land cover: *land* as a physical resource, *land cover* as the bio-physical attributes and human structures of a part of the Earth's surface, and *land use* as operations or activities carried out on land.

'Land cover' and 'land use' are often used interchangeably and/or without clear definitions but it is important to distinguish between the two. Land cover is largely concerned with the bio-physical characteristics of the land and cannot necessarily tell us what the land is used for, particularly if there are multiple uses made of a specific area of land. Also there are feedback effects that cannot be ignored, as land use effects land cover, perhaps permanently. To give an extreme example, the use of land for the production of turf, or even topsoil, clearly has long-term, probably permanent implications on bio-physical properties and therefore on land cover. The specification of any land cover mapping exercise itself reflects policy priorities and the cultural norms of agencies involved, hence the importance of the socio-economic, cultural and the political in any attempt to *know the land*. According to Owens, although land is a resource, 'it's different, it's peculiar, and it's not the same as other resources that support our society and economy. Land provides a material basis for the economy of course, but it also has powerful cultural meanings – it gives us a sense of place, and a sense of history' (Owens, 2007). Hence for Lynch

(1960) land is 'a vast mnemonic system for the retention of group history and ideals'.

This does not mean that we should adopt what Comber et al (2005a) describe as 'pure relativism', rather there is a 'middle way' which accepts that different interpretations of reality are 'meanings' rather than competing truths and that the real world is filtered through such meanings. The implication is that what is thought important to measure about land cover and land use, and the values and interpretation placed on such data will change over time. If we are at the confluence of a set of interconnected drivers relating to climate change and food security, the question that therefore arises is how well equipped are we with the data and information we need in order to judge the land use and land management implications of new and sometimes competing uses of land?

Potentially the land cover/land use distinction is useful and attempts have often been made to preserve it. For example, the standard way in which that most basic constituent of land, the soil, is classified and portrayed attempts to preserve the distinction. The Soil Survey in the UK produces for each locality *two* maps (and corresponding sets of descriptions): one of *soil type* and one of *land use capability classes* based on those soil types. The problem, of course, as indicated above, is that land use may change the underlying edaphic characteristics of the land, thereby often rendering the distinction blurred and problematic. There are many such examples: urban development, rainforest clearance, and the example of turf already given. But one further will suffice here: the British uplands are characterized by various types of heather or grass assemblages of vegetation maintained by grazing. If grazing pressure is reduced, the heather – and ultimately other shrubs – tend to dominate. There is much debate about how long, if ever, it would take, if grazing ceased, for land to return to the mixed oak forest that dominated many such areas before agriculture. The reason for the problem is that the nutrient status of the land has been much reduced after centuries of nutrient removal on the hoof as meat and wool. In short, land use affects the land.

The upland example is useful in another respect beyond that of reminding us that the land use/land cover distinction may be blurred. It also helps us to think about the nature of land as a resource or a factor of production in economic terms. On the one hand, land can be seen as a renewable or 'flow' resource (like water, wind, solar energy). Its productive potential is renewed if managed 'sustainably' – farms and forests as systems that adapt and mimic self-perpetuating ecosystems, yielding a continuous flow of output. But there are ways in which land is more akin to a 'fixed', 'stock' or 'fund' resource, akin to oil or coal. In the upland example, the agricultural use of the uplands has reduced its potential biomass yield. This may not be particularly tragic in this instance where the depletion is very slow and gradual and where grazing, should it be required, is probably sustainable for millennia to come. Moreover, the low-nutrient status of upland soils gives rise to biodiverse vegetation with amenity value. But over-exploitative farming systems can radically diminish the flow resource, using land more as a fund resource.

So land can be seen as flow and as fund. It can also be seen as 'landscape', which brings into the equation the 'values' associated with land that are not directly to do with its use as a resource. As Paul Selman (2006) puts it: 'a landscape is a relatively bounded area or unit; its recognition depends on human perception, which often is spontaneous and intuitive in its identification with a coherent tract of land; and it results from a long legacy of actions and interactions.' Selman suggests that landscape embraces three flows: energy, material and information (perceptions and values). It is these three inter-related but distinct flows which lie at the heart of that fundamental characteristic of land, and such a powerful element in policy thinking – its *multifunctionality*.

Although some writers, notably Wilson (2007), have attempted to construct a broad and a normative conceptualization of multifunctionality, our claim here is less ambitious – for us the key to understanding multifunctionality is the notion of joint production. Joint production emerges from two aspects of production: the physical production process itself and the land/capital context in which production takes place. The ubiquitous nature of joint production in the physical production process – several outputs necessarily emerging from a single production activity – may be linked to the first and second laws of thermodynamics and consequently has been proclaimed as a fundamental economic notion by Baumgärtner et al (2001). Every physical production process is a transformation of energy and matter, which can neither be created nor destroyed (first law of thermodynamics) and must generate a positive amount of entropy (second law of thermodynamics). Classically this leads to low-entropy desired goods and high-entropy waste products (Baumgärtner et al, 2001).

To that extent, agriculture is similar to many productive industries with the negative externality issue at the heart of jointness within the production process itself. But the transformation of energy and matter also takes place in space (land) and large areas of land are required for production. This extensive use of land, which also acts as an environmental, amenity and recreation resource for many people, gives jointness a special significance in agriculture. While this may not be unique to agriculture – buildings occupied by businesses, for example, may be part of an important amenity resource in a city centre – it is important in a way that is hard to imagine for many other branches of economic activity. Thus economists speak of positive economic externalities, in the sense of non-market goods, which arise from multifunctional land use. Policy analysts may speak of public goods and benefits or multiple objectives. We can even look to the post-modernists and invoke Callon's (1998) notion of 'overflowing', arising through 'the production of production' (Adkins, 2005).

Global trends in agricultural land use

The extensive use of land for farming means that agricultural land is now one of the largest ecosystems on the planet, covering some 30–40 per cent of the ice-free land surface (Turner et al, 2007; Foley et al, 2007). Recent decades have seen a significant expansion of the global agricultural area and also a marked intensification of agricultural land use associated with great leaps in land productivity and per capita food availability. Indeed, increases in agricultural output more than kept pace with population growth in the second half of the 20th century (Hassan et al, 2005). Understanding these trends is important, not least because of the growing demand for food, but also because the expansion and utilization of agricultural land is frequently at the expense of the natural environment.

Agricultural land

The global area under agricultural land management has grown steadily over the last four to five decades, and the total value of all agricultural output has roughly trebled in real terms over the same period (FAO, 2007). Agricultural expansion however, has been spatially uneven and has been much greater in developing countries, whereas the trend in developed counties has been for marginal reductions in the area devoted to agriculture. The majority of the Earth's agricultural land (69 per cent) is under pasture of various types (Smith et al, 2007) and the global importance of livestock farming is growing in association with shifting patterns of demand. Moreover, as the single largest user of land, livestock production can have profound implications for environmental management and ecosystem services. The global share of cropland has increased rapidly in recent decades but the rate of increase now appears to be slowing (FAO, 2007). Since the late 1980s Southeast Asia and parts of west and central Asia have experienced significant expansions in cropland, as have parts of East Africa, the southern Amazon Basin and Great Plains of the USA (UNEP, 2007). However, opportunities for the further expansion of cultivated land are thought to be declining given that most land that is well suited to cultivation has already been converted. Consequently, further expansion of cultivated land is likely to occur on marginal land, raising concerns that this will be associated with environmental degradation (Hassan et al, 2005; FAO, 2007). The projected increase in global population (see above) will provide a powerful driver for further agricultural expansion over the coming decades. Much will have to be accommodated through further intensification, although some 20 per cent of the associated increased agricultural production is expected to derive from the expansion of the global agricultural area, most notably in environmentally sensitive and fragile parts of South America and sub-Saharan Africa (FAO, 2007).

Agricultural output

The 40 years to 2004 saw the global output of crops increase by some 144 per cent (Hassan et al, 2005). Cereal crops are particularly important, accounting for over half of the world's harvested area and placing disproportionate demands on inputs of water, energy and agro-chemicals (Hassan et al, 2005), the use of which is likely to come under increasing scrutiny and pressure. Since the mid-1980s, when per capita cereal production peaked, cereal productivity has slowed globally (Hassan et al, 2005; FAO, 2007), while at the same time the production of oil crops has accelerated with growing demand both as feed and food in developing countries (FAO, 2007). Indeed, over the last 40 years it is the expansion of oilseed crops that has driven global cropland expansion (Hassan et al, 2005). For example, on average, global cereal outputs grew by 2.2 per cent p.a. between 1961 and 2005 compared with 4.0 per cent for oil crops (FAO, 2007). As with agricultural expansion, global figures mask significant regional variations in agricultural yields. For instance, in the last 20 years cereal yields have risen by 40 per cent in Latin America, 37 per cent in west Asia, 17 per cent in North America. The result is that whereas in the 1980s each farmer produced an annual average of 1 tonne of food and 1 ha of arable land yielded 1.8 tonnes, by 2007 this had increased to 1.4 tonnes of food per farmer and 2.5 tonnes per ha of arable land (UNEP, 2007). Thus, while the global expansion of the area under agricultural land use has been important in increasing food supply, it is this increase in the intensity of arable production that has been most important in increasing agricultural output.

Diet and food consumption

Changes in agricultural land use and production intensities have contributed to significant progress in increasing per capita food consumption from an estimated average of 2,280 kcal/person per day in the 1960s to 2,800 kcal/person per day by the early years of this century (FAO, 2007). Again, there are considerable regional variations in these figures and, while per capita food consumption in some developing countries has increased quite significantly since the 1960s, little change has occurred in sub-Saharan Africa, with an average of 2,058 kcal/person per day in the 1960s compared with 2,195 kcal/person per day at the start of the new millennium. It is also notable that per capita consumption today in developing countries is still less that that for developed countries in the 1960s (FAO, 2007). As well as the increase in food consumption in developing countries, rising incomes have been associated with a marked dietary change as the share of livestock products (meat and dairy) has increased, often at the expense of previous staples such as roots, tubers and pulses (FAO, 2007; Hassan et al, 2005; Smith et al, 2007). For instance, it has been estimated that in the 30 years to 1997 meat demand in developing countries rose from 11 kg/person per year to 24 kg/person per year (Smith et al, 2007).

The rise of bioenergy

Bioenergy consists of biofuels, biomass and other fuels produced from organic matter:

- Biofuels are liquid fuels derived from organic matter. Approximately 85 per cent of global liquid biofuel is in the form of ethanol with 90 per cent of production occurring in the USA and Brazil. The production of liquid biodiesel is largely centred on the EU (FAO, 2008b).
- Biomass is solid organic matter from crop residues, wood or short-rotation crops such as willow and miscanthus (see the chapter by Karp and colleagues) which is used to provide heat and/or electricity.
- Biogas and/or 'syngas' produced from solid biomass, food and/or animal wastes through the process of anaerobic digestion (see Chapter 5) can be used for heating, energy generation and transport.

In the context of the 'new environmentalism' of climate change mitigation bioenergy production appears to have the potential to perform a totemic role, contributing to energy security, supporting farm incomes and rural development (House of Commons, 2008; Henniges and Zeddies, 2006) and of course contributing to climate change mitigation (although the extent of such a contribution is debated; see The Royal Society, 2008). However, the rapid expansion of bioenergy through biofuel production, competition over the use of land for food production (the so-called fuel versus food debate), and the role of biofuel production in stimulating global increases in food commodity prices means that biofuels have been at the centre of much recent controversy.

As Karp and colleagues explain in Chapter 3, international treaties and national targets and legislation have stimulated the expansion of bioenergy production in recent years. In the USA, expansion of ethanol production has been described as 'exponential' (Westhoff et al, 2007) and policy support both within OECD countries and a number of developing nations means that bioenergy growth will continue, although it seems likely that the contribution of liquid biofuels to energy transport will be limited (FAO, 2008b). Righelato and Spracklen (2007), for instance, report that just a 10 per cent substitution of liquid transport fuels would require 43 per cent of the cropland area of the USA and 38 per cent of that of Europe. Even with much smaller areas under production, there are concerns that liquid biofuels have a significant impact on global agricultural markets and on food security. von Braun (2008) points out that one-third of the US maize crop is now used for ethanol production, with a knock-on effect on agricultural prices. Indeed, it has been suggested that 'biofuels have been the single most significant driver of higher prices' (Evans, 2009, p14). Commentators have also expressed concerns regarding the environmental, landscape and biodiversity implications of bioenergy (House of Commons, 2008; FAO, 2008b). Nevertheless, bioenergy is part of the emerging bio-economy, and technological advances mean that the various forms of bioenergy are likely to become increasingly important in the near future. This raises a number of practical issues and research questions, some of which are discussed in this book.

UK land use data availability and problems

As social, economic, cultural and political imperatives change, so what we want to know about our land also changes. Anyone involved in analysing time-series data about land use – either directly or indirectly – knows this. Comber et al (2005a) illustrate this by contrasting the UK Land Cover Map 2000 (LCM2000) with the earlier 1990 Land Cover Map of Great Britain (LCMGB). The 1990 map was designed to demonstrate the utility of satellite imagery for environmental monitoring, whereas the 2000 map was designed to help meet national and international policy obligations. Consequently, the different objectives of the two surveys – one science led and the other policy driven – lead to very different conceptualizations of land classes that may be nominally similar.

But at least in the case of land cover, as the name implies, a reasonably comprehensive survey of land in its totality may be undertaken and, notwithstanding the definitional difficulties, some knowledge of both the underlying physical characteristics of all land in Britain and its overlying cover is available. The same cannot be said for the use to which the land is put and its function in both physical and social systems. This presents far more tricky issues for the data gatherer and it is at this point that our knowledge of land becomes heavily compromised by two problems: the 'legacy effect' and the 'surrogacy effect'. The legacy effect refers to the long shadow cast by historic policy or science problems and objectives, or earlier data-gathering constraints. In other words, decisions on what data to collect and how, taken decades ago or even in the 19th century, cast a long shadow on what we know of land today. The problem presents a classic methodological challenge to any researchers undertaking time series analysis – the tension between the desire for continuity, on the one hand, and the need for adaptation of data sets to reflect new understanding and new objectives, on the other.

In broad terms, there are at least three main legacy effects (with some overlap) for contemporary land scientists in Britain to contend with. First, there is the importance of collecting so much data for *agricultural* purposes. This is a legacy of either the 19th century, when agricultural and landed interests in politics were so much stronger than today (for example the June Agricultural Census commenced in 1866), or the 20th century and food shortages arising from warfare (for example the comprehensive National Farm Survey undertaken in 1941–43: Short et al, 1999). Secondly, in the 20th-century debate over levels of *planning and urbanization* prompted much academic inquiry and vigorous debate from the 1950s to the 1970s. The warnings of Alice Coleman (1961), a darling of pressure groups concerned with preserving the countryside, were opposed by the more overtly scholarly and less alarmist Robin Best (Best, 1981; Best and Coppock, 1962; Best and Rogers, 1973). In both cases they operated with a rather crude distinction between urban and non-urban land driven by the urbanization debate, but their data sources were different. Best largely relied on analysis of the agricultural census, while Coleman sought to follow in the tradition of Dudley Stamp (1948), whose 1930–38 Land Utilization Survey provided a mapped inventory of every acre of mainland Britain using seven broad classes of land use. Coleman's Second Land Utilization Survey conducted between 1961 and 1968 sadly only ever resulted in 15 per cent of the maps being published, although Coleman did some analysis of the entire data sets to compare with Stamp (see Swetnam, 2007). Thirdly, and more recently, *ecological* data has come to the fore due to widespread concerns, some driven by international treaty obligations, over biodiversity losses. The most obvious legacy here is the range of data sets derived from the Countryside Survey (Carey et al, 2008; Barr et al, 1993).

The *surrogacy* effect comes into play when attempts are made to move from land type to land use or function. The data used for understanding function are frequently aspatial and therefore at best can only serve as a surrogate for spatial land function data. The most ubiquitous example of this is provided by the annual June agricultural survey (formerly census) in which a sample of between 21 per cent and 36 per cent of registered holdings provide information on cropping, stocking, land tenure and so forth. A full census is now only carried out every ten years in fulfilment of EU legislation. The last agricultural census was conducted in 2000. Time and again these data have been translated into spatial data sets at the parish level through use of the parish summaries which may be aggregated upwards to district, county, regional and national levels (see Coppock, 1976). Many are familiar with the national maps produced by the Ministry of Agriculture, Fisheries and Food (MAFF) in the 1960s showing the geographic distribution of farm types. And yet the raw data are not spatial. Famers are not expected to provide any locational information about their holdings. The parish summaries are merely an amalgamation of all holdings with a postal address in the parish. All mapping and spatial analysis of any kind is therefore based on the assumption that farm holding boundaries coincide with parish boundaries which, of course, they do not. Clearly the problem diminishes the larger the spatial unit of analysis, but as farms have amalgamated parish-based analysis has become less reliable. Another example of surrogacy is the use sometimes made of the Farm Business Survey. The FBS has been in operation in roughly the same form since the 1940s, but methods of data collection and the farms covered in the survey change, making it hard to generate accurate time series or spatial data.

Whose land is it anyway? The danger of neglecting property and markets

There is a weakness in the selection of contributions for this book that we fully acknowledge and attempt to partially remedy in this section. Farmers and other managers of rural land are the largest group of natural resource managers on the planet (FAO, 2007). We have said too little about them, therefore we offer two excuses. First, it is rather too soon to offer serious analysis of how farmers are responding to the rapidly changing technological, market and policy possibilities of the new productivism. Secondly, and more prosaically, we have another, as yet uncompleted, programme of research on farmers and the 'social question' in sustainability. In short, it is a topic to which we will return in greater detail in future publications. But it is abundantly clear that many of the possibilities discussed in this book, whether driven by markets, policies or technology, have implications for land occupiers. Maximum benefits for environment and society will accrue only with the co-operation and active engagement of farmers and land managers; and land occupiers' actions are driven by market possibilities (consumer demand), personal aspirations, perceptions and technical abilities, availability of labour and of capital, all in the context of regulatory constraints and possibilities. The literature on these topics is far too voluminous to cover here (but for a recent overview see Brookfield and Parsons, 2007) and in any case these issues are not to do with land per se. We highlight them to avoid any criticism that we have ignored the reality that farmers have agency, that how they manage the land will be determined by what consumers want to consume, what citizens want to regulate, and what they themselves want to do with the land and resources at their disposal.

However, there is one aspect of this complex of social and economic drivers that warrants some closer attention here because it is so intrinsic to the land itself, namely land occupancy arrangements. The nature of land occupancy has such obvious implications for land studies that it is surprising how little research has attempted to explicitly link occupancy to land management practices.

Contrasting tenurial arrangements have implications for the range of social and economic relationships between different groups of people. Thus, Whatmore et al (1990) classified agricultural property rights on a continuum from simple owner-occupation to contract farming with forms of secure and insecure tenancy in between. They identified three main rights to land – ownership, occupation and use – and the distribution of these rights is reflected in contrasting tenurial arrangements. Under simple owner-occupier ownership, occupation and use are all combined within the same firm or person. But as we move along the spectrum landed capital assumes responsibility for certain rights. Under an insecure tenancy, for example, landed capital has some owner and occupier rights with the tenant farmer having some user rights. Under contract farming landed capital retains all land property rights, with the contract farmer responsible only for non-land inputs such as labour and capital.

However, as our own work on tenure has shown (Winter, 2007; Winter and Butler, 2008) such a schema can be misleading as the precise content of a particular arrangement is all-important in determining its nature. There is a danger that methodological problems analogous to those surrounding de jure versus de facto ownership will appear unless the precise contents of tenancy arrangements are analysed very carefully. For example, contract farming may de jure vest all the property rights outside the hands of the farmer, but de facto the terms of the agreement *may* place considerable rights with the farmer. What our work shows is the extent to which unconventional forms of tenure – share farming, contract farming, short tenancies – have arisen which have, as yet, largely under-researched implications for land management practices. For example, contract farming now accounts for nearly 20 per cent of the land area in England's most prosperous farming region of East Anglia.

Currie (1981) has undertaken work that remains helpful in this respect. His distinction between ownership and operating structures gives rise to a classification based on differentiation according to ownership of land, ownership of labour and the provision of entrepreneurship. Currie is particularly interested in the implications of alternative tenurial arrangements for farm decision-making and contends that implications for decisions and management cannot be simply read off from tenure without considering the role of labour and capital. Thus, pure owner-occupation *may* produce highly efficient and productive farming on capitalist principles based on the use of hired labour *or* it may give rise to a form of peasant proprietorship, in which decisions on the allocation of family labour will depend upon factors other than the maximization or even the optimization of profit. It is essential to add a third category to that of capitalist producer and peasant farmer: residential proprietor, whose decisions on agricultural land use are likely to be based on a view of property as an item of consumption rather than production, a positional good, as noted by Offer (1991) for the 19th century as well.

What is the relevance of this rather arcane meander into property rights? First, it is clear that we now have a more and more complex array of actors involved with making decisions about the land. The determinants of land management and the objectives of those engaged in different aspects of land management no longer reside solely with landlord and tenant or owner-occupier. Secondly, shorter-term occupancy arrangements are now more common. This implies an increased risk of environmental asset-stripping as multiple short-term arrangements, whether formal or informal, are not necessarily the best suited to long-term stewardship. It has also been demonstrated that occupancy *change* is often a trigger for management change (Munton and Marsden, 1991) and with shorter-term arrangements this inevitably occurs more frequently. Finally, there is another issue of particular relevance to the increasing tendency to characterize the task of the land manager as the provision of ecosystem services – the fact that occupancy units do not align perfectly with natural units.

Conclusion

In this chapter we have sought to set out some of the key issues that are relevant to the new land use debate and to set the scene for the wide-ranging material presented in this book. We have a commitment to both interdisciplinarity and policy application. The book is essentially about knowledge of the land. But of course there are many different types of knowledge and the book as a whole contains contributions from a range of scientific disciplines that display different ways of knowing land and the issues relevant to the management and use of land. Conceptual unity has been less important to us than the urgent (and prior) need to establish a baseline of evidence and ideas. Following our introduction, the book is divided into two sections. The first section covers a range of new technologies and uses for land that directly or indirectly impinge on the management of land, such as anaerobic digestion (Chapter 5) and energy crops (Chapter 3), as well as ways of using land to manage water (Chapter 6) and provide ecosystem services (Chapters 2 and 7). The aim of the section is to provide state-of-the-art reviews on key issues relevant to the role of the land in climate change adaptation and mitigation. The second section of the book picks up on some of the issues and conflicts that these emerging technologies, capacities and demands give rise to. John Hopkins explores the implications for biodiversity of climate change (Chapter 8) and in Chapter 11 some of the policy aspects of this are also covered. Chapters 9, 10 and 13 serve to remind us of the breadth and challenge of the new land use debate, which encompasses not only the natural and social sciences, but also the arts (Chapter 10) and philosophy and ethics (Chapter 13).

For society as a whole, *what* we 'know' about land is determined in part by *why* we have sought to know some things but not others. The reason for this book is our strongly held perception that the importance of land to our survival as a species cannot be underestimated. As the challenges facing us escalate, so our need grows to take stock of what we know and what more we need to know about land.

References

Adkins, L. (2005) 'The new economy, property and personhood', *Theory, Culture and Society*, vol 22, pp111–130

Ambler-Edwards, S., Bailey, K., Kiff, A., Land, T., Lee, R., Marsden, T., Simons, D. and Tibbs, H. (2009) *Food Futures: Rethinking UK Strategy*, Chatham House Report, London

Andersons (2007) 'Are you prepared for 2013?' Arable Brief 2007, Andersons, Melton Mowbray

Barr, C.J., Bunce, R.G.H., Clarke, R.T., Fuller, R.M., Furse, M.T., Gillespie, M.K., Groom, G.B., Hallam, C.J., Hornung, M., Howard, D.C. and Ness, M.J. (1993) *Countryside Survey 1990: Main Report* (Countryside 1990 vol 2), Department of the Environment, London

Baumgärtner, S., Dyckhoff, H., Faber, M., Proops, J. and Schiller, J. (2001) 'The concept of joint production and ecological economics', *Ecological Economics*, vol 36, pp365–372

Benn, H. (2009) 'Challenges and opportunities: Farming for the future', Speech to Oxford Farming Conference, 6 January 2009, Oxford, UK

Best, R.H. (1981) *Land Use and Living Space*, Methuen, London

Best, R.H. and Coppock, J.T. (1962) *The Changing Use of Land in Britain*, Faber and Faber, London

Best, R.H. and Rogers, A.W. (1973) *The Urban Countryside: The Land-Use Structure of Small Towns and Villages in England and Wales*, Faber and Faber, London

Braun, J. von (2008) 'High food prices: The what, who, and how of proposed policy actions', International Food Policy Research Institute Policy Brief, May 2008, IFPRI, Washington DC

Braun, J. von (2009) 'The food crisis isn't over', *Nature*, vol 456, p701

Bridge, J. and Johnson, N. (2009) *Feeding Britain*, The Smith Institute, London

Brookfield, H. and Parsons, H. (2007) *Family Farms: Survival and Prospect*, Routledge, London

Callon, M. (1998) 'An essay on framing and overflowing', in Callon, M. (ed) *The Laws of the Market*, Blackwells, Oxford

Carey, P.D., Wallis, S., Chamberlain, P.M., Cooper, A., Emmett, B.A., Maskell, L.C., McCann, T., Murphy, J., Norton, L.R., Reynolds, B., Scott, W.A., Simpson, I.C., Smart, S.M. and Ullyett, J.M. (2008) *Countryside Survey: UK results from 2007*, NERC/Centre for Ecology & Hydrology http://www.countrysidesurvey.org.uk/reports2007.html accessed 30 July 2009

Coleman, A. (1961) 'The second land-use survey: progress and prospect', *Geographical Journal*, vol 127, pp168–186

Comber, A., Fisher, P. and Wadsworth, R. (2005a) 'What is land cover?' *Environment and Planning B: Planning and Design*, vol 32, pp199–209

Conway, G.R. (1997) *The Doubly Green Revolution. Food for All in the 21st Century*, Penguin, London

Coppock, J.T. (1976) *An Agricultural Atlas of England and Wales*, Faber and Faber, London

Currie, J.M. (1981) *The Economic Theory of Agricultural Land Tenure*, Cambridge University Press, Cambridge

Dyson, T. (1996) *Population and Food*, Routledge, London

Evans, A. (2009) *The Feeding of the Nine Billion: Global Food Security for the 21st Century*, Chatham House Report, Royal Institute of International Affairs, London

Foley, J., Monfreda, C., Ramankutty, N. and Zaks, D. (2007) 'Our share of the planetary pie', *Proceedings of the National Academy of Sciences*, vol 104(31), pp12585–12586

Food and Agriculture Organization (2007) *The State of Food and Agriculture*, UN FAO, Rome

Food and Agriculture Organization (2008a) *The State of Food Insecurity in the World 2008*, UN FAO, Rome

Food and Agriculture Organization (2008b) *The State of Food and Agriculture 2008*, UN FAO, Rome

Food and Agriculture Organization (2009) 'World food situation: Food price indices', www.fao.org/worldfoodsituation/FoodPricesIndex/en/ accessed 19 May 2009

Goodman, D. and Watts, M.J. (1997) *Globalising Food: Agrarian Questions and Global Restructuring*, Routledge, London

Hassan, R., Scholes, R. and Ash, N. (eds) (2005) *Ecosystems and Human Well-Being: Current State and Trends*, Millennium Ecosystem Assessment, Island Press, Washington

Henniges, O. and Zeddies, J. (2006) Bioenergy and agriculture: promises and challenges, International Food Policy Research Institute, Washington DC

House of Commons Environmental Audit Committee (2008) *Are Biofuels Sustainable?* First Report of Session 2007–08, vol 1, The Stationery Office, London

Lynch, K. (1960) *The Image of the City*, MIT Press, Cambridge, MA

Munton, R.J.C. and Marsden T.K. (1991) 'Occupancy change and the farmed landscape: an analysis of farm-level trends, 1970–85', *Environment and Planning A*, vol 23, pp499–510

Nellemann, C., MacDevette, M., Manders, T., Eickhout, B., Svihus, B., Prins, A.G., Kaltenborn, B.P. (eds). (2009) *The environmental food crisis – The environment's role in averting future food crises*. A UNEP rapid response assessment. United Nations Environment Programme, GRID-Arendal, www.grida.no

Offer, A. (1991) 'Farm tenure and land values in England *c.* 1750–1950', *Economic History Review*, vol 44, pp1–20

Owens, S. (2007) 'Response to "a land fit for the future"', in *20:26 Vision. What future for the Countryside?*, Campaign to Protect Rural England, London

Righelato, R. and Spracklen, D. (2007) 'Carbon mitigation by biofuels or by saving and restoring forests?' *Science*, vol 317, p9902

Selman, P. (2006) *Planning at the Landscape Scale*, Routledge, London

Short, B., Watkins, C., Foot, W. and Kinsman, P. (1999) *The National Farm Survey, 1941–43. State Surveillance and the Countryside in England and Wales in the Second World War*, CABI, Wallingford

Smith, P., Martino, D., Cai, Z., Gwary, D., Janzen, H., Kumar, P., McCarl, B., Ogle, S., O'Mara, F., Rice, C., Scholes, B. and Sirotenko, O. (2007) 'Agriculture', in Metz, B., Davidson, O.R., Bosch, P.R., Dave, R. and Meyer, L.A. (eds) *Climate Change 2007: Mitigation*, Contribution of Working Group III to the Fourth Assessment Report of the Intergovernmental Panel on Climate Change, Cambridge University Press, Cambridge/New York

Stamp, L.D. (1948) *The Land of Britain, Its Use and Misuse*, Longmans, London

Swetnam, R.D. (2007) 'Rural land use in England and Wales between 1930 and 1998: Mapping trajectories of change with a high resolution spatio-temporal dataset', *Landscape and Urban Planning*, vol 81, pp91–103

The Royal Society (2008) *Sustainable Biofuels: Prospects and Challenges*, The Royal Society, London

Turner, B., Labmin, E. and Reenberg, A. (2007) 'The emergence of land change science for global environmental change and sustainability', *Proceedings of the National Academy of Sciences*, vol 104 (52), pp20666–20671

United Nations Environment Programme (2007) *Global Environmental Outlook*: *GEO-4*, UNEP, Kenya

Westhoff, P., Thompson, W., Kruse, J. and Meyer, S. (2007) 'Ethanol Transforms Agricultural Markets in the USA', *EuroChoices*, vol 6, pp14–21

Whatmore, S., Munton R. and Marsden T. (1990) 'The rural restructuring process: Emerging divisions of agricultural property rights', *Regional Studies*, vol 24, pp235–245

Wilson, G.A. (2007) *Multifunctional Agriculture: A Transition Theory Perspective*, CABI, Wallingford

Winter, M. (2007) 'Revisiting land ownership and property rights' in Clout, H. (ed) *Contemporary Rural Geographies, Land, Property and Resources in Britain: Essay in Honour of Richard Munton*, UCL Press, London

Winter, M. and Butler, A. (2008) *Agricultural Tenure 2007*, CRPR Report 24, University of Exeter, Exeter

Part 1

New Uses of Land: Technologies, Policies, Tools and Capacities

2

Strategic Land Use for Ecosystem Services

Philip Lowe, Alan Woods, Anne Liddon and Jeremy Phillipson

The changing policy context for rural land

The policy context for rural land in the UK has changed markedly over the past half century. During and immediately after World War II the overriding imperative was expansion of food production. Whatever land that could be brought into production was cultivated. Roadside verges sprouted crops, gardens were turned over to vegetables rather than flowers, and heathland and downland were ploughed. There was little concern for the costs or efficiency of production – supply of human labour was seen as the only limiting factor (Whetham, 1952). The experience reinforced a view that UK farm land was a precious national asset that warranted long-term protection. In the words of a war-time report on land utilization, 'every agricultural acre counts' (Scott Report, 1942).

Subsequently, in the post-war period, as part of the move away from a planned, war-time economy, there was increasing stress on productivism – the economically efficient expansion of food and timber production. The focus through the 1950s and 1960s was on boosting the productivity of the land and labour, in particular through mechanization, intensification and specialization. There was substantial publicly supported investment in land drainage and in the improvement, ploughing up or afforestation of extensive areas of grassland and moorland.

Despite the increasing attention given to the efficiency of the so-called factors of production – of rural land and labour – there was little concern about whether the most effective use was being made of natural resources such as water or soils. Moreover, raising productivity relied heavily on cheap energy from fossil fuel. Farming turned from being a net generator of energy to a net consumer.

Global food supplies expanded enormously, eclipsing fears of food shortage in the UK, but the imperative to expand domestic production gained a new lease of life in the 1970s, aided by the extension of the Common Agricultural Policy to the UK when it joined the European Community (later the EU) in 1973. The productivism message was reinforced by a UK Government White Paper entitled 'Food from our own resources' in 1979. In the 1980s the success of this approach rapidly led to over-production and the consequent need to dispose of surpluses by dumping them on world markets. There was growing disquiet and scandal over the resultant costs, waste and trade tensions. Agricultural policy makers were obliged in the mid-1980s to take what at the time seemed the extreme step of introducing measures to limit farm output – including milk quotas and the compulsory idling of land through arable set-aside.

At the time, it was estimated that about three million hectares of land in Britain could be taken out of farming to bring food production and consumption into balance (Blunden and Curry, 1988, p35). Other claims on rural land could therefore be pressed. It is no coincidence then that the 1980s also saw the more confident assertion of long-standing concerns about the impact of intensive farming and forestry practices on wildlife, natural resources and landscape beauty (Mabey, 1980; Shoard, 1980; Lowe et al, 1986). The fuel crisis of the mid-1970s had given rise to concern that the economic efficiency of post-war agriculture might have been based on an increasingly inefficient use of natural resources (Centre for Agricultural Strategy, 1980).

Environmentalists began to have some purchase on agricultural policy. Dedicated schemes introduced in the mid-1980s targeted geographical areas with the aim of safeguarding valued habitats and landscapes. A complex system of regulations and rewards emerged through the 1990s to safeguard and promote what came to be known as 'multifunctional' farming. Commentators referred wishfully to the dawning of post-productivism (Wilson, 2007). Farmers were given incentives to maintain biodiversity, landscapes and access to the countryside and to protect water resources. What many farmers saw as encouragement of untidy landscapes provided room for functions other than food production. There was a similar change of emphasis in forestry policy, with state forests becoming valued in policy terms much more for their recreational and wider environmental benefits than their timber outputs.

At the same time, farmers were encouraged to seek rewards less from expanding production of low-cost commodities in surplus and more from consumer demands for high-quality, speciality and value-added products and services. Farmers' care for the environment was seen as another potential source of product differentiation that could be realized through 'green marketing'. Hence consumers could express their environmental concerns through discretionary food purchases. They were actively encouraged to 'Eat the View' (Natural England, 2009).

Government for its part took a relaxed view of where the UK sourced its food supplies in a period of relatively low commodity prices. The Department for Environment, Food and Rural Affairs (Defra) Strategy for Sustainable Farming and Food of 2002 stated that:

> The UK is now 75 per cent self-sufficient in food production – a higher figure than in the 1950s. But in an increasingly globalized world the pursuit of self-sufficiency for its own sake is no longer regarded as either necessary or desirable ... the Government will continue to assert ... that the best way of ensuring food security is through improved trading relationships, rather than a drive for self-sufficiency.

There seemed more pressing concerns, including some arising from a surfeit of food, with the Government, for example, conducting a Foresight project on obesity between 2005 and 2007.

Are we now entering a new and more volatile era marked by concerns over food supplies? Market prices for wheat doubled in the year to February 2008, and the price of milk went up by more than a quarter, prompting headlines such as:

- 'End looms for era of cheap food', *The Times, 31 July 2007.*
- 'Global food crisis looms as climate change and fuel shortages bite', *The Guardian, 3 November 2007.*
- 'Millions of families face soaring food bills' *Daily Mail, 12 August 2008.*

Yet food prices have since eased significantly, as the global economic downturn has intensified. By October 2008, the UN FAO's Food Price Index stood at 164, the same level as in August 2007 and 25 per cent lower than the Index's high of 219 in June 2008 (Evans 2009). This about-turn prompted the President of the National Farmers' Union to warn that, while demand for food should remain steady, 'the production of home-grown foods could plummet in 2009' as farmers react to the credit squeeze (NFU, 2009).

Some of the concerns and issues are redolent of the Limits to Growth period of the early 1970s with their combination of hikes in commodity prices at the end of a long period of global economic growth and the onset of stagflation. 'Food security' is once again an issue ('Echoes of Britain's wartime Dig for Victory as community gardens gain ground', *The Observer, 10 August 2008*). The Government held a 'Food Summit' at 10 Downing Street on 22 April 2008 and in November 2008 launched a Foresight project to 'examine the long-term global challenges of a growing world population and increasing pressures on food production'. It will consider 'how policy in the UK can contribute to alleviating hunger and sustaining a well-fed world in the future' (http://www.foresight.gov.uk/OurWork/ActiveProjects/LandUse/LandUse.asp). The World Bank estimates that global food supplies will have to increase by 50 per cent by 2030 to keep up with population growth (World Bank, 2008).

Food security was an important issue for many contributors to the online Land Use Debate hosted by the Relu Programme in March 2008. One commented:

> The consequences of food shortages are hard to imagine for those in the current and second generation but are well remembered by the third generation who grew up with rationing and state control of virtually all food production. There is no reason why these times of shortage should not return.

Does this mean a return to production at any cost? Many farmers and some scientists think that the balance has been tipped too recklessly towards environmental sustainability and away from food production. They urge a renewed focus on productivism. Environmentalists, however, fear that this risks forfeiting the improvements to the countryside that they have seen over the past 10–20 years. For example, the rate of decline in farmland bird numbers has been halved in recent years, thanks mainly to the policy of set-aside. With such production controls now being abandoned, conservation organizations fear the worst.

A narrow focus on productivism would be too exclusive of other goals, too fossil-fuel dependent and too wasteful of natural resources. Moreover, the countryside contributes to many aspects of quality of life which cannot be casually dismissed. The need to move away from a carbon-based economy demands that food and other vital resources must increasingly be derived from the sustainable exploitation of agricultural, forestry and marine ecosystems. Boosting food production is possible, but must be done in ways that respect environmental limits and minimize damaging trade-offs with other ecosystem services if system functionality is to be maintained.

In Relu's Land Use Debate, horticulturalist Mark Tinsley set out his view: 'We continue to reduce our commercial competitiveness by allowing the environment to dominate decision-making.' This was countered by the RSPB's Mark Avery with 'Land management can, and should, have a net positive environmental gain. Productive agriculture is a vital part of that, but the key is to recognize other land management objectives, which governments have, up to now, failed to value sufficiently.' A number of contributors fell squarely into one of these camps but others emphasized the need for multifunctionality.

Land managers in these uncertain times need to be able to plan and make decisions about the use and management of land which will realize their own business and personal expectations. At the same time, it is important that those decisions are aligned with the wider needs of society from land within a strategic land-use policy framework. Our experience of agri-environment policy over the last 20 years usefully indicates how we might balance the public and the commercial interest in the management of land, in an era of more complex and intense demands.

Multifunctional agriculture, which in the past relied upon farmers tolerating, or being paid to maintain, sub-optimal production, must now emphasize ecological efficiency as much as economic efficiency – what we might term 'smart production'. A way of taking forward such an integrated approach is provided by the 'ecosystem services' framework.

Integrated land management and the ecosystem services approach

This ecosystem services framework provided the basis for the United Nations' Millennium Ecosystem Assessment (United Nations 2005 http://www.maweb.org/; also see Figure 2.1) and has been endorsed in the UK Sustainable Development Strategy (HM Government, 2005). An ecosystems approach is essentially about

> adopting a new way of thinking and working, by shifting the focus of our policymaking and delivery away from looking at natural environment policies in separate 'silos' – e.g. air, water, soil, biodiversity – and towards a more holistic or integrated approach based on whole ecosystems (Defra, 2007).

An 'ecological system' comprises living and non-living elements within an interdependent system. Ecosystem services are the valuable benefits that a healthy environment provides for people, either directly or indirectly. The term is deliberately inclusive. Four broad types of ecosystem service are recognized in the Millennium Ecosystem Assessment: these include '*provisioning services* such as food, water, timber, and fibre; *regulating services* that affect climate, floods, disease, wastes, and water quality; *cultural services* that provide recreational, aesthetic, and spiritual benefits; and *supporting services* such as soil formation, photosynthesis, and nutrient cycling'.

The ecosystem approach has several valuable features (Defra, 2007). It emphasizes the need for holistic policymaking and delivery, recognizing that land provides economic, social and environmental services alike. Hence land should not be conceived as existing simply 'to provide food' or 'to protect our environment'. It serves diverse purposes which should be fully reflected in decision-making. The approach also recognizes that real environmental limits must be respected if the ability of land to deliver a wide range of ecosystem services is to be safeguarded. Finally, the ecosystem approach promotes adaptive

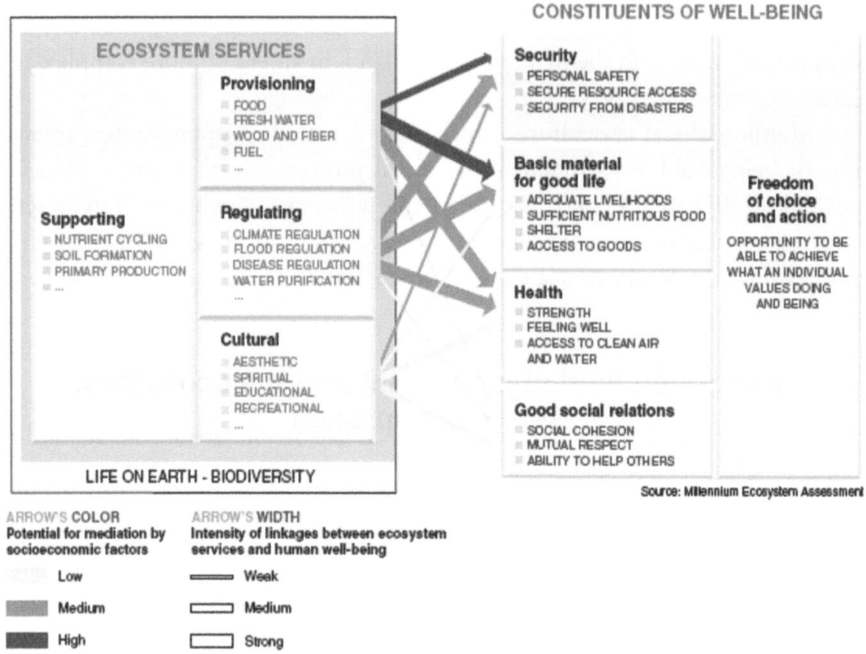

Source: http://www.maweb.org/, 2005, pvi

Figure 2.1 *Ecosystem services*

management of the environment to respond to changing conditions, including climate change.

Taking a 'more holistic approach' requires a distinct mindset – one which will be new to many of those concerned with the use of land. Can farmers, used to seeing themselves as 'food producers', conceive themselves instead as 'integrated land managers' who provide ecosystem services? The ecosystem approach means accepting the possibility that, regardless of its current main use or uses, any area of land has the potential to deliver a very wide range of services (such as food, flood management, biodiversity or recreation). It also may mean accepting a higher degree of 'multifunctional' land use than hitherto. However, there are limits to the extent to which multifunctionality can be pursued without impairing the delivery of one or more of the services involved. For example, there are well-known trade-offs to be made between crop productivity and diversity of wildlife.

The challenge lies in enabling the pattern of land use to change to create a landscape which provides a wider range of ecosystem services than hitherto. The sorts of shifts which can be envisaged include:

- Changing the extent and mix of specific land uses (e.g. the balance between agriculture, forestry, wetlands and other uses).

- Adjusting management practices (e.g. more unsprayed field margins) to foster increased biodiversity and reduce threats to water from diffuse pollution.
- Adopting new technology (e.g. 'precision' farming techniques and new varieties of crops, grasses or trees produced through traditional or novel breeding techniques).
- Introducing crops or livestock which are relatively new to the UK (e.g. novel biomass crops, soya, tea, new breeds of livestock).
- Restoring or 're-wilding' habitats (e.g. by converting intensively managed arable to extensively grazed chalk downland, or removing grazing livestock from upland grasslands to enable regeneration of heather moorland and woodland).
- Bringing previously developed, derelict or idled (set-aside) land back into productive use – to deliver a range of services, not necessarily food alone.

The shifts in attitudes required by adopting the ecosystem approach pose significant challenges for some long-standing policy concepts and mechanisms. These are often sectoral, focus on one interest in land to the exclusion of others, and emphasize land use segregation rather than multifunctionality. For example, local development plans for many years treated all land other than that zoned for development as undifferentiated 'white land', regardless of its value for food, forestry, biodiversity or other services. Attempts to distinguish between different grades of agricultural land have led to the planning concept of 'Best and Most Versatile Land'. Yet this focuses purely on the food-producing potential of land, ignoring the other ecosystem services which farmland can provide.

To counteract these partial perspectives, environmental legislation has added various designations to the planners' maps, including Areas of Outstanding Natural Beauty for landscape, National Parks for landscape and recreation, and Sites of Special Scientific Interest for biodiversity. These designations have in turn been supported by their own sets of single-purpose policies and delivery mechanisms. The literature contains many articles on the failure of policymakers to take a more integrated approach to these designations, recognizing that each will delineate areas which are of interest from diverse environmental, as well as cultural and economic perspectives. Recent legislation has added wider purposes to many of these designations, but still in a piecemeal way. It is difficult to characterize any as 'holistic'.

Identifying the potential for land to deliver different ecosystem services to society is essentially a technical exercise. It means stepping back from current or recent land use and considering what range of services could realistically be delivered, given appropriate management. Determining priorities for any one area – which services should actually be delivered there to meet society's needs – is quite another matter. Any such decision is the sum of interactions between the market, land managers and policymakers. The decisions can often be contested. For example, land managers often argue that their freedom to exercise their property rights is being infringed unduly by environmental restrictions. Equally, others argue that land managers are not being sufficiently encouraged, or required, to change management practices to deliver valued services more effectively.

Complications arise because priorities will alter over time, in response to new pressures: priorities agreed today may be far from appropriate in 2020, 2050 or 2100. For example, should the UK now invest more strongly in food production, given global population growth and the diversion of arable land from food crops to the production of biofuels? Equally, should tree crops be encouraged as sources of biomass and sustainable construction materials or discouraged because they will add stress to water resources in a dryer climate?

The challenge of reconciling public and commercial interests, short-term considerations and long-term requirements, and local and global concerns, is underlined by the imperative to respond to climate change. This is not just an additional public objective in the management of land, but is one of overriding importance; it not only re-orders existing objectives, but also exacerbates other pressures.

The challenge that climate change poses for land use

Already Earth surface temperatures are up by about 0.6°C. Most of this increase has occurred since 1970. UK warming has been higher than average, with the 1990s the warmest decade on record. The number of frost-free days, hot summer days and the growing season have all increased, meaning a significantly earlier spring and later winter. The projection for 2080 is that spring will be one to two weeks earlier and winter one to three weeks later. We are also projected to have wetter winters and drier summers, as well as a greater frequency of extreme weather events. Another important effect for the UK and many other countries is sea level rise. Global mean sea level is rising by about 1.8 mm p.a. The effects are being unevenly experienced within the UK, with lower increases for Scotland and bigger ones for South East England.

The science and politics of climate change have so far focused on mitigation. Increasingly, however, the focus is expanding to include the steps needed to adjust our economy and society to unavoidable changes in climate and their consequences. The Climate Change Act 2008, for example, establishes a National Climate Change Adaptation Framework, requiring public bodies to detail their response strategies and how these will be monitored.

The mitigation and adaptation agendas interact, with potential synergies and frictions. In simple terms, the scale of the adaptation challenge will be set by the effectiveness of efforts to meet the mitigation challenge. However, the two involve different time scales, different levels of decision-making and different actors. Efforts to mitigate climate change are science-framed and focused on governmental and intergovernmental action and regulation. Adaptation efforts are society-framed, as people, markets and institutions respond to their perceptions of resultant environmental instabilities and vulnerabilities. Both demand that we learn to manage under conditions of uncertainty (Figure 2.2).

How we use land is central to both mitigation and adaptation. On the one hand, land is both a source of emissions and a means for decreasing them. Land can produce low-carbon energy – from wind-farms, solar power, biomass crops and anaerobic digestion of waste. Equally, forests and peatlands have potential to 'lock up' substantial amounts of carbon.

On the other hand, especially as space, land is central to our capacity to adapt and adjust to the effects of climate change. Flood management areas, changing cropping zones and shifts in the geographical ranges of species are examples of this. Much of the medium-term growth in greenhouse gas emissions is already in the pipeline. So adaptation is a necessity.

It is important to ensure that short-term adaptations do not add to the long-term problem. Shifts in land use happen over divergent time scales ranging from months (e.g. an arable crop rotation) to many years (e.g. afforestation) and may be more or less reversible, which means that much of our decision-making over the use and management of land is quite path dependent. The deployment of land must therefore seek to reconcile the short and long-term perspectives.

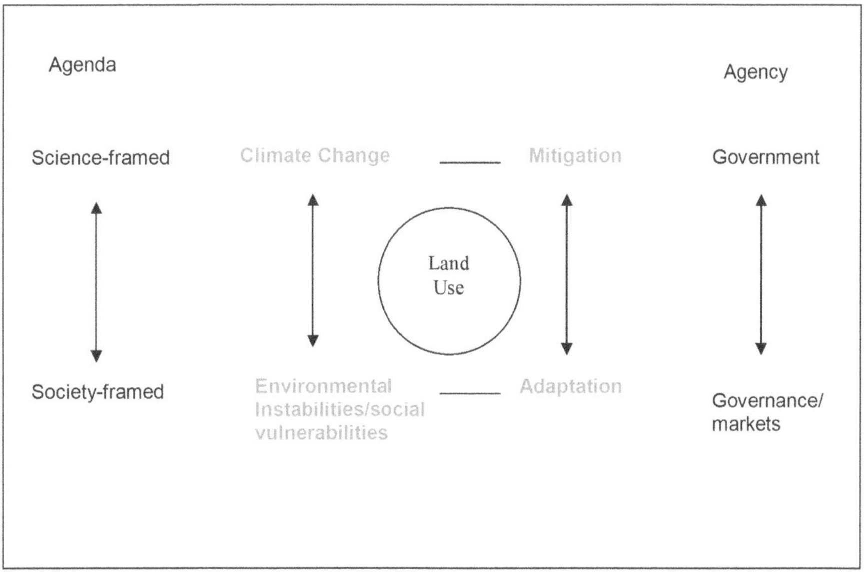

Figure 2.2 *Land use at the centre of climate change mitigation and adaptation*

What does this mean for land uses such as agriculture, biodiversity and renewable energy?

Agriculture

Agriculture is responsible for 7 per cent of UK greenhouse gases (specifically 2.7 per cent of carbon dioxide emissions, 37 per cent of methane and 66 per cent of nitrous oxide). Potential mitigation measures include (Smith et al, 2008):

- Reducing direct energy use (fuel, electricity, heating) and indirect energy (e.g. fertilizers).
- Substituting fossil energy through producing biomass for the generation of heat and power or biofuels, and anaerobic digestion of manure and organic wastes.
- Increasing carbon storage in soils through higher inputs (straw incorporation, manure, cover crops, grass in rotation) and reduced soil organic matter turnover (no-till).
- Reducing methane emissions through improved diets for ruminant animals and improved handling and storage of manures (including anaerobic digestion).
- Reducing nitrous oxide emissions by managing nitrogen cycles and technical measures to reduce emissions from manure stores, and from manures and fertilizers applied to soil.
- Developing renewable feedstocks and products to substitute for those derived from fossil fuels (e.g. for the construction industry).

Land managers are already being encouraged to pay greater attention to carbon accounting through voluntary initiatives such as 'Carbon Aware Land Management' (CALM) (CLA, 2006). The New Zealand government has formally committed itself to introduce a system of carbon accounting for agricultural commodities for 2013. Carbon accounting could be critical in stabilizing peatland and grassland as carbon stores. Packets of Walkers potato crisps already carry a carbon footprint statement. In January 2007 Tesco announced its aspiration to carbon-label all products it sells.

Agriculture will also have to adapt to climate change and its consequences, including more extreme weather events. Growing seasons will change – already the growing season for much of the UK has increased by about a month since 1990. Summers are likely to be drier, and farmers may face water shortages. Agriculture will have to cope with different pests and diseases. There will also be a shift in the geography of crops – with new opportunities as well as losses.

Biodiversity

Of course, wildlife has no option: it must adapt or perish; and adapt not only to the direct effects of climate change, but also to the indirect effects from human reactions to climate change. For example, growing new crops, increases in summer watering and geographical shifts in arable and livestock production could well occur. Regarding the direct effects, the greater frequency of extreme weather events is likely to have as much impact upon biodiversity as overall trends in temperature and precipitation. The main direct effects will be as follows:

- Changes in the timings of seasonal events, leading to loss of synchrony between species and the availability of food.
- Shifts in suitable climate conditions for individual species, leading to change in their abundance and range and altering the composition of plant and animal communities.
- Changes to habitats, such as altered water regimes, increased rates of decomposition in bogs and higher growth rates in forests.

A majority of at-risk species – birds such as song thrush, plants such as twinflower – are likely to experience shifts in the location and/or extent of areas where they can survive (Berry et al, 2007). Change is likely to be too rapid for evolutionary adaption, though genetic variation in species may give added resilience. Current habitats are fragile and may be impossible to preserve as climate changes. It will be important to facilitate movement of species, largely northwards, to enable them to survive.

The historical strategy of protecting 'islands' of suitable habitat within a more hostile 'sea' of unsuitable land uses is likely to become untenable (although see the chapter by Hopkins in this volume). New strategies will be needed to enable specialized species, in particular those that are relatively immobile, to disperse and establish in new locations as the climate changes. Action could include creating significant new areas of habitat to improve links between existing fragments, and actively introducing species to habitats which become suitable for them outside their current geographical range.

Climate change raises the challenge of dealing with invasive non-native species. Some could present substantial problems for crop management, livestock husbandry, navigation and native fauna and flora. Much depends upon changing attitudes (a key aspect of adaptation): will society perceive all non-native species as bad news or welcome their role in enhancing biodiversity? It will be important to discriminate between extensions of species ranges which are benign (e.g. little egrets) and those which are malignant (e.g. bluetongue disease).

Biofuels

An instructive case of a damaging clash between short- and long-term objectives is provided by biofuels. Replacing fossil fuels with renewable biofuels would seem in theory to offer benefits in mitigating climate change. However, short-term considerations of national energy security and farm income diversification have so far been more significant drivers. 'First generation' biofuels (grains, vegetable oils) have diverted resources from food production, forcing up food commodity prices (accounting for about a third of the increase in cereal prices experienced in 2008) while consuming considerable amounts of fossil fuel in their production. Significant losses of grassland and forest to arable cropping for biofuels, or displaced food crops, have also created alarm among environmentalists.

Second generation biofuels (i.e. dedicated energy crops or waste by-products) may be a more attractive prospect. They include biomass from wood chippings, straw, miscanthus, short-rotation willow coppice, manures and food waste. However, setting targets for renewable energy production without fully considering the implications for other ecosystem services has been a source of great contention (RSPB, 2008). For example, crops such as willow tend to use a lot of water, and targeting 'marginal land' for this purpose could damage biodiversity (Gallagher, 2008). The potential carbon-reduction benefits of second generation energy crops need to be considered in the wider context of what they replace, the demands they place on natural resources and the implications for other ecosystem services. These matters will need to be handled much more sensitively and strategically if energy crops are to have a place in the future land-use mix (Haughton et al, 2009).

A strategic approach to land use policy and delivery

There are thus multiple demands on land from different sources and directions. The key question is how to guide these demands effectively in ways that are both flexible, to allow people and businesses to adjust to environmental change, and strategic, to ensure that the long-term public good is pursued.

The following would seem to be critical needs in achieving a more flexible and strategic approach to land use policy in the UK:

- Recognizing the ecological capacities of land.
- Valuing the full range of ecosystem services.
- Promoting precision farming, to support economic and ecological efficiency.
- Reorientating production incentives to support integrated land management.

- Establishing mechanisms for locally adapting management of land, to facilitate flexible responses by people, businesses and communities.
- Promoting a co-operative environment to ensure coordinated, landscape-level action.

Finally it is important to develop a long-term strategic vision for land use which integrates diverse objectives for ecosystem services.

Recognizing the ecological capacities of land

Boosting production in a sustainable fashion involves careful trade-offs with other ecosystem services if overall ecological capacity is to be maintained. To operationalise this approach we need studies of the functionality of agro-ecosystems – not a fashionable topic among academic ecologists, who in recent years have been too preoccupied with population and behavioural ecology.

There is a clear need, for example, for experimentation and demonstration in how to integrate multiple agri-environment objectives in a whole-farm or even landscape approach. To date, studies have concentrated on one or two dimensions (e.g. water quality or supply; soil erosion, run-off and drainage; and aquatic or terrestrial biodiversity) and on techniques such as buffer zones, minimal tillage or rotational fallow. There have been few efforts to integrate these approaches, but such an understanding is needed to inform future policies for the best management of land for both high-output production (e.g. of food, fibre and energy) and critical environmental objectives (e.g. carbon storage, flood risk management and conservation of biodiversity).

Applied systems ecology here must meet up with economic and institutional analysis of the governance of land and natural resources if we are to establish clear principles and procedures for making sustainable trade-offs between desired ecosystem services. Interdisciplinary researchers are looking, for example, at how to achieve multiple objectives in floodplains, including food production, nature conservation, flood regulation, maintenance of rural livelihoods and enjoyment of the countryside (Roquette et al, 2009). An integrated analysis of ecosystems is emerging which explicitly recognizes that the different streams of environmental and productive services serve diverse stakeholder interests. Figure 2.3 provides an example of modelling land and water scenarios for Beckingham Marsh on the River Trent in Nottinghamshire and reveals the possible synergy and trade-offs between ecosystem functions and services.

Another example of important ecological capacity is peat. More carbon is stored in Britain's soils – mostly in blanket peat – than in the forests of France and Germany combined, and it is vital to safeguard their storage potential. Disturbances such as overgrazing, drainage or ploughing greatly increase

		Scenarios		
		Arable farming	Flood storage	Wetlands
Function	Use	Rapid drainage, low flood frequency	Rapid drainage, high flood frequency	Slow drainage, high flood frequency
Production	Agricultural production	High	Medium	Low
	Bio-fuel crops	H	L	M
Regulation	Flood water storage	M	H	L
Habitat	Biodiversity targets	L	M	H
Carrier	Road networks/industry	H	M	L
Information	Recreation	L	L	H
	Education	L	L	H

Note: 'high', 'medium' and 'low' refer to the degree to which land fulfils 'production', 'regulation', habitat', 'carrier' and 'information' functions under the three scenarios

Source: http://www.relu.ac.uk/research/projects/SecondCall/Morris.htm

Figure 2.3 *Example of floodplains*

carbon dioxide emissions. Much of the damage and erosion is due to extensive drainage ditches dug in the post-war period, in an unsuccessful attempt to increase the productivity of peat lands. Researchers have gone as far as proposing a scheme to allow consumers to offset their carbon footprint by paying for upland regeneration. They estimate that investing some £200/ha in blocking drains across one hectare (ha) of peatland would not only help the peat to regenerate but could also benefit downstream flood control, water quality and biodiversity (Worrall et al, 2007; Liddon, 2007).

Valuing ecosystem services

The ecosystem approach cuts across 'services' which are fully tradable (e.g. food, fibre, fuel), tradable to some extent (e.g. game and agri-environment services purchased by the state) or not-tradable (e.g. landscape beauty, and biodiversity). Some services are co-products of land management practices driven by the market (e.g. intensive farming creates a productive and highly managed landscape). However, many services are unlikely to be delivered without some public intervention (e.g. farmers without livestock may be unwilling to continue to maintain hedgerows). There are practical challenges in determining the relative roles of the market and public intervention in securing the desired level of any specific service.

Those ecosystem services that can be traded, such as producing food, fibre and energy, have an 'explicit' economic value reflected in their market price. Many other services, however, although also valued by society, cannot readily be priced; their value is 'implicit'. If it is accepted that there is market failure in delivering the service, and the political will exists to support action to correct it, these implicit values can be made explicit through government intervention. This could include: regulation to internalize costs in production systems and hence reduce undesirable impacts; or incentives to reward the active provision of the desired service.

The Water Framework Directive provides one example of significant government intervention at EU level to require markets to value more highly a resource – water – that has a largely implicit value. It does this by setting demanding targets for improving the wider water environment and requiring Member States to develop programmes of measures to implement these. The measures can include: designating 'Water Protection Zones' within which land management activities may be strictly controlled to protect water quality; promoting the adoption of voluntary good practice to minimize soil erosion; or the provision of incentives to restore heavily modified watercourses or create new wetland habitats.

Deciding what absolute or relative value to place on specific non-marketable ecosystem services presents challenges to policymakers in making explicit their implicit values to society. For example, to what extent can the services purchased through agri-environment schemes be valued financially in their own right, rather than in comparison with food production? EU legislation currently requires agri-environment payments to be calculated on the basis of the income foregone by adopting specified land management practices. This seems unbalanced, given that food is valued by the market without necessarily taking account of the resulting environmental costs and benefits. It also seems out of step with the ecosystem approach, which values all services from land in their own right, not as adjuncts to or competitors with food production alone.

Promoting 'precision farming' to support economic and ecological efficiency

Precision farming, aided by developments in agricultural engineering and technology, could play a critical role in enabling high-output production with reduced inputs. The capacity of land to deliver goods and services can be increased many times by the application of engineering, as evidenced over the centuries by drainage, terrace construction and irrigation. The technical potential exists to further enhance or adjust the capacity of most agricultural land. This applies even without genetic improvements in crops and livestock – whether through traditional or novel methods.

Better management of water could be a key target in enhancing the productive capacity of land. Innovative systems for water management could combine sensors and informatics to control soil water much more precisely, allowing greater efficiency in water use and curbing nutrient run-off. Another prospective development also promising tremendous efficiency gains is the use of micro-robots within field crops for highly targeted seed and fertilizer placement, weed control and pest management.

Precision farming does not necessarily depend upon high-tech gadgetry and is required not just to boost production but to do so in ways that are ecologically efficient and conserve resources. The experimental and demonstration work of bodies such as the RSPB and Game Conservancy Trust is showing how techniques such as field margins and headlands, skylark patches and beetle banks can be integrated into commercial farming in ways that enhance both production and biodiversity.

Much depends on the attitudes and motivations of land managers. A research team led by Professor Bill Sutherland at Cambridge is investigating management options for biodiverse arable farming. They are determining why arable farmers deviate from the optimal land management practices defined by economic models. They are also using ecological models to predict how weed and bird populations will respond to changes in these land management practices. Their work will help improve understanding of what approaches will best deliver multiple ecosystem services on arable farms.

Reorientating production incentives to support integrated land management

There are potentially large-scale public resources for integrated land management available from the continuing reform of the Common Agricultural Policy (CAP). Annual public expenditure under the CAP is over €70 billion (€6 billion within the UK), but three-quarters of it is skewed towards farm income support (Figure 2.4). Its pattern and level of distribution reflect historical production levels and previous budget allocations. The forthcoming EU budget review provides an opportunity to reorient the CAP towards future sustainability challenges.

CAP ‘Pillar 1’ funding, covering market support and the Single Farm Payment, does little positively to encourage integrated land management. Farmers receiving CAP payments are subject to ‘cross-compliance’, which requires them to respect certain environmental and agricultural production standards. However, few countries rigorously enforce these requirements and they are insufficient to address the full range of concerns over loss of farmland

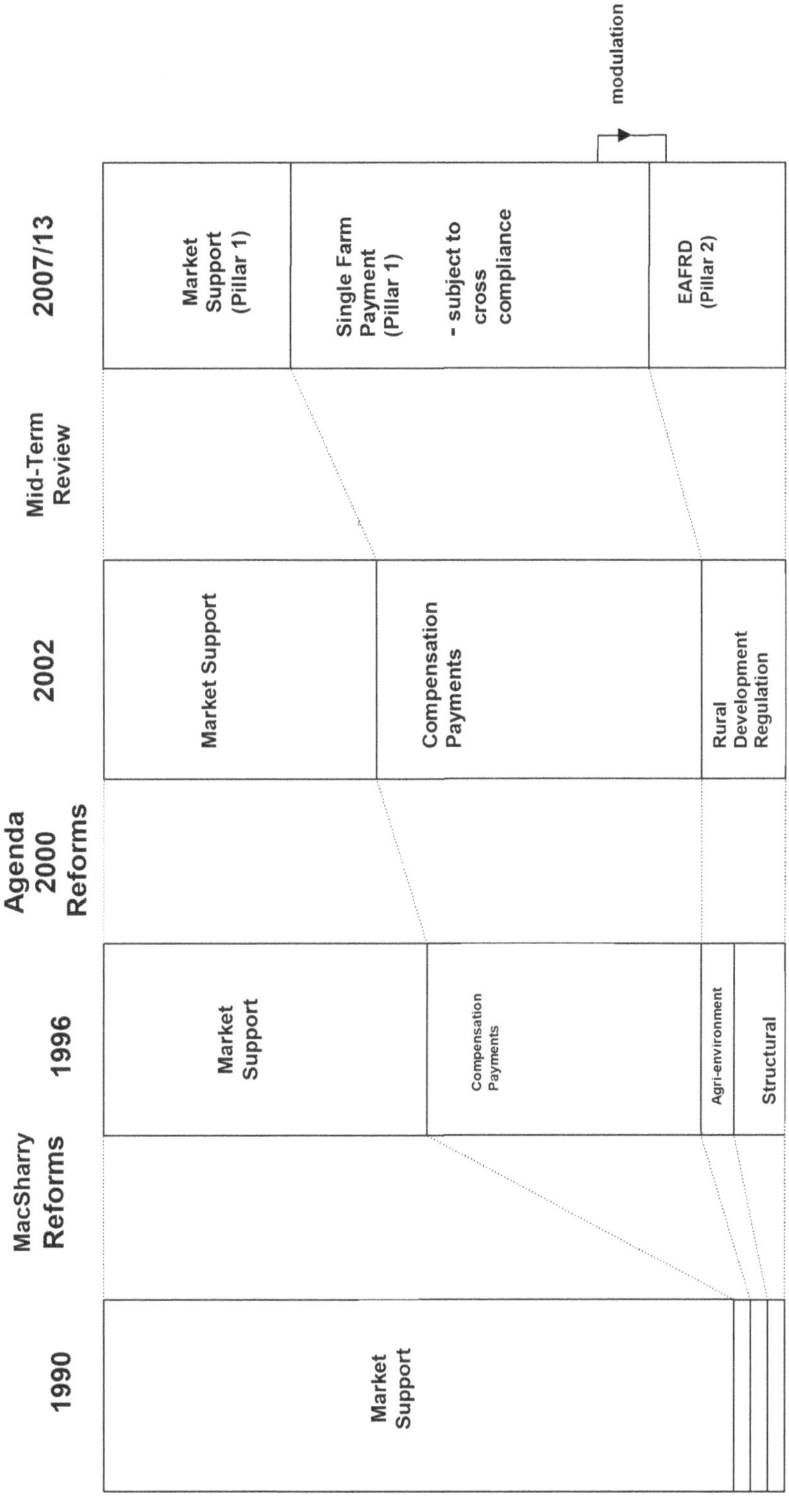

Figure 2.4 *The changed architecture of the CAP*

biodiversity, sustainable use of water and accelerating climate change. If European farmers are to justify continuing receipt of payments on anything like this scale beyond the current budget period, further greening of Pillar 1 will be necessary.

The architecture of the CAP would be significantly different if sustainable development was at its core. That would suggest a major shift of resources to 'Pillar 2' – the European Agricultural Fund for Rural Development. This supports agri-environment, farm development and rural diversification schemes. If it is to be enlarged, it will need to be reworked as an integrated land and resource management fund, with an emphasis on investment in the full range of ecosystem services from land. At the moment its scope is too restricted, it is too centralized and rigid, and it is not goal-oriented – it stifles the local initiative, problem-solving and innovative behaviour that need to be encouraged among land managers if adaptability and resilience are to be fostered.

Mechanisms for locally adapting management of land

A critical need in adopting a more strategic approach to land use policy is to complement centralized policymaking with action to encourage and support local innovation in tackling land use challenges, whether by individuals, businesses, or local communities. For example it is unrealistic to expect a single agri-environment scheme, developed on a national basis, to deliver the specific services required in any one locality. Devolution to UK territories, and within England to Regional Development Agencies, suggests that the need for better integration between top-down and bottom-up approaches is recognized. Participatory budgeting, to enable local communities to identify and fund locally determined priorities, offers one possibility.

New approaches and policy mechanisms must balance the public and private interest in land use and facilitate coordinated action. Land managers need a clear framework of incentives and responsibilities, and better structures for engaging with local communities. A new framework could be developed by rationalizing the existing piecemeal framework of legal measures for securing appropriate land management (including management agreements, cross-compliance and codes of practice) into a generic stewardship obligation on rural land managers that would allow for property rights to be adapted progressively in step with ecological conditions.

As the timescales for planning the delivery of various environmental services lengthen, the need for measures to secure long-term benefits will intensify. For example, the Water Framework Directive sets milestones for improving water management through to 2027. An agri-environment agreement which secures benefits between 2015 and 2021, but not thereafter, will be inadequate. Equally, the Climate Change Act sets targets for reductions in greenhouse gas emissions by 2050: is it prudent to use public resources to invest in carbon-reduction measures now if the benefits cannot be guaranteed beyond five years?

International experience suggests that there are various ways to secure long-term benefits through land management agreements (Green Balance, 2008). For example, the USA Grassland Reserve Program purchases easements to protect grasslands while maintaining production. Owners receive an annual payment based on the length of the easement and the value of the land. Another option might be for the state, or voluntary bodies such as the National Trust, to purchase land and lease it back to the former owner (or new tenants) under restrictive conditions at appropriate rents. This might also be an option for eroding coastlines, to ease adjustment for existing property owners who face losing their land and/or homes to the sea.

Promotion of a cooperative environment to ensure coordinated landscape-level action

It is important to recognize that different spatial scales are relevant for planning and securing the delivery of different services. While marsh fritillary butterflies require sensitive management of vegetation at a field scale, golden eagles require a mix of habitats at a landscape scale. River catchments are an appropriate context for safeguarding services related to water, but less so for food production or biodiversity: the local authority areas often used for Local Biodiversity Action Plans do not always fit well with the distribution of habitats. A 'landscape-scale approach' is being urged to facilitate adaptation to climate change (East of England Biodiversity Forum, 2008).

Whatever the ideal spatial scale for setting objectives for different land services, within the current property rights framework, the scale at which many of those services will be delivered is essentially that of the land holding. Collaborative action between land managers will be essential where action on one holding alone, or scattered holdings across any area, will not in itself deliver desired ecosystem services. For example, the health of a river will depend on the practices adopted by thousands of riparian owners. There might be a role for a 'collaboration' bonus, where more than a defined minimum proportion of land or of land managers within a targeted area is entered into an agri-environment scheme.

Environmental cooperatives such as those found in the Netherlands (Franks and McGloin, 2007) may offer one way of achieving coordinated responses. These emerged in the 1990s from within the farming community, responding to environmental concerns in their local areas. These initiatives demonstrate the value not only of local commitment but also of a supportive government framework which, in the Dutch case, includes covering start-up and overhead costs, funding training and conservation schemes, and being open to co-operative representatives in the development of policy.

Another approach would be the local or regional targeting of agri-environmental incentives. This is important to get the most out of measures such as the Entry Level Stewardship scheme (ELS), introduced in the UK in 2005, which is open to all farmers and land managers and encourages them to take up simple environmental management from a menu of options. By 2007 there were over 28,000 ELS agreements in England, covering more than four million hectares or approximately 40 per cent of the total farmed area. The ELS establishes a flexible framework within which land management incentives may be altered across a broad scale and yet at a very detailed level. It would be possible to offer regionally differentiated menus of eligible management measures and payment rates, to target the scheme towards local environmental priorities (Hodge, 2009).

The need for greater integration in policymaking and delivery is also central to any new strategic approach. Current rural land use policy is often fragmented, inadequately targeted and dominated by short-term thinking. For example, Catchment Sensitive Farming Initiatives have been focused on areas affected by high phosphate levels, without exploiting links to Nitrate Vulnerable Zone initiatives. Similarly, an initial focus in agri-environment schemes on conservation benefits meant that opportunities to tackle diffuse pollution were missed; 'if the focus had been on diffuse pollution, it is likely that the conservation goals could also have been met' (Harris et al, 2008). An integrated approach to policymaking should maximize synergies between policy areas and avoid unintended impacts. This is fully consistent with the more holistic approach to policymaking and delivery of the ecosystem approach (Defra, 2007).

'Habitat banking' – the restoration, creation or enhancement of habitats to compensate for development impacts elsewhere – may be one means of using a market mechanism to secure long-term benefits, encourage collaboration between land managers or secure action at a landscape scale. Developers purchase credits from a 'bank', which uses these to fund the purchase and management of land (or agreements with land managers) to yield long-term environmental gains. Habitat banking has been operating for some 30 years in the USA, where the market may be worth more than $1.5 billion annually. It is now being promoted in the EU (Wilkinson and Hill, 2008).

Conclusions: Developing a long-term strategic vision for land use

While the requirements of integrated land management are challenging, significant opportunities currently present themselves to realize this more flexible and strategic approach.

First, growing public consciousness of the enormity of climate change has propelled a willingness to alter personal behaviour and support concerted political and governmental action. Secondly, although the current global recession will undoubtedly test such commitments, it may serve to reinforce a sense of global interdependence and collective responsibility at various levels, and provide impetus for institutional reform. Much will depend on the terms whereby governments renegotiate the world's economic order, as they seek paths to recovery. That will set a framework in which competing national concerns for food, water and energy security are played out, as well as the balancing of priorities for economic development and environmental protection.

Thirdly, and specifically salient to the use of rural land across Europe, the continuing reform of the CAP provides opportunities to reassign resources – human, financial and land – to integrated land management. Fourthly, within the UK, decades of experimentation mean that we have a great diversity of instruments and means for what one might call 'strategic environmental planning' – such as designated areas, agri-environmental payments, Nitrate Vulnerable Zones, Catchment Management Plans, National Park Management Plans, Biodiversity Action Plans, Regional Rural Delivery Frameworks and Integrated Regional Strategies. In other fields, strategies are drawn up but resources often fail to flow; in land use policy we have the inverse problem: the resources and initiatives are many, but the coherent strategy is absent.

A new strategic approach to land use policy and delivery should: develop a long-term vision for securing integrated land management using an ecosystems approach; reconcile competing demands between different ecosystem services, at different spatial and temporal scales; better integrate policymaking and delivery; and overcome diverse risks which could hinder or prevent progress.

That requires a long-term vision which integrates diverse objectives for ecosystem services. The Government's Foresight Project on Land Use Futures was set up in April 2008 to explore how land use in the UK could change over the next 50 years. It will identify scenarios to prompt further thinking about future challenges and potential responses. It is being complemented by a further Foresight Project on Farming and Food. The breadth and intensity of the work being undertaken for these projects is immense. While they will not in themselves deliver a new 'vision', they should provide, together with ongoing research, a sound basis from which politicians and other players can do so.

A new strategic approach must also reconcile the competing demands for security in the delivery of the full range of ecosystem services, whether these are 'provisioning', 'regulating', 'cultural' or 'supporting' services (where 'security' implies a national responsibility, but not necessarily 'self-sufficiency'). We have international obligations as well as legitimate national interests. A strategic approach must facilitate land-based mitigation of climate change and adaptation steps (e.g. flood control, shifts in wildlife ranges). It should also set a framework for multifunctional land use, embracing both rural and urban land. Land use is all about coordinating the long-term and the short-term, the small-scale and the large-scale, the public interest and private interests, the economy and the environment. Land use policy must be well informed, but ultimately it is a matter not of technocratic determination but of democratic choice. The future use and management of land merits wider public debate.

Acknowledgements

Relu is a collaboration between the Economic and Social Research Council, the Natural Environment Research Council and the Biotechnology and Biological Sciences Research Council, with additional funding from Defra and the Scottish Government. Thanks to Terry Carroll, Jeremy Franks, Mark Kibblewhite, Andrew Lovett, Joe Morris, Chris Rodgers and Peter Sutton for assistance and comments on this chapter.

References

Berry, P.M., O'Hanley, J., Thomson, C.L., Harrison, P.A., Masters G.J. and Dawson, T.P. (2007) *Modelling Natural Resource Responses to Climate Change: Monarch 3 Contract Report*, Technical Report, United Kingdom Climate Impacts Programme, Oxford

Blunden, J. and Curry, N. (1988) *A Future for Our Countryside*, Basil Blackwell, Oxford

Centre for Agricultural Strategy (1980) *The Efficiency of British Agriculture*, CAS, Reading University

CLA (2006) *Climate Change and the European Countryside: Impacts on Land Management and Response Strategies*, www.cla.org.uk/policy accessed 30 July 2009

Defra (2002) *Strategy for Sustainable Farming and Food. Facing the Future*, Defra, London

Defra (2007) *Securing a Healthy Natural Environment: An Action Plan for Embedding an Ecosystems Approach*, Defra, London

East of England Biodiversity Forum (2008) *East of England Biodiversity Delivery Plan 2008–2015*

Evans, A. (2009) *The Feeding of the Nine Billion: Global Food Security for the 21st Century*, Chatham House Report, Royal Institute of International Affairs, London

Franks, J. and McGloin, A. (2007) 'Environmental cooperatives as instruments for delivering across-farm environmental and rural policy objectives: Lessons for the UK' *Journal of Rural Studies*, vol 23, pp472–489

Gallagher, E. (2008) *The Gallagher Review of the Indirect Effects of Biofuels Production*, Renewable Fuels Agency, St Leonards-on-Sea

Green Balance (2008) *The Potential of Conservation Covenants*, National Trust, London

Harris, R., Surridge, B., Holt, A. and Lerner, D. (2008) 'Rising to the land use challenge', Feedback from the Catchment Science Centre, University of Sheffield, Sheffield

Haughton, A.J, Bond, A.J., Lovett, A.A., Dockerty, T., Sünnenberg, G., Clark, S.J., Bohan, D.A., Sage, R.B., Mallott, M.D., Mallott, V.E., Cunningham, M.E., Riche, A.B.,

Shield, I.F., Finch, J.W., Turner, M.M. and Karp, A. (2009) 'A novel, integrated approach to assessing social, economic and environmental implications of changing rural land-use: A case study of perennial biomass crops', *Journal of Applied Ecology*, vol 46 (2), pp315–322

HM Government (2005) *UK Sustainable Development Strategy. Securing the future*, The Stationery Office, London

Hodge, I. (2009) 'The future development of agri-environmental schemes', in Brouwer, F. and Martijn van der Heide, C. (eds) *Multifunctional Rural Land Management*, pp33–52, Earthscan, London

Liddon, A. (2007) 'Uplands under pressure', *The Land Journal*, November/December

Lowe, P., Cox, G., MacEwen, M., O'Riordan, T. and Winter, M. (1986) *Countryside Conflicts*, Gower, London

Mabey, R. (1980) *The Common Ground: A Place for Nature in Britain's Future?* Hutchinson/Nature Conservancy Council

National Farmers' Union (2009) 'Farm output could plummet in 2009', Press Release, 1 January 2009

Natural England (2009) 'Eat the View – promoting sustainable local products', Countryside Agency Archive, http://p1.countryside.gov.uk/LAR/archive/ETV/index.asp accessed 30 July 2009

Rouquette, J.R., Posthumus, H., Gowing, D.J.G., Tucker, G., Dawson, Q.L., Hess, T.M. and Morris, J. (2009) 'Valuing nature-conservation interests in agricultural flood-plains', *Journal of Applied Ecology*, vol 46 (2), pp289–296

Royal Society for the Protection of Birds (2008) *A Cool Approach to Biofuels*, RSPB, Sandy

Scott Report (1942) *Report of the Committee on Land Utilisation in Rural Area*, Cmd. 6378, HMSO, London

Shoard, M. (1980) *The Theft of the Countryside*, Temple Smith, London

Smith, P. et al (2008) 'Greenhouse gas mitigation in agriculture' *Philosphical Transactions of the Royal Society B*, vol 363, pp789–813

United Nations (2005) Millennium Ecosystem Assessment, http://www.maweb.org accessed 30 July 2009

Whetham, E.H. (1952) *British Farming 1939–1949*, Nelsons Agriculture Series, London

Wilkinson, D. and Hill, D. (2008) 'Banking on it', in Country Land and Business Association, *Land and Business*, December, p47. See also www.environmentbank.com accessed 30 July 2009

Wilson, G.A. (2007) *Multifunctional Agriculture: A Transition Theory Perspective*, CAB International, Wallingford

World Bank (2008) *Agriculture for Development*, World Development Report, World Bank, Washington

Worrall, F., Burt, T.B., Adamson, J., Reed, M.S., Warburton, J., Armstrong, A. and Evans, M. (2007) 'Predicting the future carbon budget of an upland peat catchment', *Climatic Change*, vol 85, pp139–158

3

Perennial Energy Crops: Implications and Potential

Angela Karp, Alison J. Haughton, David A. Bohan, Andrew A. Lovett, Alan J. Bond, Trudie Dockerty, Gilla Sünnenberg, Jon W. Finch, Rufus B. Sage, Katy J. Appleton, Andrew B. Riche, Mark D. Mallott, Victoria E. Mallott, Mark D. Cunningham, Suzanne J. Clark and Martin M. Turner

Introduction

Policy interest in perennial energy crops, such as short-rotation coppice (SRC) willow and miscanthus grass, is firmly established in Europe, and the UK government has strongly supported the use of biomass crops as a source of electricity, heat and even transport fuel. The UK Biomass Strategy envisages a major expansion in both the supply and use of biomass, which is seen as playing a central role in meeting the EU target for renewable energy. There are various drivers of this policy interest, notably the potential of biomass as a low-carbon energy source in response to the challenge of climate change, the need to improve energy security and the desire to strengthen rural economic development in the context of agricultural decline. From a land use perspective, the prospect of diverting a significant proportion of farmland principally from food production to energy production would represent the most fundamental change in land use since the decline of horses as the primary source of power in agriculture, and a significant source of power for general transport, during the first half of the 20th century.

This chapter considers the range of potential impacts of increasing rural land use under perennial energy crops, drawing on environmental, social and economic research to provide a broad-based assessment. The use of farmland for perennial energy crop production on the scale envisaged will have potentially far-reaching implications for biodiversity, hydrology, landscape and the rural economy. This chapter not only explores the research evidence for the nature and scale of these effects, but does so in the context of the use of sustainability appraisal (SA) as a tool for land use planning. In particular, this review will identify the scientific tools that should underpin the conduct of Environmental Impact Assessments (EIAs), Strategic Environmental Assessments (SEAs) or SAs where strategic decisions on the planting of such crops have to be made.

Background

Global warming and energy security are high on the agendas of nations worldwide. The impacts of climate change, rising fuel costs and concerns over future energy security already affect the everyday lives of people and technological solutions are urgently sought. Producing feedstock from crops as renewable sources of carbon for conversion into heat and power (bioenergy) and liquid transport fuels (biofuels) has the potential to provide promising solutions from the agricultural sector. Biofuels, in particular, have been heralded as 'green gold', providing a renewable resource that could mitigate global climate change, promote energy security and support agricultural producers around the world (FAO, 2008).

The UK is a signatory of the Kyoto Protocol (United Nations, 1998), which aims to reduce global CO_2 emissions, and in 2000 set a target that 10 per cent of UK electricity supply should come from renewable sources by 2010. The most recent Energy White Paper included an aspiration to double this proportion to 20 per cent by 2020 (DTI, 2007). Several recent strategy and policy documents confirmed the potential of biomass as an energy source (RCEP, 2004; Defra, 2004; DTI and Carbon Trust, 2004) and a number of practical measures are now in place to support expansion; including planting grants, the UK Government's Defra bioenergy infrastructure scheme and the renewables obligations for licensed electricity suppliers (ROs) and transport fuels (RTFO). The UK has also committed itself to reducing greenhouse gases by 80 per cent, compared with 1990 levels, by 2050; (UK Parliament, 2008a), with a strategy to achieve this that includes support for small-scale renewables and encouraging renewable heat (UK Parliament, 2008b).

Within a short time frame of recognizing the potential of bioenergy and biofuel crops, however, there have been increasing counter-concerns over the potential implications for two other globally recognized future challenges: food security and water availability. Agricultural commodity prices have risen rapidly and governments have been alerted to the indirect consequences of growing energy crops on land currently used for food production, as well as the failure of some biofuel chains to achieve positive carbon balances and significant greenhouse gas (GHG) reductions (The Royal Society, 2008; Gallagher, 2008; Searchinger et al, 2008; FAO, 2008).

The UK Government and the EU remain committed to including energy from crops in the renewable energy portfolio, but the emphasis has become sharply focused onto the development of sustainable bioenergy and biofuel chains, which includes producing energy from crops that can be grown on sub-prime arable land to reduce competition with food production. Perennial biomass crops have the potential to offer sustainable bioenergy production that fit these criteria (Karp and Shield, 2008). As perennials, they recycle the major proportion of their nutrients, require fewer inputs and thus have the potential to be grown on lower-grade land. Life-cycle analyses of fuel chains for these crops also indicate higher energy gains and greater GHG reductions than those associated with 'first generation bioenergy crops' (Rafaschieri et al, 1999; Adler et al, 2007; von Blottnitz and Curran, 2007).

There are two types of perennial energy crops currently envisaged for UK farmland: coppiced trees and grasses. Of these, the most advanced agronomically and, in terms of plantings, the most common are short-rotation coppice (SRC) willow (*Salix* spp) and miscanthus grass (*Miscanthus x giganteus*). SRC willow and miscanthus have particular advantages as energy crops. Unlike annual crops, there is no intensive cultivation cycle, which improves the energy and greenhouse gas balance, and compared with other perennials they are fast growing, with the potential to produce large yields from low inputs of fertilizers and pesticides. Government incentives have been introduced to encourage establishment of biomass crops (e.g. the Energy Crops Scheme, ECS, www.naturalengland.org.uk/planning/grants-funding/energy-crops/default.htm). These have led to a recent increase in the area of land under SRC willow and miscanthus; for example, between 2001 and 2007 approximately 7,450 ha were established under this scheme in England alone (see statistics at www.nnfcc.co.uk).

While the potential benefits of growing perennial biomass crops for renewable energy are clear, conversion of large areas of land to SRC willow and miscanthus could constitute a major land use change, depending on the land use they replace. In particular, they are physically different and are managed differently to arable crops currently grown in the UK. They are in the ground for 7–25 years; although miscanthus is harvested annually in early spring, harvesting cycles for willow are every two to three years in winter. The crops are also very tall (three to four metres) and dense, and deeper rooting than traditional arable crops. These factors have potential implications for the appearance of the rural landscape (Figure 3.1), tourist income, farm income, hydrology and biodiversity.

A number of research investigations have been conducted to assess the impacts of perennial biomass crops (see Rowe et al, 2009, for recent review, and below). Most studies have concentrated on specific economic or environmental impacts, such as on economic viability, hydrology or biodiversity. Some have

Figure 3.1 *SRC willow (top) and miscanthus (bottom) are very different to crops traditionally grown in the UK*

suggested how assessments of impacts should be made (Firbank, 2008, for ecological impacts). Very few, however, have tackled how the different social, economic and environmental impacts could be assessed in an integrative way or how evidence relating to negative or positive impacts should be utilized in decision-making and land use planning.

The Relu-Biomass project (www.Relu-Biomass.org.uk) funded as part of the UK Relu programme recognized that targeted scientific research was needed to provide a science-based integrated evaluation of the social, economic and environmental impacts of perennial biomass crops. To achieve this, during three years (1 January 2006 to 31 December 2008) Relu-Biomass brought together scientists of the main disciplines to review existing information and also conduct new specific research activities to help fill knowledge gaps. Social science research, including Geographical Information Systems (GIS)-based approaches, was carried out by the University of East Anglia, economic studies by the University of Exeter, hydrological studies by the Centre for Ecology and Hydrology (CEH) and biodiversity studies by Rothamsted Research and the Game and Wildlife Conservation Trust. Defra funded a complementary project on biodiversity, which enabled size of field, age of crops and also bird use of miscanthus to be investigated. Two contrasting farming systems, typical of different regions of England (Figure 3.2), were chosen as the main study areas: (i) the arable cropping-dominated system as represented by the East Midlands; and (ii) a grassland-dominated system typified by the South West of England.

These two regions also have contrasting environmental characteristics (e.g. more rainfall in the South West) and, with Yorkshire and the Humber, contain the main concentrations of the 15,000-plus hectares of energy crops currently planted in England.

Two basic approaches to integration were taken: a GIS-based constraint mapping exercise and the development of a SA framework as a decision tool for land use planning. The latter involved direct stakeholder engagement and the project also benefited greatly from an advisory committee of principal stakeholders. Specific objectives were to: (i) develop an integrated scientific framework of the medium- and long-term conversion of land to energy crops; (ii) evaluate the implementation of the SA framework and identify the most appropriate planting scenarios; (iii) apply an integrated scientific approach to update existing best practice guides for planting SRC willow and miscanthus at any scale; and (iv) provide the scientific tools to underpin the conduct of EIAs, SEAs or SAs involving projects, policies or programmes where increased planting of energy crops is proposed or anticipated.

In the sections that follow, we first describe the approach taken in the different disciplines in the Relu-Biomass project and provide some of the results achieved so far. We then describe how these data are being integrated into tools for decision-making and land use planning. Finally, we conclude by

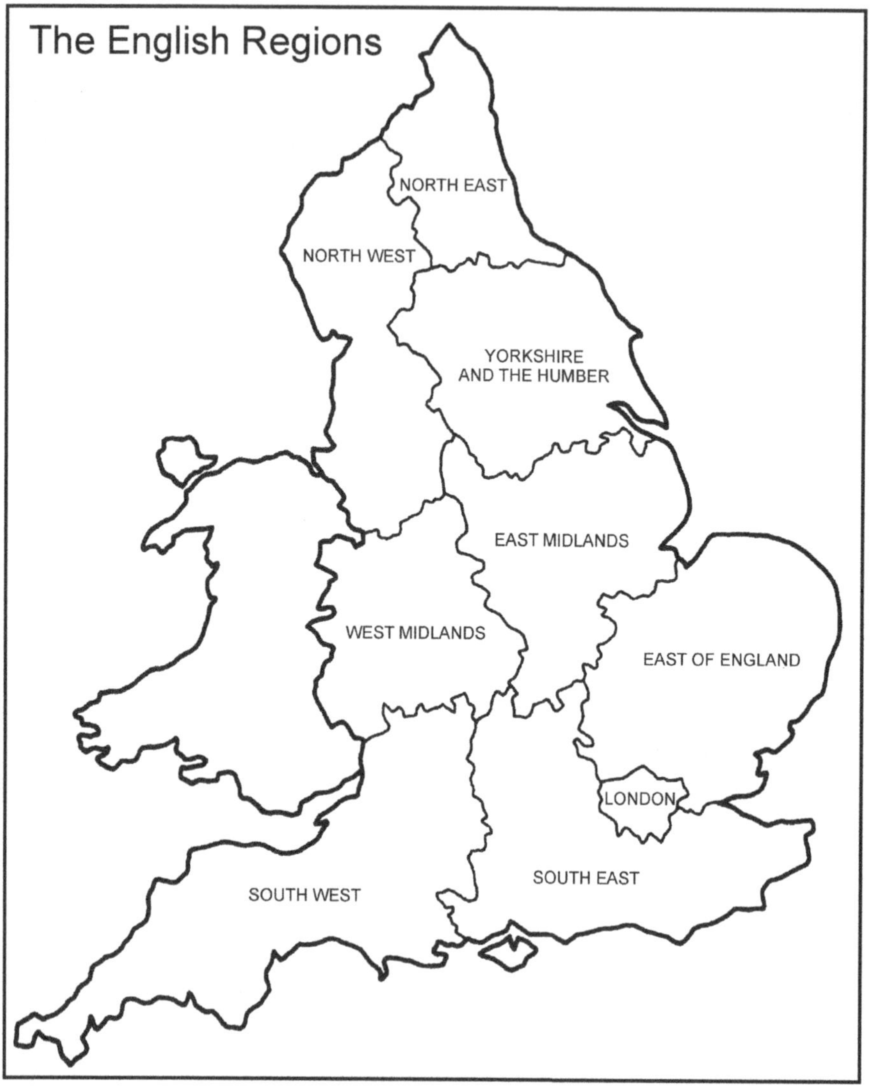

Figure 3.2 *English regions*

commenting on the implications and future potential of perennial biomass crops in the light of the findings so far.

Filling in evidence gaps: disciplinary research on impacts in the Relu-Biomass project

Public attitudes to renewable energy and visual impact

Best practice guides and guidance notes are available for both energy crops, SRC willow and miscanthus, which include siting considerations and plantation management protocols that are designed to address visual impact (as well as biodiversity and hydrological considerations), but these are usually at site level (DTI, 1999; Defra, 2001, 2002; Bell and McIntosh, 2001). In the Relu-Biomass project, it was considered important to move from site- to landscape-scale assessments of planting schemes and evaluate different means of communicating information on potential impacts to the public and other stakeholders.

One valuable framework for assessing the visual impacts of increased energy crop planting in different parts of England is provided by the set of Joint Character Areas (JCAs), defined by the Countryside Agency, English Heritage and English Nature in the mid 1990s. There are 159 JCAs, each with a description of what makes it distinctive, how this character arose and potential future management issues. These areas have since been used for various planning and management purposes, including judging the capacity of landscapes to absorb different amounts or types of change. Several recent studies have used this framework to assess the suitability of areas for planting of energy crops, including work in the South West by Capener et al (2004) and the national opportunities mapping exercise for energy crops in England sponsored by Defra (Defra, 2006). These issues of landscape sensitivity were incorporated into the GIS constraint mapping study described below in the section on integrative approaches.

A questionnaire survey was conducted during the summer of 2007 in town centres within the East Midlands and the South West, where SRC and miscanthus are grown in the surrounding areas, to assess current awareness and opinions of the general public regarding energy crops. Two urban centres (one larger and one smaller in size) were selected for survey in each region. These were Lincoln and Retford in the East Midlands and Taunton and Bridgwater in the South West. The content of the questionnaire was designed to be conducted in the street and therefore needed to take no more than about five minutes. Most of the questions involved selection from a set of possible responses and sets of photographs were used to show the appearance of SRC, miscanthus and a biomass power station. The answer choices and photographs were contained in a booklet which was given to respondents while the survey was conducted. These questions and the booklet layout were street tested in a pilot exercise in advance of the main survey. Overall, 490 completed questionnaires were obtained, with an acceptable age and gender balance compared with census statistics for each centre.

The results of this questionnaire survey indicated that there were many positive public attitudes towards planting of biomass crops as a renewable source of energy. For instance, over three-quarters of respondents said they thought SRC and miscanthus would fit very or reasonably well into the landscape of their local area. Over 60 per cent of survey participants also said that they would not mind seeing the crops within the view from their home.

However, showing participants a photo of a biomass power station and stating that this would need to be located within, for example, 25 miles had a distinct impact on responses, with the percentage saying they would not mind having the crops within the view from their house dropping below 30 per cent. This suggests that it is issues of infrastructure, particularly the scale of energy generation, rather than landscape or other concerns, that will need to be addressed if public support for biomass crops is to be developed. These matters are currently being examined further through a number of focus groups and discussions with regional stakeholders.

A third approach used in this part of the project was the development of GIS-based computer visualizations of energy crop plantings in specific locations within the study regions for use as a visual tool in stakeholder and public consultations. For example, Figure 3.3 shows a simulated view along a road in Lincolnshire before and after the planting of miscanthus. Interactive real-time models have also been created to run on an Elumens VisionStation and have been used at a number of public events (Figure 3.4). These visualizations were used at the focus group meetings to assess the visual impact of different management options (e.g. size of field margins) and opinions on the public acceptability of various scales and distributions of energy crop planting in different landscape settings.

Figure 3.3 *Visualizations showing view (a) prior to miscanthus planting and (b) with a mature crop*

Figure 3.4 *Visualization display at the British Association Festival of Science, York, September 2007*

Biodiversity

Only limited research has been carried out in the UK, or elsewhere in Europe, on wildlife use of miscanthus (Semere and Slater, 2007; Bellamy et al, 2009). In contrast, for SRC willow, several studies have been undertaken. Early, non-commercial plantings in the UK were found to provide new habitat opportunities for a variety of wildlife. Plant diversity consisted of a mixture of pre-existing and colonizing plants (Sage, 1998; Cunningham et al, 2004). Although some research has been directed at ground-dwelling species (Coates and Say, 1999), invertebrate studies have mostly focused on the crop canopy, where a variety of herbivorous and phytophagous invertebrates have been recorded, including some pests (Sage and Tucker, 1997; Sage, 2001). For birds, assessments suggest SRC supports some species not normally found in intensively managed arable crops and others that are. Comparisons indicated that fields of harvested SRC contain arable and grassland birds in addition to scrub and woodland species, leading to a net conservation gain (Sage and Robertson, 1996; Sage, 2001).

In the Relu-Biomass project, an extensive study has been carried out of biodiversity in 16 fields each of established miscanthus and willow SRC plantations using the approaches developed for the farm scale evaluations (FSEs) of genetically modified, herbicide-tolerant crops (Firbank et al, 2003; Haughton et al, 2003; Heard et al, 2005; Bohan et al, 2005). Although the two study regions did not provide sufficient sites, sampling was achieved across both the 'easterly' and 'westerly' lowland Environmental Zones of Great Britain (see Haines-Young et al, 2000). The complementary Defra-funded project extended the total number of fields sampled to 24 per crop. As with the FSEs, sampling was limited to weed plant and invertebrate species that can be sampled relatively easily (Firbank et al, 2003). At each site, twelve 32-m-long transects were evenly distributed around the edges of the crops, and ran perpendicular to the crop edge into the crop. Weed species were counted within quadrats at five sampling points along each transect, on two standardized dates over the season, with samples of weed biomass also being taken at the final species count. Monthly weed-seed rain samples were collected from two points along four of the transects. Vortis suction sampling was conducted for plant and soil-surface dwelling invertebrates, at three points along four of the transects on two dates within the season. Sticky traps were placed within the crop and in the canopy at three points along four transects once during the season to assess flying invertebrates. Bees and butterflies were counted along four line-transects of 100 m that ran along the external edges of the crop (headlands) monthly during the season to assess foraging behaviour.

A rich data source is being generated from this project, only a portion of which has been analysed at the time of writing. Together with existing data on the abundance and diversity of plants and invertebrates in conventional arable management from the FSEs and existing long-term invertebrate and plant species lists for arable (from LINK-IFS projects), grassland and woodland systems, these data will allow: (i) the risks for species diversity and abundance to be evaluated and appropriate criteria for power and sample size follow-up testing to be set; (ii) the biodiversity associated with socially/aesthetically acceptable cropping to be estimated; and (iii) predictions for the optimum balance between biodiversity and the costs of cropping to be made for each crop and environment.

Here we present data for the butterflies in headlands of miscanthus and SRC willow crops compared with the arable crops of the FSEs. For butterfly family groups and total butterflies, year totals per kilometre of transect walked were calculated for each FSE site (0.3 km per visit, and up to seven sampling visits) for miscanthus or SRC willow fields (each 0.4 km per visit, and from two to five visits). Following $\log_{10}$ transformation, after adding an offset of one to allow for zeros, the mean and variance of the logged counts was computed over all sites for each crop. For each group the difference between the miscanthus or SRC willow and FSE mean logged counts was then computed along with a 95 per cent confidence interval. The mean and confidence limits were then back-transformed to the ratio scale.

The abundance of total butterflies was significantly greater in field margins surrounding both miscanthus and SRC willow than in the headlands of arable crops. There were 60 per cent and 132 per cent more butterflies in headlands of miscanthus and SRC willow, respectively, than in arable headlands. The abundance of families of butterflies varied between headlands of biomass crops and the arable crops, where the abundance of the Pieridae in miscanthus and SRC willow field margins was significantly lower than in arable field margins at 56 per cent and 64 per cent, respectively. Except for Lycaenidae in miscanthus headlands, all other families of butterfly were significantly more abundant in headlands of biomass crops than in those surrounding arable crops. The Satyrinae showed the largest differences of 370 per cent and 620 per cent in miscanthus and SRC willow respectively, (Figures 3.5 and 3.6) (Haughton et al, 2009).

While we already know that commercial SRC can be good for certain groups of farmland and scrub type birds on UK farmland (Sage et al, 2006), relatively little work has been undertaken in miscanthus. Semere and Slater (2007) and Bellamy et al (2009) did find a variety of farmland birds using miscanthus

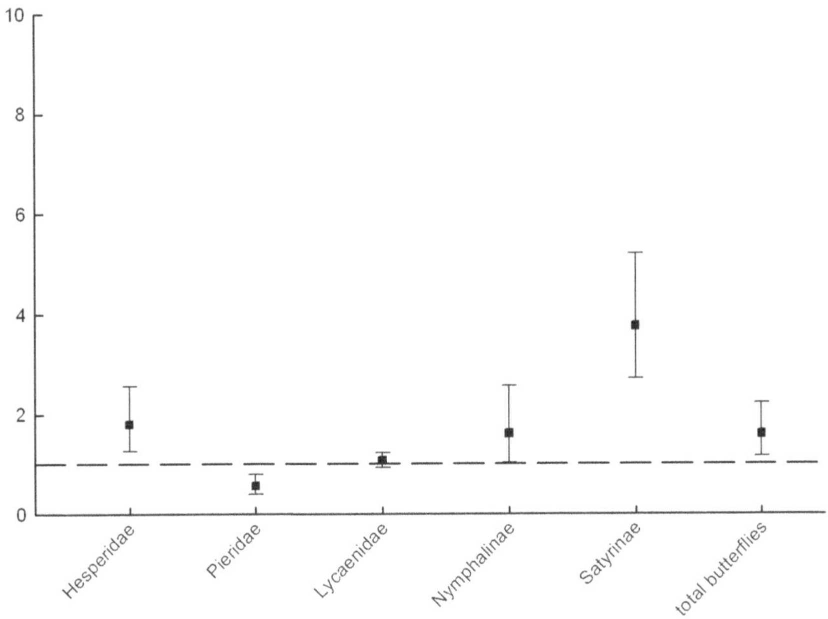

Note: Dashed line is line of unity. Error bars around mean are 95 per cent confidence limits, back-transformed to the ratio scale (hence asymmetry)

Source: Adapted from Haughton et al (2009)

Figure 3.5 *Mean ratio (R) of families of butterfly in field margins around miscanthus crops to arable crops*

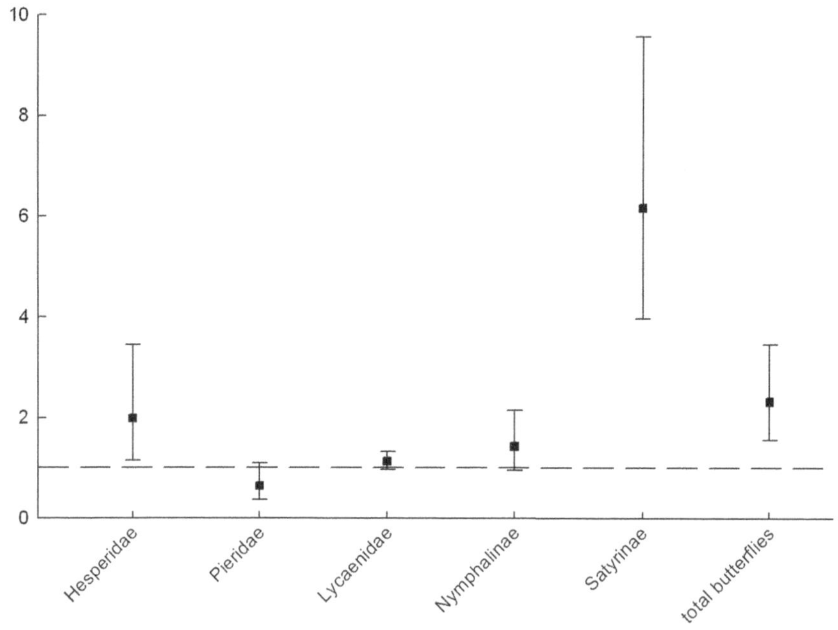

Note: Dashed line is line of unity. Error bars around mean are 95 per cent confidence limits, back-transformed to the ratio scale (hence asymmetry)

Source: Adapted from Haughton et al (2009)

Figure 3.6 *Mean ratio (R) of families of butterfly in field margins around SRC willow crops to arable crops*

plots. However, the fields available to these studies tended to be agronomically poor (excessively patchy cropping and weedy, as documented by the authors) and hence commercially unrepresentative, which may have heavily influenced bird use. In the Defra-funded complementary project on biodiversity, some simple bird-monitoring work was therefore included at a subsample of 16 recent commercial miscanthus sites in winter and summer to address this imbalance. Comparisons were made with adjacent fields of grass or cereals and, in summer, only with a handful of SRC fields that had been cut the previous winter (because meaningful comparisons with past data were not possible due to substantial methodological differences).

Here we present summary data for birds in summer. In May and June miscanthus crop height was low enough to allow birds to be surveyed by systematically walking transects through the crop (as in the controls). In July, the height and density of the miscanthus (more than two metres) meant a novel technique was required, involving two surveyors, one watching from a high vantage point (usually a deer seat) while the other walked through the crop flushing birds. We found that encounters with territorial males in miscanthus in May and June were very low compared with the cut SRC and, in June, slightly lower even than the (much larger) cereal control fields (Figure 3.7). In July, encounters with post-breeding individuals, primarily blackbirds and reed buntings, in miscanthus increased overall densities but again were much lower than in the recently cut SRC.

Hydrology

As miscanthus and SRC willow are deeper rooting than other arable crops and have a longer growing period, land conversion to energy cropping may have implications for water resources. Howes et al (2002) concluded that the effects of energy crop production on water quality were likely to be beneficial due to the reduced requirement for nitrogen fertilizer and pesticide inputs. However, there is serious concern about amounts of water needed by energy crops and the possible implications for stream flow and groundwater recharge. With the introduction of the Water Framework Directive (WFD), this concern relates not only to the direct impacts on resource availability, but also to the implications of lower flows for the ecology of water courses. There have been two major studies into the potential hydrological impacts of energy crops in the UK, both carried out by the CEH. In the most recent, Finch et al (2004) concluded that there was no simple answer since the balance of impacts depended on the type of energy crop, the land cover replaced, soil characteristics and climatic

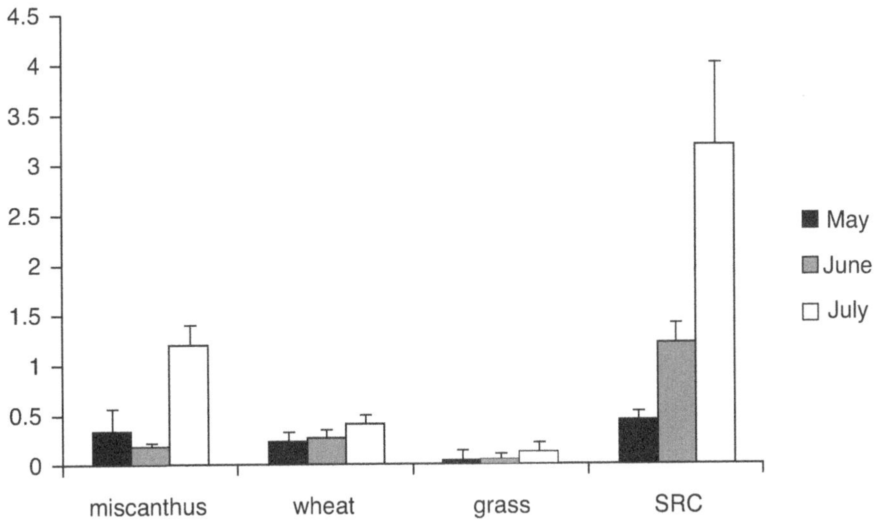

Note: While SRC is not harvested every year, data on SRC here are from recently cut plots only

Figure 3.7 *Mean number of birds per hectare (excluding corvids) recorded during May, June and July in 16 miscanthus fields compared with control plots*

variables. Moreover, these studies were based on limited numbers of measurements, particularly for miscanthus, so there were significant uncertainties in the predictions that could be generated for different rainfall scenarios.

In the Relu-Biomass project, the significant body of previous work on the water use of SRC willow was used to provide parameter values and data for the calibration of a numerical model of the land surface water and energy balance, based on Finch (2001). In addition, two full years of measurements of the land-surface fluxes of water and energy were carried out to test the outputs of the model. The collected dataset was used to quantify the uncertainties in the modelled output. Previous work showed that the differences in water use of the different willow varieties studied were less than the uncertainties in the measurements and model predictions, so predictions for a 'typical' willow SRC type can be made.

Although there is only one variety of miscanthus currently grown in the UK, there is comparatively little information on its water use. Recently published studies suggest that the water efficiency is atypical of plants that use the C_4 photosynthetic pathway. In particular, it appears that, in southern England, it is able to maintain high quantum yield assimilations of CO_2 in the early growing season despite low temperatures, that is, less than 12°C. This has major implications for predictions of both the growth and the annual water use which are likely to be higher than estimated by Finch et al (2004). In the Relu-Biomass project, measurements have been made at two sites in order to provide parameter values for and to test the numerical model.

A range of automatically logged and manually operated instruments were used at sites of both crops to measure: (i) net radiation, soil heat, latent heat, sensible heat fluxes (the latter two using eddy covariance); (ii) soil water content using Profile Probes and a neutron probe; (iii) soil water potentials using pressure transducers; and (iv) forcing variables using an automatic weather station (AWS). Additional measurements were made at the miscanthus site: (i) stomatal conductance using an infra red gas analyser (IRGA); (ii) leaf area index (LAI) using a Sunfleck Ceptometer, and canopy height; and (iii) throughfall using a system of troughs and tipping bucket raingauges.

Although the analysis is not fully complete, the results so far appear to confirm the hypothesis that the water use of miscanthus is higher than that given by Finch et al (2004), because of a longer growing period. As a consequence of the deeper rooting depth of miscanthus, compared with the majority of other agricultural crops, water use (and by implication yield) are only likely to be limited by the availability of soil water during exceptionally dry summers, such as occurred in 2003. About 55 per cent of the total annual water use is made up by transpiration of the crop, the remainder being due to evaporation from the soil and interception (rainfall caught on the leaves and evaporating directly back into the atmosphere). Although leaf fall is mainly complete by the end of the

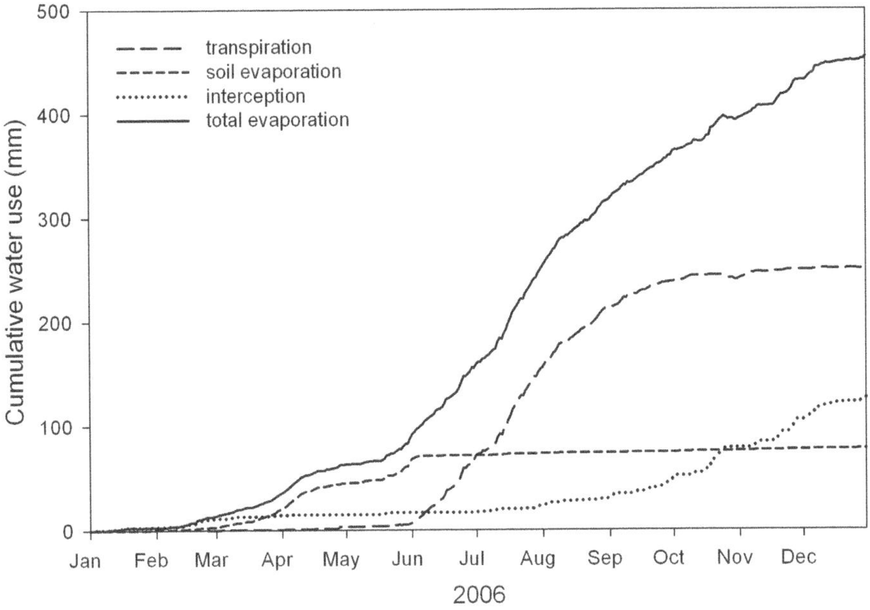

Figure 3.8 *Simulated cumulative total water use of miscanthus and its components: crop transpiration, soil evaporation and crop interception loss*

year, there is a significant water loss due to interception in the autumn (Figure 3.8). As a result, around two-thirds of the water use occurs in the second half of the year, very different to the situation with cereals such as winter wheat.

Rural economics

The uptake of energy crops by UK farmers will depend on a range of economic factors. While there is already considerable experience of the issues involved in policy-driven attempts to substantially modify traditional and conventional farming systems, for example through the introduction of various agri-environment schemes, the adoption of energy crops such as SRC and miscanthus will require major changes in farm planning horizons (Turner et al, 1999). Coupled with this, the requirement for incentive payments to facilitate uptake by farmers needs to be balanced with wider issues, including a clear identification of the rationale for Government intervention and an assessment in terms of value-for-money criteria. Once an energy-crop supply chain infrastructure is established, returns equivalent to some arable crops could be achieved by UK farmers. However, the initial investment required to establish the crop and secure a market can be prohibitive, introducing risk. In addition, the low-input nature of the crops may have implications for the wider rural economy as large-scale changes of land use to energy crops could reduce demand for seeds, fertilizer, pesticides, machinery and labour, so negatively affecting the agricultural supply trade. There has been no previous attempt to quantify this possible effect.

In the Relu-Biomass project, the basic farm-level economics of energy crop production was studied by focusing on the following aspects. First, an examination was undertaken of the establishment of definitive baseline costs of energy crop production, expressed on an annual equivalent value basis, inclusive of short-run and long-run cost profiles. Second, a review of the impacts on farm business organization and management was carried out, including its implications for cash flows (both short and long term), capital investment cycles (e.g. requirements for specialized machinery), and farm structures and systems (including the potential use of agricultural contractors). Third, an investigation was undertaken to identify the effects on farm resource utilization, including seasonal/cyclical implications in the demand for labour. The medium-term implications of the recent 'decoupling' reform of farm support under the Common Agricultural Policy on farmers' attitudes to long-term crops such as miscanthus and SRC were then examined. Finally an attempt was made to gain insight into the likely dynamics of energy crop uptake through estimation of the energy supply curve, under different farm system scenarios, at different prices and under competing alternative cropping assumptions.

The economic appraisal was based on modelling of the impact on, and scope for integration with, a number of alternative farming systems, assuming the replacing of both arable crops and grassland. The research drew on published economic data and involved focused surveys of farmers and farm managers. In addition, aspects of farmers' decision-making were addressed through identifying those issues of significance in investing for long-term production patterns.

While the full analyses of the study results have not yet been completed, it is clear that plantings of these biomass crops has taken place within a very wide range of farming systems and farmer decision-making has been driven by diverse factors. Similarly, the survey has uncovered a considerable range in terms of planted areas per holding, both in absolute terms (as land area) and as a proportion of the farmed area. So one tentative conclusion is that, on the evidence to date, the adoption of biomass as an agricultural enterprise is such that the traditional English patchwork of cropping has so far not been challenged seriously, simply modified by the addition of a novel crop. For largely historical reasons, the majority of miscanthus has been grown in the South West and, more recently, the Midlands, while most SRC has been in the North East, but this pattern is changing quite rapidly. In terms of production economics it is evident that, although the commercial attractiveness of biomass crops looked highly questionable in the context of 2007 cereal prices, the situation in the autumn of 2008 looks very different.

The contrasting experiences provided by the volatile returns from cereal growing over the past two seasons, not the only alternative to biomass by any means, but nevertheless a benchmark comparison for many, underline both the weakness and strength of biomass and point to its likely future role on many farms. Its low returns mean that it is unlikely to be the dominant enterprise on most farms, except in special circumstances such as where the farmer is looking for a reduced commitment of time and effort; but the predictability of those returns may look very attractive as part of a risk-management strategy in an era of greater market volatility. In terms of the wider rural economy, the conclusions are already clear: biomass production provides less employment and requires a smaller supply infra-structure than conventional food crops. The implications of this depend largely on the scale of plantings in the years to come.

Integrative approaches in the Relu-Biomass project

GIS-based constraints mapping

GIS techniques have been used to perform a variety of data assembly and integrative roles during the research in the Relu-Biomass project. A constraint mapping exercise (see Lovett et al, 2009) has been carried out based on a review of previous policy documents and studies (Capener et al, 2004; Defra, 2006; Centre for Sustainable Energy, 2007). A series of data layers were compiled at 100 m resolution for the whole of England with the following 11 factors incorporated into overlays:

- Unsuitable soil types
- Biodiversity Action Plan priority and semi-natural habitats
- Existing woodland
- Slope steepness
- Urban areas
- Major rivers
- Lakes
- Designated areas (e.g. Sites of Special Scientific Interest, SSSIs)
- Cultural heritage
- Landscape sensitivity
- Improved grassland.

A total area for England was calculated in the GIS as 13,039,461 ha. The first nine factors listed above were regarded as absolute constraints and, excluding the areas covered by any of these, reduced the available land to 7,770,628 ha (59.6 per cent of England). Two additional factors; environmentally sensitive landscapes and improved grassland, are not absolute constraints to the planting of perennial energy crops such as miscanthus but represent types of areas where there would be reasons for avoiding extensive planting. Including these constraints as well reduced the available land by more than 3 million ha to 4,719,756 ha (36.2 per cent of England). When this was then further restricted to Grade 3 or 4 land (intermediate to poor categories where ECS data show that energy crops are currently being planted), the resulting distribution of potentially suitable land was as shown in Figure 3.9 and totalled 3,120,173 ha (23.9 per cent of England).

This analysis was combined with a yield mapping exercise through collaboration with another RCUK-funded project, TSEC-Biosys (www.tsec-biosys.ac.uk). An empirical yield model (Richter et al, 2008) was combined with the GIS data to produce yield estimates for the land highlighted in Figure 3.9 and the results were compared with agricultural census data on the distributions of currently grown food crops. This analysis indicated that areas with the highest biomass yields do co-locate with important food-producing areas. Nevertheless, investigation of a scenario involving energy crop planting on 350,000 ha (a target in the UK Biomass Strategy for 2020) (DTI et al, 2007) suggested that this could be achieved without requiring higher-grade land and so would not necessarily greatly impact on UK food security. GIS-based yield and constraint mapping can thus help identify important issues in bioenergy generation potentials and land use trade-offs at regional or finer spatial scales that would be missed in analyses at the national level (Lovett et al, 2009).

GIS analyses were also used to help identify appropriate locations for the public questionnaire surveys and focus groups as described above. In addition, detailed GIS maps for smaller areas (based on Ordnance Survey MasterMap) provided the framework for the landscape visualizations. All the suitability maps will be updated as additional data becomes available from other disciplines in the project and from integration of the research on hydrology, biodiversity and economics that is currently being investigated.

SA framework

SA is an objectives-driven approach that relies on the derivation of aspirational sustainability objectives, against which different plan performances can be compared. In the SA approach measurable empirical objectives are identified by engaging stakeholders with relevant scientists. Targets and indicators of these empirical objectives are then used to assess the performance of alternative plans. Essentially, SA can be seen as an analytic-deliberative process which fuses quantitative, expert-derived data with stakeholder concerns and values (see Petts, 2003; Wiklund, 2005; Chilvers, 2007). SA has been criticized as being a process which facilitates trade-offs whereby environmental impacts can be

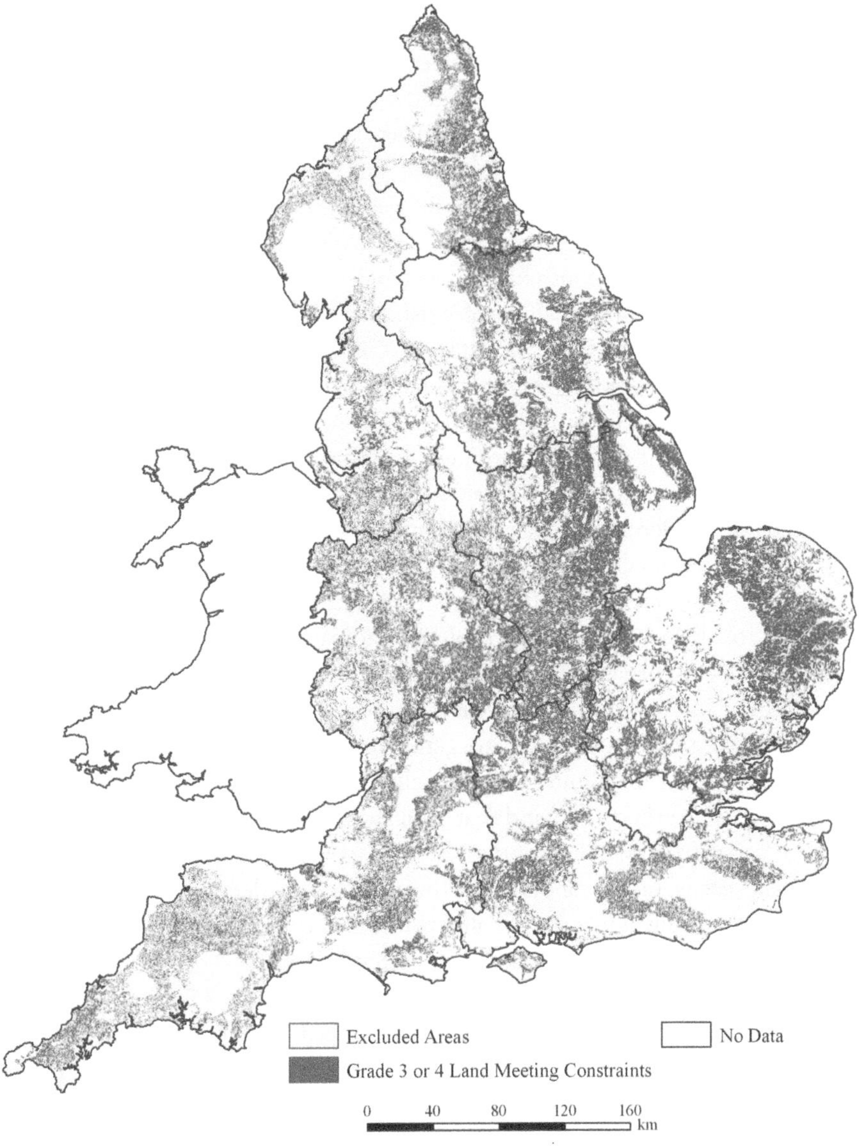

Figure 3.9 *Land outside the 11 planting constraints and classed as Grades 3 or 4*

offset by economic benefits (Morrison-Saunders and Fischer, 2006; Gibson, 2006) (this suggests, for example, that planting may be seen as appropriate in environmentally designated areas or Areas of Outstanding Natural Beauty). In the Relu-Biomass project, the potential for such trade-offs has been avoided through the combination of SA with constraints mapping, whereby the SA is applied only to the land area not screened out under any of the other constraints.

At the onset of the Relu-Biomass project, a scoping exercise involving a literature review and consultation was used to identify the impacts that needed to be assessed. This involved discussion and collaboration between all the partners from the different disciplines in the project. Initial objectives, targets and indicators were identified for each discipline and existing work reviewed to see if suitable measures existed or if new ones need to be derived.

It was clear that a separate SA framework was required for each study region to take into account existing regional plans and programmes, tensions with other policies and analysis of existing baseline sustainability conditions (Office of the Deputy Prime Minister, 2005). Stakeholder lists were drawn up for each region, including: farmers; energy companies; central, regional and local government; NGOs (e.g. National Farmers' Union); Natural England; Environment Agency; English Heritage; Forestry Commission; SRC willow and miscanthus businesses; and academics. One-day stakeholder meetings were held to brainstorm (Donnelly et al, 2006) and legitimize (see Owens et al, 2004) appropriate sustainability objectives which, through an iterative process of development by scientists and reviews by stakeholders, were then accepted as the basis for the SA framework. Table 3.1 indicates the objectives selected and demonstrates the broad scope of stakeholder concerns which need to be taken into account in decision-making related to planting decisions.

A similar process was used to generate indicators which were evaluated against criteria for examining the suitability of indicators for SA (see Donnelly et al, 2007). The criteria demand that indicators should be policy relevant, cover a range of environmental receptors, be relevant to the plan in question, show trends; be easily understandable to the public and decision-makers, be well founded technically and scientifically, prioritize key issues, provide early warning and be adaptable to reflect differing circumstances.

In developing the SA framework it became clear that some of the existing indicators are not suitable and that better indicators could be derived from the results of the research activities in the project. In particular, a specific challenge in relation to biomass crops is to identify more appropriate measures to examine the biodiversity implications of changing a proportion of land from conventional arable crops to biomass crops. The SA framework stakeholder discussions revealed that, while biodiversity was of concern, current indicators of biodiversity are not appropriate. The initial results of Relu-Biomass biodiversity studies have shown that butterflies are sensitive to the change to biomass cropping (Haughton et al, 2009). Other research and researchers have also shown that butterflies respond quickly and consistently to various changes in environmental factors. We have proposed, therefore, that butterflies might prove more appropriate and reliable than other more commonly used indicators of environmental change, such as birds (Haughton et al, 2009). As more of the results from the biodiversity studies in the project become known, as well as the

Table 3.1 *Stakeholder-derived sustainability objectives for the East Midlands and South West regions*

East Midlands objectives	South West objectives
Minimize transport movements	Minimize additional vehicle movements
Enhance rural quality of life	Enhance rural quality of life
Increase water availability	Maintain water availability
Improve public enjoyment of the countryside	Improve public connection with the countryside
Safeguard the historic environment	Safeguard the historic environment
Reduce energy costs to the consumer	Reduce energy costs
Increase amount of energy produced locally	Increase amount of energy produced and used locally
Increase the viability of local economies	Increase the viability of local economies
Enhance tourism potential	Maintain tourism resource
Enhance viability of farming	Enhance viability of farming
Maximize waste management opportunities	Maximize waste management opportunities
Enhance employment	Enhance rural employment
Enhance local landscape character	Enhance local landscape character
Improve water quality	Improve water quality
Protect soil resources	Protect and improve soil resources
Improve air quality	Improve air quality
Protect and enhance biodiversity	Protect and enhance biodiversity
Reduce greenhouse gas emissions	Reduce greenhouse gas emissions
Maintain food security	

hydrology and economic research, it is anticipated that further indicators will be identified that are more appropriate for biomass crops compared with the existing ones. Ultimately, use of better indicators will enhance the analytic component of SA, which could have far-reaching implications for the level of understanding of the social, economic and environmental consequences of future decision-making in many sectors.

In the latter phases of the Relu-Biomass project, the SA framework is being used to assess the impact of different planting scenarios. These include differences in: total land cover taken up by biomass crops; biomass end use; crop management in terms of field distribution/pattern and crop management in terms of headland size.

For each scenario the wider stakeholder community will be consulted and the scenarios will be compared and modified through an iterative process to identify the most appropriate one for each type and region relative to sustainability objectives. Results will be written in the SA report, which will also examine the practicalities of undertaking SAs for energy crops and the key quality assurance issues that need to be addressed. Our findings will be discussed with stakeholders to produce authoritative, science-based recommendations for planners and other professionals involved in conducting environmental assessments of energy crop planting schemes.

Conclusions

The potential for energy crops to contribute to the future energy supplies through providing a source of renewable carbon for heat and power and liquid transport fuels has been recognized by governments throughout the world. However, policy changes, rising concerns over the economic viability of energy cropping and increasing awareness of possible impacts on food security, water and other environmental issues, have checked the growth of the bioenergy and biofuel industries. Scientifically based approaches which enable policymakers and decision-makers to examine the possible impacts of different scenarios of energy crop plantings are urgently needed. For this to be satisfied, a robust evidence base and holistic approaches which enable integration across the different social, economic and environmental disciplines are also required. The UK interdisciplinary Relu-Biomass project has focused on filling in knowledge gaps though targeted research on the impacts of SRC willow and miscanthus on visual appearance, social acceptability, biodiversity, hydrology and rural economics and has developed two approaches to provide integration and decision-making tools: GIS-based constraints mapping and SA. Continual engagement with stakeholders has been an essential component of the project, and in future the expectation is that the tools will be utilized in planning expansion of energy crop plantings in the UK in ways which maximize positive impacts and minimize negative ones.

Acknowledgements

The Relu-Biomass project (www.Relu-Biomass.org.uk) was funded under the Rural Economy and Land Use (Relu) programme of the Economic and Social Research Council (ESRC), Biotechnology and Biological Sciences Research Council (BBSRC) and Natural Environment Research Council (NERC). Rothamsted Research is an Institute of the BBSRC of the UK. We thank Bical, ADAS, Coppice Resources Ltd (CRL), Strawson's Energy Group, Renewable Energy Growers and TV Energy for assisting us in finding study sites, and the miscanthus and SRC willow growers for allowing access to their farms. We thank the project Advisory Committee (see project web site) and all stakeholders who attended the meetings for their contribution to the development of the

SA framework. Natural England provided details of approved ECS agreements. The FSEs were funded by the Department for Environment, Food and Rural Affairs (Defra) and the Scottish Executive.

References

Adler, P.R., Del Grosso, S.J. and Parton, W.J. (2007) 'Life-cycle assessment of net greenhouse-gas flux for bioenergy cropping systems', *Ecological Applications*, vol 17, no 3, pp675–691

Bell, S. and McIntosh, E. (2001) 'Short rotation coppice in the landscape', Guideline Note, Forestry Commission, Wetherby, West Yorkshire

Bellamy, P.E., Croxton, P.J., Heard, M.S., Hinsley, S.A., Hulmes, L., Hulmes, S., Nuttall, P., Pywell, R.F. and Rothery, P. (2009) 'The impact of growing miscanthus for biomass on farmland bird populations', *Biomass and Bioenergy*, vol 33, no 2, pp191–199

Blottnitz, H. von, and Curran, M.A. (2007) 'A review of assessments conducted on bioethanol as a transportation fuel from a net energy, greenhouse gas, and environmental life cycle perspective', *Journal of Cleaner Production*, vol 15, no 7, pp607–619

Bohan, D.A., Boffey, C.W.H., Brooks, D.R., Clark, S.J., Dewar, A.M., Firbank, L.G., Haughton, A.J., Hawes, C., Heard, M.S., May, M.J., Osborne, J.L., Perry, J.N., Rothery, P., Roy, D.B., Scott, R.J., Squire, G.R., Woiwod I.P. and Champion, G.T. (2005) 'Effects on weed and invertebrate abundance and diversity of herbicide management in genetically modified herbicide-tolerant winter-sown oilseed rape', *Proceedings of the Royal Society B – Biological Sciences*, vol 272 (1562), pp463–474

Capener, P., Scholes, H., Ward, S. and Evans, N. (2004) 'Revision 2010 – Establishing county/sub regional targets for renewable electricity development to 2010', in Capener P. and Consultants, B. (eds) *Report to the Government Office for the South West and the South West Regional Assembly*, Government Office for the South West and the South West Regional Assembly, Bristol, Avon, UK

Centre for Sustainable Energy (2007) 'Devon *Miscanthus* and Woodfuels Opportunities Statement', in Centre for Sustainable Energy B, *Report to Devon Wildlife Trust*, Centre for Sustainable Energy, Bristol, UK

Chilvers, J. (2007) Towards analytic-deliberative forms of risk governance in the UK? Reflecting on learning in radioactive waste. *Journal of Risk Research*, vol 10, pp197–222

Coates, A. and Say, A. (1999) 'Ecological assessment of short rotation coppice report and appendices', ETSU B/W5/00216/REP/1, Didcot, Oxfordshire, UK

Cunningham, M.D., Bishop, J.D., McKay, H.V. and Sage, R.B. (2004) *ARBRE Monitoring – Ecology of Short Rotation Coppice*. Contract no. B/U1/00627/REP. DTI, publication no URN 04/961. Department of Trade and Industry, London

Department for Environment, Food and Rural Affairs (2001) 'Planting and growing *Miscanthus* – Best practice guidelines', Defra, London

Department for Environment, Food and Rural Affairs (2002) 'Growing SRC – Best practice guidelines', Defra, London

Department for Environment, Food and Rural Affairs (2006) 'Opportunities and optimum sitings for energy crops', www.defra.gov.uk/farm/crops/industrial/energy/opportunities/index.htm accessed July 2009.

Department for Environment, Food and Rural Affairs and Department of Trade and Industry (2004) 'A strategy for non-food crops and uses – Creating value from renewable materials', Defra, London

Department of Trade and Industry (1999) 'SRC for energy production – good practice guidelines', DTI, London

Department of Trade and Industry (2007) 'Meeting the energy challenge: A White Paper on energy', DTI, The Stationery Office, Norwich

Department of Trade and Industry and The Carbon Trust (2004) 'Renewables innovation review', DTI, London

Department of Trade and Industry, Department for Transport and Department for Environment Food and Rural Affairs, Department for Trade and Industry and Department for Transport (2007), *UK Biomass Strategy* (Department for Environment, Food and Rural Affairs, London)

Donnelly, A., Jennings, E., Mooney, P., Finnan, J., Lynn, D., Jones, M., O'Mahony, T., Thérivel, R. and Byrne, G. (2006) 'Workshop approach to developing objectives, targets and indicators for use in SEA', *Journal of Environmental Assessment Policy and Management*, vol 8, no 2, pp135–156

Donnelly, A., Jones, M., O'Mahony, T. and Byrne, G. (2007) 'Selecting environmental indicators for use in strategic environmental assessment', *Environmental Impact Assessment Review*, vol 27, no 2, pp161–175

Finch, J.W. (2001) Estimating change in direct groundwater recharge using a spatially distributed soil water balance model. *Quarterly Journal of Engineering Geology and Hydrogeology*, vol 34, pp71–83

Finch, J.W., Hall, R.L., Rosier, P.T.W., Clark, D.B., Stratford, C., Davies, H.N., Marsh, T.J., Roberts, J.M., Riche, A.B. and Christian, D.G. (2004) 'The hydrological impacts of energy crop production in the UK', *Final Report to ETSU* B/CR/000783/00/00, CEH, Wallingford

Firbank, L.G. (2008) 'Assessing the ecological impacts of bioenergy projects', *BioEnergy Research*, vol 1, no 1, pp12–19

Firbank, L.G., Heard, M.S., Woiwod, I.P., Hawes, C., Haughton, A.J., Champion, G.T., Scott, R.J., Hill, M.O., Dewar, A.M., Squire, G.R., May, M.J., Brooks, D.R., Bohan, D.A., Daniels, R.E., Osborne, J.L., Roy, D.B., Black, H.I.J., Rothery, P. and Perry, J.N. (2003) 'An introduction to the Farm-Scale Evaluations of genetically modified herbicide-tolerant crops', *Journal of Applied Ecology*, vol 40, pp2–16

Food and Agriculture Organization (2008) 'Biofuels: prospects, risks and opportunities', Rome, 2008, ISBN 978-92-5-105980-7, ISSN 0081-4539, TC/P/I0100/E FAO109530, http://www.earthprint.com/productfocus.php?id=FAO109530 accessed July 2009

Gallagher, E. (2008) 'Review of the indirect effects of biofuels', Renewable Fuels Agency, www.dft.gov.uk/rfa/reportsandpublications/reviewoftheindirecteffectsofbiofuels/executivesummary.cfm accessed July 2009

Gibson, R. B. (2006) 'Beyond the pillars: Sustainability assessment as a framework for effective integration of social, economic and ecological considerations in significant decision-making', *Journal of Environmental Assessment Policy and Management*, vol 8, no 3, pp259–280

Haines-Young, R.H., Barr, C.J., Black, H.I.J., Briggs, D.J., Bunce, R.G.H., Clarke, R.T., Cooper, A., Dawson, F.H., Firbank, L.G., Fuller, R.M., Furse, M.T., Gillespie, M.K., Hill, R., Hornung, M., Howard, D.C., McCann, T., Morecroft, M.D., Petit, S., Sier,

A.R.J., Smart, S.M., Smith, G.M., Stott, A.P., Stuart R.C. and Watkins J.W. (2000) 'Accounting for nature: Assessing habitats in the UK countryside', www.countrysidesurvey.org.uk/archiveCS2000/report_pdf.htm accessed 15 October 2008

Haughton, A.J., Champion, G.T., Hawes, C., Heard, M.S., Brooks, D.R., Bohan, D.A., Clark, S.J., Dewar, A.M., Firbank, L.G., Osborne, J.L., Perry, J.N., Rothery, P., Roy, D.B., Scott, R.J., Woiwod, I.P., Birchall, C., Skellern, M.P., Walker, J.H., Baker, P., Browne, E.L., Dewar, A.J.G., Garner, B.H., Haylock, L.A., Horne, S.L., Mason, N.S., Sands, R.J.N. and Walker, M.J. (2003) 'Invertebrate responses to the management of genetically modified herbicide-tolerant and conventional spring crops. II. Within-field epigeal and aerial arthropods', *Philosophical Transactions of the Royal Society B*, vol 358, pp1863–1878

Haughton, A.J., Bond, A.J., Lovett, A.A., Dockerty, T., Sünnenberg, G., Clark, S.J., Bohan, D.A., Sage, R.B., Mallott, M.D., Mallott, V.E., Cunningham, M.D., Riche, A.B., Shield, I.F., Finch, J.W., Turner, M.M. and Karp, A. (2009) 'A novel, integrated approach to assessing social, economic and environmental implications of changing rural land-use: a case study of perennial biomass crops', *Journal of Applied Ecology*, vol 46, pp323–333

Heard, M.S., Rothery, P., Perry, J.N. and Firbank, L.G. (2005) 'Predicting longer-term changes in weed populations under GMHT management', *Weed Research*, vol 45, pp331–338

Howes, P., Barker, N., Higham, I., O'Brien, S., Talvitie, M., Bates, J., Adams, M., Jones, H. and Dumbleton, F. (2002) 'Review of power production from renewable and related sources', R&D Technical Report P4–097/TR, Environment Agency, Bristol

Karp, A. and Shield, I. (2008) Bioenergy from plants and the sustainable yield challenge. *New Phytologist*, vol 179, no 1, pp15–32

Lovett, A.A., Sünnenberg, G.M., Richter, G.M., Dailey, A.G., Riche, A.B. and Karp, A. (2009), 'Land Use Implications of Increased Biomass Production Identified by GIS-Based Sustainability and Yield Mapping for Miscanthus in England', *Bioenergy Research*, vol 2(1), pp17–28

Morrison-Saunders, A. and Fischer, T.B. (2006) 'What is wrong with EIA and SEA anyway? A sceptic's perspective on sustainability assessment', *Journal of Environmental Assessment Policy and Management*, vol 8, no 1, pp19–39

Office of the Deputy Prime Minister (2005) 'Sustainability appraisal of regional spatial strategies and local development documents', http://www.communities.gov.uk/documents/planningandbuilding/pdf/142520.pdf, accessed 14 October 2008

Owens, S., Rayner, T. and Bina, O. (2004) 'New agendas for appraisal: reflections on theory, practice, and research', *Environment and Planning A*, vol 36, pp1943–1959

Petts, J. (2003) 'Barriers to deliberative participation in EIA: Learning from waste policies, plans and projects', *Journal of Environmental Assessment Policy and Management*, vol 5, pp269–293

Rafaschieri, A., Rapaccini, M. and Manfrida, G. (1999) 'Life cycle assessment of electricity production from poplar energy crops compared with conventional fossil fuels', *Energy Conversion and Management*, vol 40, no 14, pp1477–1493

Royal Commission on Environmental Pollution (2004) 'Biomass as a renewable energy source', RCEP, Department for Environment, Food and Rural Affairs, London

Richter, G.M., Riche, A.B., Dailey, A.G., Gezan, S.A. and Powlson, D.S. (2008) 'Is UK biofuel supply from *Miscanthus* water-limited?' *Soil Use and Management*, vol 24, pp235–245

Rowe, R., Street, N. and Taylor, G. (2008) 'Identifying potential environmental impacts of large-scale deployment of bioenergy crops in the UK', *Renewable & Sustainable Energy Reviews*, vol 13, pp271–290

Sage, R.B. (1998) 'Short rotation coppice for energy: towards ecological guidelines', *Biomass and Bioenergy*, vol 15, pp39–47

Sage, R.B. (2001) 'The ecology of short-rotation coppice crops: wildlife and pest management', PhD thesis, University of Hertfordshire at Hatfield, UK

Sage, R.B. and Robertson, P.A. (1996) 'Factors affecting songbird communities using new short rotation coppice habitats in spring', *Bird Study*, vol 43, pp201–213

Sage, R.B. and Tucker, K. (1997) 'Invertebrates in the canopy of willow and poplar short rotation coppices', *Aspects of Applied Biology*, vol 49, pp105–111

Sage, R.B., Cunningham, M. and Boatman, N. (2006) 'Birds in short rotation coppice compared with other arable crops in central England and a review of bird census data from energy crops in the UK', *Ibis*, vol 148, pp184–197

Searchinger, T., Heimlich, R., Houghton, R.A., Dong, F.X., Elobeid, A., Fabiosa, J., Tokgoz, S., Hayes, D. and Yu, T.H. (2008) 'Use of US croplands for biofuels increases greenhouse gases through emissions from land-use change', *Science*, vol 319 (5886), pp1238–1240

Semere, T. and Slater, F.M. (2007) 'Ground flora, small mammal and bird species diversity in miscanthus (*Miscanthus x giganteus*) and reed canary-grass (*Phalaris arundinacea*) fields', *Biomass & Bioenergy* (31) pp20–29

The Royal Society (2008) *Sustainable Biofuels: Prospects and Challenges*, The Royal Society, London

Turner, M.M., Naish, R.W., Barr, D. and Fogerty, M.W. (1999) *The South West Forest: a Review of the Economic Factors Affecting Decision Making by Farmers.* Report commissioned by the Countryside Commission. University of Exeter

United Nations (1998) *Kyoto Protocol to the United Nations Framework Convention on Climate Change*, UN, Geneva

United Kingdom Parliament (2008a), 'Climate Change Act', available at <http://www.opsi.gov.uk/acts/acts2008/pdf/ukpga_20080027_en.pdf>, last accessed July 29th 2009.

United Kingdom Parliament (2008b), 'Energy Act', available at <http://www.opsi.gov.uk/acts/acts2008/pdf/ukpga_20080032_en.pdf>, last accessed July 29th 2009

Wiklund, H. (2005) 'In search of arenas for democratic deliberation: A Habermasian review of environmental assessment', *Impact Assessment and Project Appraisal*, vol 23, pp281–292

4

Soaking Up the Carbon

Pete Smith

Overview

Soils contain about three times the amount of carbon (C) as the atmosphere, and vegetation contains a slightly smaller amount than the atmosphere. Land management has a large impact on soil and vegetation carbon levels and converting forest land or grassland to croplands has caused significant loss of terrestrial carbon. Historically, soils have lost between 40 and 90 Pg (1 Pg = 1 Gt = 10^{15} g = thousand million tonnes) C globally through cultivation and disturbance, with total historical losses (including vegetation) of 180 Pg C. Biological carbon sequestration is the process whereby land use and land management are used to restore soil and vegetation carbon stocks.

There are a number of practices, under the broad activities of cropland management, grazing-land management, restoration of cultivated organic soils and restoration of degraded lands, that have been shown to promote soil carbon sequestration. The global potential for soil organic carbon (SOC) sequestration is estimated to be around 0.4, 0.6 and 0.7 Pg C $year^{-1}$ at carbon prices of USD 0–20, 0–50 and 0–100 t CO_2-$equivalent^{-1}$, respectively, similar to the economic potential for carbon sequestration in forestry. Carbon sequestration potential in agriculture is similar to the greenhouse gas (GHG) mitigation potential in the energy, transport, industry and forestry sectors; is higher than the potential in the waste sector; but is lower than the potential in the buildings sector.

When considering biological carbon sequestration, account must be taken of impacts on other GHGs and the limited duration and potential reversibility of the carbon sinks. Soil carbon sequestration is currently encouraged more by non-climate policy than by climate policy. Policies encouraging soil carbon sequestration are most often implemented for other reasons, including improvement of water and air quality, improvement of soil fertility and improvement of biodiversity.

There are a number of barriers that mean that the economic potential for carbon sequestration might not be reached. These barriers, divided into five categories – economic, risk-related, political/bureaucratic, logistical and educational/societal barriers – may prevent best management practices from being implemented. The most significant barriers to implementation of mitigation measures in developing countries (and for some economies in transition) are economic. These are mostly driven by poverty and in some areas these are exacerbated by a growing population. To begin to overcome these barriers, global sharing of innovative technologies for efficient use of land resources and agricultural chemicals, to eliminate poverty and malnutrition, will significantly help to remove barriers that currently prevent implementation of mitigation measures in agriculture. Capacity building and education in the use of innovative technologies and best management practices would also serve to reduce barriers.

More broadly, macroeconomic policies to reduce debt and to alleviate poverty in developing countries, through encouraging sustainable economic growth and sustainable development, would serve to lower or remove barriers. Mitigation measures that also improve food security and profitability (such as improved use of fertilizer) would be more favourable than those which have no economic or agronomic benefit. Such practices are often referred to as 'win-win' options, and strategies to implement such measures can be encouraged on a 'no regrets' basis; that is, they provide other benefits even if the mitigation potential is not realized.

The role of soils and vegetation in the global carbon cycle

Globally, soils contain about three times the amount of carbon in vegetation and twice the amount in the atmosphere (IPCC, 2000a); that is, about 1500 Pg of organic C (Batjes, 1996), with 560 Pg in vegetation (IPCC, 2000a). The annual fluxes of CO_2 from atmosphere to land (global net primary productivity, NPP) and land to atmosphere (respiration and fire) are each of the order of 60 Pg C $year^{-1}$ (IPCC, 2000a). During the 1990s, fossil fuel combustion and cement production emitted 6.3 ± 1.3 Pg C $year^{-1}$ to the atmosphere, while land-use change emitted 1.6 ± 0.8 Pg C $year^{-1}$ (Schimel et al, 2001; IPCC, 2001). Atmospheric carbon increased at a rate of 3.2 ± 0.1 Pg C $year^{-1}$; the oceans absorbed 2.3 ± 0.8 Pg C $year^{-1}$ with an estimated residual terrestrial sink of 2.3 ± 1.3 Pg C $year^{-1}$ (Figure 4.1) (Schimel et al, 2001; IPCC, 2001).

The size of the pools of SOC and vegetation carbon are therefore large compared with gross and net annual fluxes of carbon to and from the terrestrial biosphere (Smith, 2004a). Small changes in the SOC or vegetation pools could have dramatic impacts on the concentration of CO_2 in the atmosphere. The response of SOC and vegetation carbon to global warming is therefore of critical importance. One of the first examples of the potential impact of increased release of terrestrial carbon on further climate change was given by Cox et al (2000). Using a climate model with a coupled carbon cycle, Cox et al (2000)

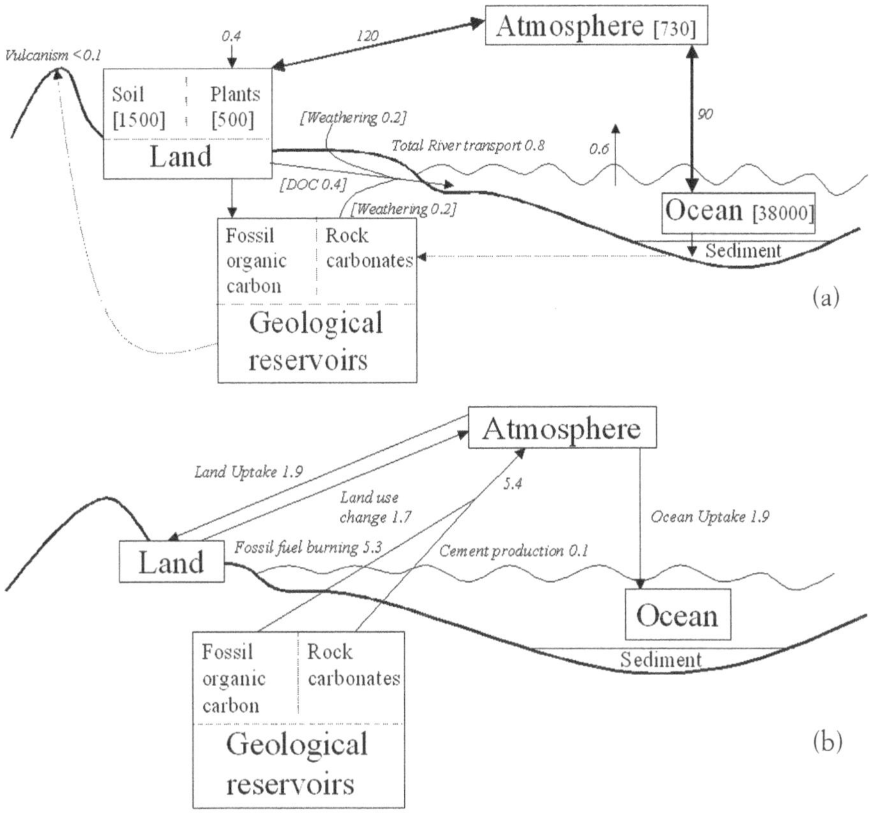

Note: DOC dissolved organic carbon

Source: Redrawn from IPCC, 2001; Smith, 2004a

Figure 4.1 *The global carbon cycle for the 1990s (Pg C): (a) the natural carbon cycle and (b) the human perturbation*

showed that release of terrestrial carbon under warming would lead to a positive feedback, whereby carbon release would result in increased global warming. Since then, a number of coupled climate carbon cycles (so-called C4 models) have been developed. However, there remains considerable uncertainty concerning the extent of the terrestrial feedback, with the difference between the models amounting to 200 ppm CO_2-C by 2100 (Friedlingstein et al, 2006). This difference is of the same order as the difference between fossil fuel carbon emissions under the IPCC Special Report on Emissions Scenarios (SRES) (IPCC, 2000b). It is clear that better quantifying the response of terrestrial carbon, a large proportion of which derives from the soil, is essential for understanding the nature and extent of the earth's response to global warming. Understanding interactions between climate and land-use change will also be critically important.

Historically, soils have lost between 40 and 90 Pg C globally through cultivation and disturbance (Houghton, 1999; Houghton et al, 1999; Schimel, 1995; Lal, 1999). The total historical system loss (including vegetation) is closer to 180 Pg C (Houghton et al, 1999; De Fries et al, 2000).

Land use impacts on terrestrial carbon stocks

The main method for increasing vegetation carbon stocks is through aforestation or through improving vegetation management to increase the standing stock of carbon (Nabuurs et al, 2007). The main method for increasing soil carbon stocks is through improved soil management (Smith et al, 2007b).

Land-use change significantly affects soil carbon stock (Guo and Gifford, 2002). Most long-term experiments on land-use change show significant changes in SOC (Smith et al, 1997, 2000, 2001a, 2002). This is likely to continue into the future; in a recent modelling study examining the potential impacts of climate and land-use change on SOC stocks in Europe, land-use change was found to have a larger net effect on SOC storage than projected climate change (Smith, J.U. et al, 2005).

In a meta-analysis of long-term experiments, Guo and Gifford (2002) have shown that converting forest land or grassland to croplands causes significant loss of SOC; whereas conversion of forestry to grassland does not result in SOC loss in all cases. Total ecosystem carbon (including above-ground biomass) does, however, decrease due to loss of the tree biomass carbon. Similar results have been reported in Brazil, where total ecosystem carbon losses are large, but where soil carbon does not decrease (Veldkamp, 1994; Moraes et al, 1995; Neill et al, 1997; Smith et al, 1999), though other studies have shown a loss of SOC upon conversion of forest to grassland (Allen, 1985; Mann, 1986; Detwiller and Hall, 1988). In the most favourable case, only about 10 per cent of the total ecosystem carbon lost after deforestation (due to tree removal, burning, etc.) can be recovered (Fearnside, 1997; Neill et al, 1997; Smith et al, 1999).

The largest per-area losses of SOC occur where the carbon stocks are largest, for example in highly organic soils such as peatlands, either through drainage, cultivation or liming. Organic soils hold enormous quantities of SOC, accounting for 329–525 Pg C or 15–35 per cent of the total terrestrial carbon (Maltby and Immirizi (1993), with about one-fifth (70 Pg) located in the tropics. Studies of cultivated peats in Europe show that they can lose significant amounts of SOC through oxidation and subsidence; between 0.8 and 8.3 t C ha^{-1} $year^{-1}$ (Nykänen et al, 1995; Lohila et al, 2004; Maljanen et al, 2001, 2004). The potential for SOC loss from land-use change on highly organic soils is therefore very large.

In short, SOC tends to be lost when converting grasslands, forest or other native ecosystems to croplands, or by draining, cultivating or liming highly organic soils. SOC tends to increase when restoring grasslands, forests or native vegetation on former croplands or by restoring organic soils to their native condition. Where the land is managed, best management practices that increase carbon inputs to the soil (e.g. improved residue and manure management) or reduce losses (e.g. reduced-impact tillage, reduced residue removal) help to maintain or increase SOC levels. Management practices to increase SOC storage are discussed in the next section.

The most effective mechanism for reducing SOC loss globally would be to halt land conversion to agriculture, but with the population growing and diets changing in developing countries (Smith et al, 2007b; Smith and Trines, 2007) more land is likely to be required for agriculture. To meet growing and changing food demands without encouraging land conversion to agriculture will require productivity on current agricultural land to be increased (Vlek et al, 2004). In addition to increasing agricultural productivity, there are a number of other management practices that can be used to prevent SOC loss.

Agricultural management to sequester SOC

Smith et al (2007a, 2008) have reviewed the GHG mitigation potential of agricultural management practices. About 90 per cent of total GHG mitigation potential in agriculture stems from soil carbon sequestration (Smith et al, 2007a,b; 2008). Smith et al (2007a, 2008) examined practices under the broad activities of cropland management, grazing-land management, restoration of cultivated organic soils and restoration of degraded lands. The effects of these management practices on SOC are summarized below.

Cropland management

Mitigation practices in cropland management include the following partly overlapping categories

Agronomy

Improved agronomic practices that increase yields and generate higher inputs of carbon residue can lead to increased soil carbon storage (Follett, 2001). Examples of such practices include: using improved crop varieties; extending crop rotations, notably those with perennial crops that allocate more carbon below ground; and avoiding or reducing use of bare (unplanted) fallow (West and Post, 2002; Smith, 2004b,c; Lal, 2003, 2004a; Freibauer et al, 2004). Adding more nutrients, when the soil is deficient, can also promote soil carbon gains (Alvarez, 2005), but the benefits from nitrogen fertilizer can be offset by higher N_2O emissions from soils and CO_2 from fertilizer manufacture (Schlesinger, 1999; Robertson, 2004; Gregorich et al, 2005). Another group of agronomic practices are those that provide temporary vegetative cover between successive agricultural crops or between rows of tree or vine crops. These 'catch' or 'cover' crops add carbon to soils (Barthès et al, 2004; Freibauer et al, 2004).

Tillage/residue management

Advances in weed control methods and farm machinery now allow many crops to be grown with minimal tillage (reduced tillage) or without tillage (no-tillage). These practices are now increasingly used throughout the world (Cerri et al, 2004). Since soil disturbance tends to stimulate soil carbon losses through enhanced decomposition and erosion, reduced- or no-tillage agriculture often results in soil carbon gain, but not always (West and Post, 2002; Ogle et al, 2005; Gregorich et al, 2005; Alvarez, 2005). Adopting reduced- or no-tillage may also affect N_2O, emissions but the net effects are inconsistent and not well quantified globally (Smith and Conen, 2004; Helgason et al, 2005; Li et al, 2005; Cassman et al, 2003). The effect of reduced tillage on N_2O emissions may depend on soil and climatic conditions. In some areas, reduced tillage promotes N_2O emissions, while elsewhere it may reduce emissions or have no measurable influence (Marland et al, 2001). Further, no-tillage systems can reduce CO_2 emissions from energy use (Marland et al, 2003; Koga et al, 2003). Systems that retain crop residues also tend to increase soil carbon, because these residues are the precursors for soil organic matter, the main carbon store in soil.

Water management

About 18 per cent of the world's croplands now receive supplementary water through irrigation (Millennium Ecosystem Assessment, 2005). Expanding this area (where water reserves allow) or using more effective irrigation measures can enhance carbon storage in soils through enhanced yields and residue returns (Follett, 2001; Lal, 2004a). But some of these gains may be offset by CO_2 from energy used to deliver the water (Schlesinger 1999; Mosier et al, 2005) or from N_2O emissions from higher moisture and fertilizer nitrogen inputs (Liebig et al, 2005). The latter effect has not been widely measured. Drainage of croplands in humid regions can promote productivity (and hence soil carbon) and perhaps also suppress N_2O emissions by improving aeration (Monteny et al, 2006).

Rice management

Increasing rice production can also enhance SOC stocks (Pan et al, 2006). Methane (CH_4) emissions can be reduced by adjusting the timing of organic residue additions (e.g., incorporating organic materials in the dry period rather than in flooded periods; Xu et al, 2003; Cai and Xu, 2004), by composting the residues before incorporation or by producing biogas for use as fuel for energy production (Wang and Shangguan, 1996; Wassmann et al, 2000).

Agro-forestry

Agro-forestry is the production of livestock or food crops on land that also grows trees for timber, firewood or other tree products. It includes shelter belts and riparian zones/buffer strips with woody species. The standing stock of carbon above ground is usually higher than the equivalent land use without trees, and planting trees may also increase soil carbon sequestration (Oelbermann et al, 2004; Guo and Gifford, 2002; Mutuo et al, 2005; Paul et al, 2003); but the effects on N_2O and CH_4 emissions are not well known (Albrecht and Kandji, 2003).

Land-use change

One of the most effective methods of reducing emissions is often to allow or encourage the reversion of cropland to another land use, typically one similar to the native vegetation. The conversion can occur over the entire land area ('set-aside') or in localized spots such as grassed waterways, field margins or shelterbelts (Follett, 2001; Freibauer et al, 2004; Lal, 2004b; Falloon et al, 2004; Ogle et al, 2003). Such land-use change often increases carbon storage. For example, converting arable cropland to grassland typically results in the accrual of soil carbon because of lower soil disturbance and reduced carbon removal in harvested products. Compared with cultivated lands, grasslands may also have reduced N_2O emissions from lower nitrogen inputs and higher rates of CH_4 oxidation, but recovery of oxidation may be slow (Paustian et al, 2004). Similarly, converting drained croplands back to wetlands can result in rapid accumulation of soil carbon (removal of atmospheric CO_2). This conversion may stimulate CH_4 emissions because waterlogging creates anaerobic conditions (Paustian et al, 2004). Planting trees can also reduce emissions. Because land-use conversion comes at the expense of lost agricultural productivity, it is usually an option only on surplus agricultural land or on croplands of marginal productivity.

Grazing-land management and pasture improvement

Grazing lands occupy much larger areas than croplands (FAOSTAT, 2008) and are usually managed less intensively. The following are examples of practices to reduce GHG emissions and to enhance removals.

Grazing intensity

The intensity and timing of grazing can influence the removal, growth, carbon allocation and flora of grasslands, thereby affecting the amount of carbon accrual in soils (Conant et al, 2001, 2005; Freibauer et al, 2004; Conant and Paustian, 2002; Reeder et al, 2004). Carbon accrual on optimally grazed lands is often greater than on ungrazed or overgrazed lands (Liebig et al, 2005; Rice and Owensby, 2000). The effects are inconsistent, however, owing to the many types of grazing practices employed and the diversity of plant species, soils and climates involved (Schuman et al, 2001; Derner et al, 2006).

Increased productivity (including fertilization)

As for croplands, carbon storage in grazing lands can be improved by a variety of measures that promote productivity. For instance, alleviating nutrient deficiencies by fertilizer or organic amendments increases plant litter returns and, hence, soil carbon storage (Schnabel et al, 2001; Conant et al, 2001). Adding nitrogen, however, often stimulates N_2O emissions (Conant et al, 2005), thereby offsetting some of the benefits. Irrigating grasslands, similarly, can promote soil carbon gains (Conant et al, 2001). The net effect of this practice, however, depends also on emissions from energy use and other activities on the irrigated land (Schlesinger, 1999).

Fire management

Burning can affect the proportion of woody versus grass cover, notably in savannahs, which occupy about an eighth of the global land surface. Reducing the frequency or intensity of fires typically leads to increased tree and shrub cover, resulting in a CO_2 sink in soil and biomass (Scholes and van der Merwe, 1996). This woody-plant encroachment mechanism saturates over 20–50 years, whereas avoided CH_4 and N_2O emissions continue as long as fires are suppressed.

Species introduction

Introducing grass species with higher productivity, or carbon allocation to deeper roots, has been shown to increase soil carbon. For example, establishing deep-rooted grasses in savannahs has been reported to yield very high rates of carbon accrual (Fisher et al, 1994), although the applicability of these results has not been widely confirmed (Conant et al, 2001; Davidson et al, 1995). In the Brazilian savannah (Cerrado Biome), integrated crop–livestock systems using *Brachiaria* grasses and zero tillage are being adopted (Machado and Freitas, 2004). Introducing legumes into grazing lands can promote soil carbon storage (Soussana et al, 2004), through enhanced productivity from the associated nitrogen inputs and perhaps also reduced emissions from fertilizer manufacture if biological N_2 fixation displaces applied nitrogen fertilizer (Sisti et al, 2004; Diekow et al, 2005). Ecological impacts of species introduction need to be considered.

Restoration of cultivated organic/peaty soils

Organic or peaty soils contain high densities of carbon accumulated over many centuries, because decomposition is suppressed by absence of oxygen under flooded conditions. To be used for agriculture, these soils are drained, which aerates the soil, favouring decomposition and therefore high CO_2 and N_2O fluxes. CH_4 emissions are usually suppressed after draining, but this effect is far outweighed by pronounced increases in N_2O and CO_2 (Kasimir-Klemedtsson et al, 1997). Emissions from drained organic soils can be reduced to some extent by practices such as avoiding row crops and tubers, avoiding deep ploughing and maintaining a shallower water table. But the most important mitigation practice is avoiding the drainage of these soils in the first place or re-establishing a high water table (Freibauer et al, 2004).

Restoration of degraded lands

A large proportion of agricultural lands has been degraded by excessive disturbance, erosion, organic-matter loss, salinization, acidification or other processes that reduce productivity (Batjes, 1999; Foley et al, 2005; Lal, 2003, 2004b). Often, carbon storage in these soils can be partly restored by practices that reclaim productivity, including: re-vegetation (e.g., planting grasses); improving fertility by nutrient amendments; applying organic substrates such as manures, biosolids and composts; reducing tillage and retaining crop residues; and conserving water (Lal, 2004b; Bruce et al, 1999; Olsson and Ardö, 2002; Paustian et al, 2004). Where these practices involve higher nitrogen amendments, the benefits of carbon sequestration may be partly offset by higher N_2O emissions.

Carbon sequestration in vegetation

The main methods to increase carbon stocks in vegetation occur through afforestation/reforestation (planting trees to increase the per-area NPP of a unit of land), forest management (fire suppression, thinning, fertilization, other techniques to improve per-area NPP of a forested unit of land), re-vegetation (increasing vegetation carbon stocks using plants other than trees) and cropland and grazing land management (increasing the above-ground standing stock of carbon in agricultural systems; Smith, 2004b).

Comparing biological carbon sequestration with other forms of carbon sequestration

Compared with abiotic carbon sequestration, biological carbon sequestration potential is small. Abiotic sequestration can take the form of oceanic sequestration through deep-ocean injection of CO_2 (5,000–10,000 Pg C potential), geological sequestration through the capture, liquefaction, transport and injection of CO_2 into coal seams, old oil wells, stable rock strata or saline aquifers, or scrubbing of CO_2 and mineral carbonation at point of CO_2 emission (Lal, 2008). However, Lal (2008) points out that abiotic technologies are expensive, have leakage risks and may not be available for routine use until 2025 and beyond, whereas soil carbon sequestration is natural, cost-effective, with ancillary benefits and is immediately applicable (Lal, 2008).

Soil carbon sequestration can be achieved by increasing the net flux of carbon from the atmosphere to the terrestrial biosphere by increasing global carbon inputs to the soil (via increasing NPP), by storing a larger proportion of the carbon from NPP in the longer-term carbon pools in the soil or by reducing carbon losses from the soils by slowing decomposition. Many reviews have been published recently discussing options available for soil carbon sequestration and mitigation potentials (IPCC, 2000a; Cannell, 2003; Metting et al, 1999; Smith et al, 2000, 2007a; Lal, 2004a; Lal et al, 1998; Nabuurs et al, 1999; Follett et al, 2000; Freibauer et al, 2004). For soil carbon sinks, the best options are to increase carbon stocks in soils that have been depleted in carbon, that is, agricultural soils and degraded soils or to halt the loss of carbon from cultivated peatlands (Smith et al, 2007a).

Early estimates of the potential for additional soil carbon sequestration varied widely. Based on studies in European cropland (Smith et al, 2000), US cropland (Lal et al, 1998), global degraded lands (Lal, 2001) and global estimates (IPCC, 2000a), an estimate of global soil carbon sequestration potential of 0.9 ± 0.3 Pg C year^{-1} was made by Lal (2004a,b), between one-third and one-quarter of the annual increase in atmospheric carbon levels. Over 50 years, the level of carbon sequestration suggested by Lal (2004a) would restore a large part of the carbon lost from soils historically.

The most recent estimate (Smith et al, 2007a) is that the technical potential for SOC sequestration globally is around 1.3 Pg C year^{-1}, but this is very unlikely to be realized. Economic potentials for SOC sequestration estimated by Smith et al (2007a) were 0.4, 0.6 and 0.7 Pg C year^{-1} at carbon prices of USD 0–20, 0–50 and 0–100 t CO_2-equivalent^{-1}, respectively. At reasonable carbon prices, then, global soil carbon sequestration seems to be limited to around 0.4–0.7 Pg C year^{-1}. Even then, there are barriers (e.g. economic, institutional, educational, social) that mean the economic potential may not be realized (Trines et al, 2006; Smith and Trines, 2007). The estimates for carbon sequestration potential in soils are of the same order as for forest trees, which have a technical potential to sequester approximately 1–2 Pg C year^{-1} (IPCC, 2000c; Trexler, 1988, cited in Metting et al, 1999), but economic potential for carbon sequestration in forestry is similar to that for soil carbon sequestration in agriculture (IPCC WGIII, 2007).

Comparison between soil and vegetation carbon sequestration and GHG mitigation potential in other sectors

Soil carbon sequestration potential in agricultural soils is of a similar size to that available through forest carbon sequestration and prices up to USD 100 t CO_2^{-1}. Figure 4.2 shows the findings from the IPCC Fourth Assessment Report on global economic mitigation potential (IPCC WGIII, 2007). At all carbon prices, the greatest mitigation potential is in the buildings sector. At low carbon prices (USD 20 t CO_2-equivalent^{-1}), agricultural mitigation potential (of which 90 per cent is due to carbon sequestration; Smith et al, 2007a, 2008) is similar to the potential in the energy and transport sectors and is higher than that in the industry, forestry and waste sectors. At medium carbon prices (USD 50 t CO_2-equivalent^{-1}), the mitigation potential for soil carbon sequestration is lower than the potential in the buildings, energy and industry sectors, but is higher than the potential in the transport, forestry and waste sectors. At high carbon prices (USD 100 t CO_2-equivalent^{-1}), the mitigation potential from agricultural soil carbon sequestration is similar to the industry and energy supply sectors, lower than the buildings sector, but higher than the transport, forestry and

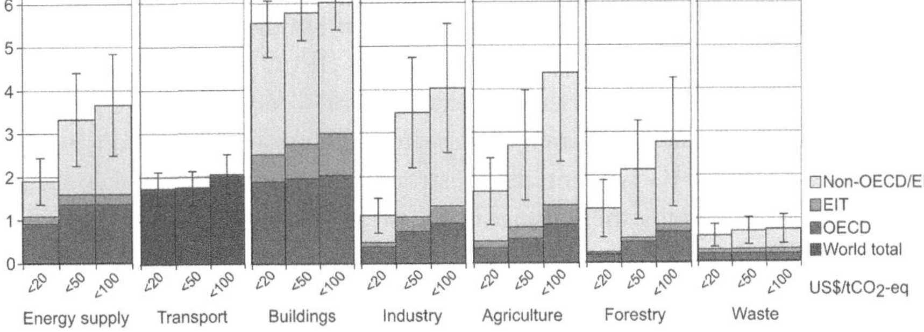

Source: IPCC WGIII, 2007

Figure 4.2 *Estimated sectoral economic potential for global mitigation (gigatonnes of CO_2-equivalent year^{-1}) for different sectors as a function of carbon price in 2030 from bottom-up studies*

waste sectors. It should be noted that there is considerable uncertainty (denoted by error bars in Figure 4.2) associated with the mitigation potential in all sectors, but especially in the energy supply, industry, agriculture, forestry and waste sectors (IPCC WGIII, 2007).

To put the figures for soil carbon sequestration potential in the context of global annual carbon emissions and the annual rise in atmospheric CO_2-carbon concentration, at USD 100 t CO_2-equivalent^{-1}, 0.7 Pg C year^{-1} can be sequestered in agricultural soils (Smith, 2008). The current annual emission of CO_2-carbon to the atmosphere is 6.3 ± 1.3 Pg C year^{-1}. Carbon emission gaps by 2100 could be as high as 25 Pg C year^{-1}, meaning that the carbon emission problem could be up to four times greater than at present. The maximum annual global carbon sequestration potential is about 0.7 Pg C year^{-1} (Smith et al, 2007a), meaning that, even if these rates could be maintained until 2100, soil carbon sequestration would contribute a maximum of about 1–3 per cent towards reducing the carbon emission gap under the highest emission scenarios. When we also consider the limited duration of carbon sequestration options in removing carbon from the atmosphere, we see that carbon sequestration could play only a minor role in closing the emission gap by 2100. It is clear from these figures that if we wish to stabilize atmospheric CO_2 concentrations by 2100, the increased global population and its increased energy demand can only be supported if there is a large-scale switch to non-carbon-emitting technologies in the energy, transport, building, industry, agriculture, forestry and waste sectors (IPCC WGIII, 2007).

This demonstrates that soil carbon sequestration alone can play only a minor role in closing the carbon emission gap by 2100. Nevertheless, if atmospheric CO_2 levels are to be stabilized at reasonable concentrations by 2100 (e.g. 450–550 ppm), drastic reductions in emissions are required over the next 20–30 years (IPCC, 2000b; IPCC WGIII, 2007). During this critical period, all measures to reduce net carbon emissions to the atmosphere would play an important role – there will be no single solution (IPCC WGIII, 2007). IPCC WGIII (2007) show that there is significant potential for greenhouse gas mitigation at low cost across a range of sectors but, for stabilization at low atmospheric CO_2/GHG concentrations, strong action needs to be taken in the very near future, echoing the findings of the Stern Review (Stern, 2006). Given that carbon sequestration is likely to be most effective in its first 20 years of implementation, it should form a central role in any portfolio of measures to reduce atmospheric CO_2 concentrations over the next 20–30 years while new technologies, particularly in the energy sector, are developed and implemented (Smith, 2004a, 2008).

Other considerations for biological carbon sequestration

One also needs to consider the trade-off between different sources of CO_2. For example, nitrogen fertilizer production has an associated carbon cost, and some authors have argued that the additional carbon sequestration for increased production is outweighed by the carbon cost in producing the fertilizer (Schlesinger 1999). However, other studies in developing countries suggest that, when accounting for increased production per unit of land allowed by increased fertilizer use and the consequent avoided use of new land for agriculture, there is a significant carbon benefit associated with increased fertilizer use in these countries (Vlek et al, 2004).

Biological carbon sinks are not permanent and will continue only for as long as appropriate management practices are maintained. If a land management or land-use change is reversed, the carbon accumulated will be lost, usually more rapidly than it was accumulated (Smith et al, 1996). For the greatest potential of soil carbon sequestration to be realized, new carbon sinks, once established, need to be preserved in perpetuity. Within the Kyoto Protocol, mechanisms have been suggested to provide disincentives for sink reversal; that is, when land is entered into the Kyoto process it has to continue to be accounted for and any sink reversal will result in a loss of carbon credits. This process is termed 'sink reversibility' (IPCC, 2000c).

Soil carbon sinks increase most rapidly soon after a carbon-enhancing land management change has been implemented, but soil carbon levels may decrease initially if there is significant disturbance, for example when land is afforested. Sink strength, that is, the rate at which carbon is removed from the atmosphere, in soil becomes smaller with time, as the soil carbon stock approaches a new equilibrium. At equilibrium, the sink has saturated: the carbon stock may have increased, but the sink strength has decreased to zero (Smith, 2004c). This process is termed 'sink saturation' (IPCC, 2000c).

The time taken for sink saturation (i.e. new equilibrium) to occur is variable. The period for soils in a temperate location to reach a new equilibrium after a land-use change is around 100 years (Jenkinson, 1988; Smith et al, 1996), but tropical soils may reach equilibrium more quickly. Soils in boreal regions may take centuries to approach a new equilibrium. As a compromise, current IPCC good practice guidelines for GHG inventories use a figure of 20 years for soil carbon to approach a new equilibrium (IPCC, 1997; Paustian et al, 1997).

Policies encouraging soil carbon sequestration

Smith et al (2007a,b) have recently reviewed policies that encourage agricultural GHG mitigation, including many which encourage soil carbon sequestration. They have found that soil carbon sequestration is encouraged more by non-climate policy than by climate policy. Policies encouraging soil carbon sequestration are most often implemented for other reasons, including improvement of water and air quality, improvement of soil fertility and improvement of biodiversity. Climate and non-climate policies affecting soil carbon sequestration are reviewed briefly below.

Impact of climate policies

Many recent studies have shown that actual levels of GHG mitigation are far below the technical potential for these measures. The gap between technical potential and realized GHG mitigation occurs due to barriers to implementation and cost considerations.

Globally and for Europe, Cannell (2003) has shown that the realistically achievable potential for carbon sequestration and bioenergy-derived fossil fuel offsets are less than 20 per cent of the technical potential. Similar figures have been derived by Freibauer et al (2004) and the European Climate Change Programme (2001) for agricultural carbon sequestration in Europe. Smith et al (2005) have shown recently that carbon sequestration in Europe and for four case-study countries in Europe, is likely to be negligible by the first Commitment Period of the Kyoto Protocol (2008–2012), despite significant biological/technical potential (Smith et al, 2000; Freibauer et al, 2004; Smith, 2004b). The estimates of global economic mitigation potential at different costs reported by Smith et al (2008) are 35, 43 and 56 per cent of technical potential at USD 0–20, 0–50 and 0–100 t CO_2-equivalent^{-1}.

In Europe, there is little evidence that climate policy is affecting GHG emissions from agriculture (see Smith et al, 2005), with most emission reduction occurring through non-climate policy (Freibauer et al, 2004). Non-climate policies affecting GHG emissions are discussed below in 'Impact of non-climate policies'. Some countries have agricultural policies designed to reduce GHG emissions (e.g. Belgium), but most do not (Smith et al, 2005). In Europe, the European Climate Change Programme (2001) has recommended the reduction of livestock CH_4 emissions as being the most cost-effective GHG mitigation options for European agriculture.

In North America, while the USA is not a participant in the Kyoto Protocol, it hosts multinational companies that have reduced GHG intensity as a by-product of their worldwide current Kyoto exposure or through their activities to explore options for future climate agreements. Some of this activity has involved agricultural sector activities, including pig-manure management, farm tillage and afforestation of agricultural land. In the USA, some states are imposing, or are considering imposing, policies. The USA also runs the Clear Skies Initiative, which is a voluntary programme to reduce GHG intensity per dollar of gross domestic product (GDP) by 18 per cent by 2010. A substantial sign-up has occurred on the voluntary registry. However, the programme is projected to allow emissions to increase by 12 per cent even though the intensity has been reduced, as GDP is growing. There is also a long-term diminishing trend in emissions per capita, largely caused by energy conservation, and the programme does not deviate much from a continuation of that trend.

In Canada, the agriculture sector contributes about 10 per cent to national emissions, so mitigation (removals and emission reductions) is considered to be an important contribution to achieving Kyoto targets (and at the same time reduce risk to air, water and soil quality). Examples include: the Agriculture and Agri-foods Canada (AAFC) mitigation program, which encourages voluntary adoption of GHG mitigation practices on farms; national research programmes aimed at reducing the energy intensity of crop production systems, enhancing biological sinks and enhanced bioenergy capacity (i.e. CH_4 capture); and the domestic offset trading system designed to encourage soil carbon sequestration and emission reductions.

In Oceania, vegetation management policies in Australia have assisted in progressively restricting the emissions from land-use change (mainly land-clearing for agriculture) to about 60 per cent of 1990 levels. Complementary policies that aim to foster establishment of both commercial and non-commercial forestry and agro-forestry are resulting in significant afforestation of agricultural land in both Australia and New Zealand.

In Latin America and the Caribbean, climate change mitigation has still not been considered as an issue for mainstream policy implementation. Most countries in the region have devoted efforts to capacity building for complying with obligations under the United Nations Framework Convention on Climate Change (UNFCCC), and a few of them have prepared National Strategy Studies for the gross domestic product (GDP). Carbon sequestration in agricultural soils would be the climate change mitigation option with the highest potential in the region, and its exclusion from the GDP has hindered a wider adoption of land use management practices (e.g. zero tillage).

In Asia, China and India have policies that reduce GHG emissions, but these were implemented for reasons other than climate policy. No African country has emission-reduction targets under the Kyoto Protocol, so the impacts of climate policy on agricultural emissions in Africa are small. We are unaware of any approved GDP projects in Africa related to the reduction of agricultural GHG emissions per se, although several projects are under investigation in relation to the restoration of agriculturally degraded lands, the carbon sequestration potential of agro-forestry and the reduction in sugarcane burning.

Agricultural GHG offsets can be encouraged by market-based trading schemes. Offset trading, or trading of credits, allows farmers to obtain credits for reducing their GHG emission reductions. The primary agricultural project types include CH_4 capture and destruction, and soil carbon sequestration. Although not currently included in current projects, measures to reduce N_2O emissions could be included in the future. The vast majority of agricultural projects have been focused on reducing CH_4 from livestock wastes in North America (Canada, Mexico and the USA), South America (Brazil), China and Eastern Europe. Of those projects that do exist, the majority have resulted in the production of Certified Emission Reductions (CERs) from Kyoto's Clean Design Mechanism (CDM) and other types of certificate. CERs are then bought and sold through the use of offset aggregators, brokers and traders. Although the CDM does not currently support soil carbon sequestration projects, emerging markets in Canada and the USA are considering supporting offset trading from this project type. Credits created from CH_4 capture in the USA will provide an active role in the developing Regional Greenhouse Gas Initiative (RGGI) on the East Coast and will certainly be included should any national market-based trading scheme be implemented. For soil carbon offsets, Canada's Pilot Emission Removals, Reductions and Learnings (PERRL) initiatives programme, under the direction of the Saskatchewan Soil Conservation Association (SSCA) encourages farmers to adopt no-tillage practices in return for carbon offset credits. In addition Chicago Climate Exchange (CCX) (http://www.chicagoclimatex.com/) allows GHG offsets from no-tillage and conversion of cropland to grasslands to be traded by a voluntary market-trading mechanism. These approaches to agriculturally derived GHG offset are likely to expand geographically and in scope.

Impact of non-climate policies

Many policies other than climate policies affect GHG emissions from agriculture. These include other UN conventions such as biodiversity, desertification and actions on sustainable development; macroeconomic policy such as EU Common Agricultural Policy (CAP)/CAP reform, international free trade agreements, trading blocks, trade barriers, region-specific programmes, energy policy and price adjustment; and other environmental policies, including various environmental/agro-environmental schemes. These are described further below.

Non-climate-related UN conventions

In Asia, China has introduced laws to convert croplands to forest and grassland in Vulnerable Ecological Zones under the UN Convention on Desertification. This will increase carbon storage and reduce N_2O emissions. Under the UN Convention on Biodiversity, China has initiated a programme that restores croplands close to lakes, the sea or other natural lands to conservation zones for wildlife. This may increase soil carbon sequestration and if restored to wetland could increase CH_4 emissions. In support of UN Sustainable Development guidelines, China has introduced a Land Reclamation Regulation (1988) in which land degraded by construction, mining and so on is restored for use in agriculture, increasing carbon storage in these degraded soils. In Europe (including the former Soviet Union) and North America, none of the UN conventions have had significant impacts on agricultural GHG emissions.

Macroeconomic policy

In Latin America, the burden of a high external debt triggered the adoption in the 1970s of policies designed for improving the trade balance, mainly through a promotion of exports of agricultural commodities (Tejo, 2004). This resulted in the changes in land use and management, which are still causing increases in annual GHG emissions today. In other regions, for example in the former Soviet Union and many East European countries, political changes occurring since 1990 have meant that agriculture has de-intensified, with less inputs of organic and mineral fertilizer and more land abandonment. This has led to a decrease in agricultural GHG emissions. In Africa, the cultivated area in southern Africa has increased by 30 per cent since 1960, while agricultural production has doubled (Scholes and Biggs, 2004). The macroeconomic development framework for Africa (NEPAD, 2005) emphasizes agriculture-led development. It is therefore anticipated that the cropped area will continue to increase, especially in Central, East and southern Africa, perhaps at an accelerating rate. In Western Europe, North America, Asia (China) and Oceania, macroeconomic policy has tended to reduce GHG emissions, though enlargement of the EU may intensify agriculture in the new member states and may increase GHG emissions. Smith et al (2007a,b) provide a non-exhaustive summary of various macroeconomic policies that potentially affect agricultural GHG emissions in each major world region.

Other environmental policies

In most world regions, environmental policies have been put in place to improve fertility, reduce erosion and soil loss, improve agricultural efficiency and reduce losses from agriculture. The majority of these environmental policies also reduce GHG emissions. Smith et al (2007a) provide a non-exhaustive summary of various environmental policies that were not implemented specifically to address GHG emissions but that potentially affect agricultural GHG emissions in each major world region. In all regions, policies to improve other aspects of the environment have been more effective in reducing GHG emissions from agriculture than policies aimed specifically at reducing agricultural GHG emissions.

Overcoming barriers to implementation

There are a number of barriers that mean that the economic potential for carbon sequestration might not be reached (Smith et al, 2007a,b). These barriers may prevent best management practices from being implemented. Trines et al (2006) have divided these into five categories: economic, risk-related, political/bureaucratic, logistical and educational/societal barriers. Trines et al (2006) have considered barriers preventing a range of agricultural and forestry GHG mitigation measures (including soil carbon sequestration) in developed countries, developing countries and countries with economies in transition; and Smith and Trines (2007) have considered the particular barriers prevalent in developing countries:

- Economic barriers include the cost of land, competing for land, continued poverty, lack of existing capacity, low price of carbon, population growth, transaction costs and monitoring costs.
- Risk-related barriers include the delay on returns due to slow system responses, issues of permanence (particularly of carbon sinks) and issues concerning leakage and natural variation in carbon-sink strength.
- Political and bureaucratic barriers include the slow land-planning bureaucracy and the complexity and lack of clarity in C/GHG accounting rules, resulting in a lack of political will.
- One of the logistical barriers considered by Trines et al (2006) is that land owners are often scattered and have very different interests, that large areas are unmanaged, the managed areas can be inaccessible and some areas are not biologically suitable.
- The education/societal barriers relate to the sector and legislation governing it being very new, stakeholder perceptions and the persistence of traditional practices.

Competition with other land uses is a barrier that necessitates a comprehensive consideration of mitigation potential for the land-use sector. It is important that forestry and agricultural land management options are considered within the same framework to optimize mitigation solutions. Costs of verification and monitoring could be reduced by clear guidelines on how to measure, report and verify GHG emissions from agriculture.

Transaction costs, on the other hand, will be more difficult to address. The process of passing the money and obligations back and forth between those who realize the carbon sequestration and the investors or those who wish to acquire the carbon benefits involves substantial transaction costs, which increases with the number of landholders involved. Given the large number of smallholder farmers in many developing countries, the transaction costs are likely to be even higher than in developed countries, where costs can amount to 25 per cent of the market price (Smith et al, 2007b). Organizations such as farmers' collectives may help to reduce this significant barrier by drawing on the value of social capital. Farmers in developing countries are in touch with each other, through local organizations, magazines or community meetings, providing forums for these groups to set up consortia of interested forefront players. In order for these collectives to work, regimes need to be in place already, and it is essential that the credits are actually paid to the local owner.

For a number of practices, especially those involving carbon sequestration, risk-related barriers such as delay on returns and potential for leakage and sink reversal can be significant barriers. Education, emphasizing the long-term nature of the sink, could help to overcome this barrier, but fiscal policies (guaranteed markets, risk insurance) might also be required.

Education/societal barriers affect many practices in many regions. There is often a societal preference for traditional farming practices and, where mitigation measures alter traditional practice radically (not all practices do), education and extension would help to reduce some of the barriers to implementation.

But the most significant barriers to implementation of mitigation measures in developing countries (and for some economies in transition) are economic. These are mostly driven by poverty and in some areas these are exacerbated by a growing population. In developing countries many farmers are poor and struggle to make a living from agriculture, with food insecurity and child malnutrition still prevalent in poor countries (Conway and Toenniessen, 1999). Given the challenges many farmers in these regions already face, climate change mitigation is a low priority. To begin to overcome these barriers, global sharing of innovative technologies for efficient use of land resources and agricultural chemicals, to eliminate poverty and malnutrition, will significantly help to remove barriers that currently prevent implementation of mitigation measures in agriculture (Smith et al, 2007b). Capacity building and education in the use of innovative technologies and best management practices would also serve to reduce barriers.

More broadly, macroeconomic policies to reduce debt and to alleviate poverty in developing countries, through encouraging sustainable economic growth and sustainable development, would serve to lower or remove barriers: farmers can only be expected to consider climate mitigation when the threat of poverty and hunger are removed. Mitigation measures that also improve food security and profitability (such as improved use of fertilizer) would be more favourable than those that have no economic or agronomic benefit. Such practices are often referred to as 'win-win' options, and strategies to implement such measures can be encouraged on a 'no regrets' basis (Smith and Powlson, 2003); that is, they provide other benefits even if the mitigation potential is not realized.

Maximizing the productivity of existing agricultural land and applying best management practices would help to reduce GHG emissions (Smith et al, 2007b). Ideally agricultural mitigation measures need to be considered within a broader framework of sustainable development. Policies to encourage sustainable development will make agricultural mitigation in developing countries more achievable. Current macroeconomic frameworks do not support sustainable development policies at the local level, while macroeconomic policies to reduce debt and to alleviate poverty in developing countries, through encouraging sustainable economic growth and sustainable development, are desperately needed.

Ideally policies associated with fair trade, reduced subsidies for agriculture in the developed world and less onerous interest rates on loans and foreign debt all need to be considered. This may provide an environment in which climate change mitigation in agriculture in developing countries could flourish. The UK's Stern Review (www.sternreview.org.uk; Stern, 2006) warns that unless we take action in the next 10–20 years, the environmental damage caused by climate change later in the century could cost between 5 and 20 per cent of global GDP every year. The barriers to implementation of mitigation actions in developing countries need to be overcome if we are to realize even a proportion of the 70 per cent of global agricultural climate mitigation potential that is available in these countries. Since we need to act now to achieve low atmospheric CO_2/GHG stabilization targets (IPCC WGIII, 2007), overcoming these barriers in developing countries should be a priority (Smith and Trines, 2007).

References

Albrecht, A. and Kandji, S.T. (2003) 'Carbon sequestration in tropical agroforestry systems', *Agriculture, Ecosystems and Environment*, vol 99, pp15–27

Allen J.C. (1985) 'Soil response to forest clearing in the United States and tropics: geological and biological factors', *Biotropica*, vol 17, pp15–27

Alvarez, R. (2005) 'A review of nitrogen fertilizer and conservative tillage effects on soil organic storage', *Soil Use and Management*, vol 21, pp38–52

Barthès, B., Azontonde, A., Blanchart, E., Girardin, C., Villenave, C., Lesaint, S., Oliver, R. and Feller, C. (2004) 'Effect of a legume cover crop (*Mucuna pruriens* var. *utilis*) on soil carbon in an Ultisol under maize cultivation in southern Benin', *Soil Use and Management*, vol 20, pp231–239

Batjes, N.H. (1996) 'Total carbon and nitrogen in the soils of the world', *European Journal of Soil Science*, vol 47, pp151–163

Batjes, N.H. (1999) *Management Options for Reducing CO_2-Concentrations in the Atmosphere by Increasing Carbon Sequestration in the Soil*, Dutch National Research Programme on Global Air Pollution and Climate Change Report 410-200-031 and ISRIC Technical Paper 30, International Soil Reference and Information Centre, Wageningen

Bruce, J.P., Frome, M., Haites, E., Janzen, H., Lal, R. and Paustian, K. (1999) 'Carbon sequestration in soils', *Journal of Soil and Water Conservation*, vol 54, pp382–389

Cai, Z.C. and Xu, H. (2004) 'Options for mitigating CH_4 emissions from rice fields in China', in Hayashi, Y. (ed) *Material Circulation through Agro-Ecosystems in East Asia and Assessment of Its Environmental Impact*, NIAES Series 5, National Institute for Agro-Environmental Sciences, Tsukuba, pp45–55

Cannell, M.G.R. (2003) 'Carbon sequestration and biomass energy offset: Theoretical, potential and achievable capacities globally, in Europe and the UK', *Biomass and Bioenergy*, vol 24, pp97–116

Cassman, K.G., Dobermann, A., Walters, D.T. and Yang, H. (2003) 'Meeting cereal demand while protecting natural resources and improving environmental quality', *Annual Review of Environment and Resources*, vol 28, pp315–358

Cerri, C.C., Bernoux, M., Cerri, C.E.P. and Feller, C. (2004) 'Carbon cycling and sequestration opportunities in South America: The case of Brazil', *Soil Use and Management*, vol 20, pp248–254

Conant, R.T. and Paustian, K. (2002) 'Potential soil carbon sequestration in overgrazed grassland ecosystems', *Global Biogeochemical Cycles*, vol 16, pp90–1 to 90–9

Conant, R.T., Paustian, K. and Elliott, E.T. (2001) 'Grassland management and conversion into grassland: Effects on soil carbon', *Ecological Applications*, vol 11, pp343–355

Conant, R.T., Paustian, K., Del Grosso, S.J. and Parton, W.J. (2005) 'Nitrogen pools and fluxes in grassland soils sequestering carbon', *Nutrient Cycling in Agroecosystems*, vol 71, pp239–248

Conway, G. and Toenniessen, G. (1999) 'Feeding the world in the twenty-first century', *Nature*, vol 402, ppC55–C58

Cox, P.M., Betts, R.A., Jones, C.D., Spall, S.A. and Totterdell, I.J. (2000) 'Acceleration of global warming due to carbon-cycle feedbacks in a coupled climate model', *Nature*, vol 408, pp84–187

Davidson, E.A., Nepstad, D.C., Klink, C. and Trumbore, S.E. (1995) 'Pasture soils as carbon sink', *Nature*, vol 376, pp472–473

De Fries, R.S., Hansen, M.C., Townshend, J.R.G., Janetos, A.C. and Loveland, T.R. (2000) 'A new global 1-km data set of percentage tree cover derived from remote sensing', *Global Change Biology*, vol 6, pp247–254

Derner, J.D., Boutton, T.W. and Briske, D.D. (2006) 'Grazing and ecosystem carbon storage in the North American Great Plains', *Plant and Soil*, vol 280, pp77–90

Detwiller, R.P. and Hall, A.S. (1988) 'Tropical forests and the global carbon cycle', *Science*, vol 239, pp42–47

Diekow, J.J., Mielniczuk, H., Knicker, C., Bayer, D.P. and Kögel-Knabner, I. (2005) 'Soil C and N stocks as affected by cropping systems and nitrogen fertilization in a southern Brazil Acrisol managed under no-tillage for 17 years', *Soil and Tillage Research*, vol 81, pp87–95

European Climate Change Programme (2001) *Agriculture. Mitigation potential of Greenhouse Gases in the Agricultural Sector*, Working Group 7, Final report, COMM(2000)88, European Commission, Brussels

Falloon, P., Smith, P. and Powlson, D.S. (2004) 'Carbon sequestration in arable land – the case for field margins', *Soil Use and Management*, vol 20, pp240–247

FAOSTAT (2008) 'FAOSTAT Agricultural Data', http://faostat.fao.org/faostat/collections?version=ext&hasbulk=0&subset=agriculture accessed 10 March 2008

Fearnside, P.M. (1997) 'Greenhouse gases from deforestation in Brazilian Amazonia: Net committed emissions', *Climatic Change*, vol 35, pp321–360

Fisher, M.J., Rao, I.M., Ayarza, M.A., Lascano, C.E., Sanz, J.I., Thomas, R.J. and Vera, R.R. (1994) 'Carbon storage by introduced deep-rooted grasses in the South American savannas', *Nature*, vol 371, pp236–238

Foley, J.A., DeFries, R., Asner, G., Barford, C., Bonan, G., Carpenter, S.R., Chapin, F.S., Coe, M.T., Dailey, G.C., Gibbs, H.K., Helkowski, J.H., Holloway, T., Howard, E.A., Kucharik, C.J., Monfreda, C., Patz, J.A., Prentice, I.C., Ramankutty, N. and Snyder, P.K. (2005) 'Global consequences of land use', *Science*, vol 309, pp570–574

Follett, R.F. (2001) 'Organic carbon pools in grazing land soils', in Follett, R.F., Kimble, J.M. and Lal, R. (eds) *The Potential of U.S. Grazing Lands to Sequester Carbon and Mitigate the Greenhouse Effect*, Lewis, Boca Raton, FL, pp65–86

Follett, R.F., Kimble, J.M. and Lal, R. (2000) 'The potential of U.S. grazing lands to sequester soil carbon', in Follett, R.F., Kimble, J.M. and Lal, R. (eds) *The Potential of U.S. Grazing Lands to Sequester Carbon and Mitigate the Greenhouse Effect*, Lewis, Boca Raton, FL, pp401–430

Freibauer, A., Rounsevell, M., Smith, P. and Verhagen, A. (2004) 'Carbon sequestration in the agricultural soils of Europe', *Geoderma*, vol 122, pp1–23

Friedlingstein, P., Cox, P., Betts, R., Bopp, L., Von Bloh, W., Brovkin, V., Cadule, P., Doney, S., Eby, M., Fung, I., Bala, G., John, J., Jones, C., Joos, F., Kato, T., Kawamiya, M. Knorr, W., Lindsay, K., Matthews, H.D., Raddatz, T., Rayner, P., Reick, C., Roeckner, E., Schnitzler, K. G., Schnur, R., Strassmann, K., Weaver, A.J., Yoshikawa, C. and Zeng, N. (2006) 'Climate-carbon cycle feedback analysis: Results from the (CMIP)-M-4 model intercomparison', *Journal of Climate*, vol 19, pp3337–3353

Gregorich, E.G., Rochette, P., VandenBygaart, A.J. and Angers, D.A. (2005) 'Greenhouse gas contributions of agricultural soils and potential mitigation practices in Eastern Canada', *Soil and Tillage Research*, vol 83, pp53–72

Guo L.B. and Gifford, R. M. (2002) 'Soil carbon stocks and land use change: A meta analysis', *Global Change Biology*, vol 8, pp345–360

Helgason, B.L., Janzen, H.H., Chantigny, M.H., Drury, C.F., Ellert, B.H., Gregorich, E.G., Lemke, Pattey, E., Rochette, P. and Wagner-Riddle, C. (2005) 'Toward improved coefficients for predicting direct N_2O emissions from soil in Canadian agroecosystems', *Nutrient Cycling in Agroecosystems*, vol 71, pp87–99

Houghton, R.A. (1999) 'The annual net flux of carbon to the atmosphere from changes in land use 1850 to 1990', *Tellus*, vol 50B, pp298–313

Houghton, R.A., Hackler, J.L. and Lawrence, K.T. (1999) 'The US carbon budget: contributions from land-use change', *Science*, vol 285, pp574–578

IPCC (1997) *IPCC (Revised 1996) Guidelines for National Greenhouse Gas Inventories, Workbook*, Intergovernmental Panel on Climate Change, Paris

IPCC (2000a) *Special Report on Land Use, Land Use Change, and Forestry*, Cambridge University Press, Cambridge

IPCC (2000b) *Special Report on Emissions Scenarios*, Cambridge University Press, Cambridge

IPCC (2000c) *Special Report on Land Use, Land Use Change, and Forestry*, Cambridge University Press, Cambridge, UK

IPCC (2001) *Climate Change: The Scientific Basis*, Cambridge University Press, Cambridge

IPCC WGIII (2007) 'Summary for policy makers', Working Group III contribution to the Intergovernmental Panel on Climate Change Fourth Assessment Report, *Climate Change 2007: Mitigation of Climate Change*, Cambridge University Press, Cambridge

Jenkinson, D.S. (1988) 'Soil organic matter and its dynamics', in Wild, A. (ed) *Russell's Soil Conditions and Plant Growth*, 11th edn, Longman, London, pp564–607

Kasimir-Klemedtsson, A., Klemedtsson, L., Berglund, K., Martikainen, P., Silvola, J. and Oenema, O. (1997) 'Greenhouse gas emissions from farmed organic soils: A review', *Soil Use and Management*, vol 13, pp245–250

Koga, N., Tsuruta, H., Tsuji, H. and Nakano, H. (2003) 'Fuel consumption-derived CO_2 emissions under conventional and reduced tillage cropping systems in northern Japan', *Agriculture, Ecosystems and Environment*, vol 99, pp213–219

Lal, R. (1999) 'Soil management and restoration for C sequestration to mitigate the accelerated greenhouse effect', *Progress in Environmental Science*, vol 1, pp307–326

Lal, R. (2001) 'Potential of desertification control to sequester carbon and mitigate the greenhouse effect', *Climate Change*, vol 15, pp35–72

Lal, R. (2003) 'Global potential of soil carbon sequestration to mitigate the greenhouse effect', *Critical Reviews in Plant Sciences*, vol 22, pp151–184

Lal, R. (2004a) 'Soil carbon sequestration impacts on global climate change and food security', *Science*, vol 304, pp1623–1627

Lal, R. (2004b) 'Soil carbon sequestration to mitigate climate change', *Geoderma*, vol 123, pp1–22

Lal, R. (2008) 'Carbon sequestration', *Philosophical Transactions of the Royal Society, B*, vol 363, pp815–830

Lal, R., Kimble, J.M., Follet, R.F. and Cole, C.V. (1998) The *potential of U.S. cropland to sequester carbon and mitigate the greenhouse effect*, Ann Arbor Press, Chelsea, MI

Li, C., Frolking, S. and Butterbach-Bahl, K. (2005) 'Carbon sequestration in arable soils is likely to increase nitrous oxide emissions, offsetting reductions in climate radiative forcing', *Climatic Change*, vol 72, pp321–338

Liebig, M.A., Morgan, J.A., Reeder, J.D. Ellert, B.H., Gollany, H.T. and Schuman, G.E. (2005) 'Greenhouse gas contributions and mitigation potential of agricultural practices in northwestern USA and western Canada', *Soil and Tillage Research*, vol 83, pp25–52

Lohila, A., Aurela, M., Tuovinen, J.P. and Laurila, T. (2004) 'Annual CO_2 exchange of a peat field growing spring barley or perennial forage grass', *Journal of Geophysical Research*, vol 109, D18116, doi:10.1029/2004JD004715

Machado, P.L.O.A. and Freitas, P.L. (2004) 'No-till farming in Brazil and its impact on food security and environmental quality', in Lal, R., Hobbs, P.R., Uphoff, N. and Hansen, D.O. (eds) *Sustainable Agriculture and the International Rice–Wheat System*, Marcel Dekker, New York, pp291–310

Maljanen, M., Martikainen, P.J., Walden, J. and Silvola, J. (2001) 'CO_2 exchange in an organic field growing barley or grass in eastern Finland', *Global Change Biology*, vol 7, pp679–692

Maljanen, M., Komulainen, V.M., Hytonen, J., Martikainen, P. and Laine, J. (2004) 'Carbon dioxide, nitrous oxide and methane dynamics in boreal organic agricultural soils with different soil characteristics', *Soil Biology and Biochemistry*, vol 36, pp1801–1808

Maltby, E. and Immirzi, C.P. (1993) 'Carbon dynamics in peatlands and other wetlands soils: regional and global perspective', *Chemosphere*, vol 27, pp999–1023

Mann, L.K. (1986) 'Changes in soil carbon storage after cultivation', *Soil Science*, vol 142, pp279–288

Marland, G., McCarl, B.A. and Schneider, U.A. (2001) 'Soil carbon: policy and economics', *Climatic Change*, vol 51, pp101–117

Marland, G., West, T.O., Schlamadinger, B. and Canella, L. (2003) 'Managing soil organic carbon in agriculture: The net effect on greenhouse gas emissions', *Tellus*, vol 55B, pp613–621

Metting, F.B., Smith, J.L. and Amthor, J.S. (1999) 'Science needs and new technology for soil carbon sequestration', in Rosenberg, N.J., Izaurralde, R.C. and Malone, E.L. (eds) *Carbon Sequestration in Soils: Science, Monitoring and Beyond*, Battelle, Columbus, Ohio, pp1–34

Millennium Ecosystem Assessment (2005) *Findings from the Conditions and Trend Working Group*, Island, Washington, DC

Monteny, G.J., Bannink, A. and Chadwick, D. (2006) 'Greenhouse gas abatement strategies for animal husbandry', *Agriculture, Ecosystems and Environment*, vol 112, pp163–170

Moraes, J.F.L. de, Volkoff, B., Cerri, C.C., Bernoux, M. (1995) 'Soil properties under Amazon forest and changes due to pasture installation in Rondônia, Brazil', *Geoderma*, vol 70, pp63–86

Mosier, A.R., Halvorson, A.D., Peterson, G.A., Robertson, G.P. and Sherrod, L. (2005) 'Measurement of net global warming potential in three agroecosystems', *Nutrient Cycling in Agroecosystems*, vol 72, pp67–76

Mutuo, P.K., Cadisch, G., Albrecht, A., Palm, C.A. and Verchot, L. (2005) 'Potential of agroforestry for carbon sequestration and mitigation of greenhouse gas emissions from soils in the tropics', *Nutrient Cycling in Agroecosystems*, vol 71, pp43–54

Nabuurs, G.J., Daamen, W.P., Dolman, A.J., Oenema, O., Verkaik, E., Kabat, P., Whitmore, A.P. and Mohren, G.M.J. (1999) 'Resolving issues on terrestrial biospheric sinks in the Kyoto Protocol', Dutch National Programme on Global Air Pollution and Climate Change, Report 410 200 030, RIVM: Bilthoven, NL

Nabuurs, G.J., Masera, O., Andrasko, K., Benitez-Ponce, P., Boer, R., Dutschke, M., Elsiddig, E., Ford-Robertson, J., Frumhoff, P., Karjalainen, T., Krankina, O., Kurz, W., Matsumoto, M., Oyhantcabal, W., Ravindranath, N.H., Sanz Sanchez, M.J. and Zhang, X. (2007) 'Forestry', in Metz, B., Davidson, O.R., Bosch, P.R., Dave, R. and Meyer, L.A. (eds) *Climate change 2007: Mitigation*, Chapter 9, Contribution of Working Group III to the Fourth Assessment Report of the Intergovernmental Panel on Climate Change, Cambridge University Press, Cambridge/New York

Neill, C., Melillo, J.M., Steudler, P.A., Cerri, C.C., Moraes, J.F.L. de, Piccolo, M.C. and Brito, M. (1997) 'Soil carbon and nitrogen stocks following forest clearing for pasture in the Southwestern Brazilian Amazon', *Ecological Applications*, vol 7, pp1216–1225

NEPAD (2005) 'The NEPAD Framework Document, http://www. uneca.org/nepad/

Nykänen. H., Alm, J., Lang, K., Silvola, J. and Martikainen, P.J. (1995) 'Emissions of CH_4, N_2O and CO_2 from a virgin fen and a fen drained for grassland in Finland', *Journal of Biogeography*, vol 22, pp351–357

Oelbermann, M., Voroney, R.P. and Gordon, A.M. (2004) 'Carbon sequestration in tropical and temperate agroforestry systems: A review with examples from Costa Rica and southern Canada', *Agriculture Ecosystems and Environment*, vol 104, pp359–377

Ogle, S.M., Breidt, F.J., Eve, M.D. and Paustian, K. (2003) 'Uncertainty in estimating land use and management impacts on soil organic storage for US agricultural lands between 1982 and 1997', *Global Change Biology*, vol 9, pp1521–1542

Ogle, S.M., Breidt, F.J. and Paustian, K. (2005) 'Agricultural management impacts on soil organic carbon storage under moist and dry climatic conditions of temperate and tropical regions', *Biogeochemistry*, vol 72, pp87–121

Olsson, L. and Ardo, J. (2002) 'Soil carbon sequestration in degraded semiarid agro-ecosystems – perils and potentials', *Ambio*, vol 31, pp471–477

Pan, G.X., Zhou, P., Zhang, X.H., Li, L.Q., Zheng, J.F., Qiu, D.S. and Chu, Q.H. (2006) 'Effect of different fertilization practices on crop C assimilation and soil C sequestration: A case of a paddy under a long-term fertilization trial from the Tai Lake region, China', *Acta Ecologica Sinica*, vol 26 (11), pp3704–3710

Paul, E.A., Morris, S.J., Six, J., Paustian, K. and Gregorich, E.G. (2003) 'Interpretation of soil carbon and nitrogen dynamics in agricultural and afforested soils', *Soil Science Society of America Journal*, vol 67, pp1620–1628

Paustian, K., Andrén, O., Janzen, H.H., Lal, R., Smith, P., Tian, G., Tiessen, H., Noordwijk, M. van and Woomer, P.L. (1997). 'Agricultural soils as a sink to mitigate CO_2 emissions', *Soil Use and Management*, vol 13, pp229–244

Paustian, K., Babcock, B.A., Hatfield, J., Lal, R., McCarl, B.A., McLaughlin, S., Mosier, A., Rice, C., Robertson, G.P., Rosenberg, N.J., Rosenzweig, C., Schlesinger, W.H. and Zilberman, D. (2004) *Agricultural Mitigation of Greenhouse Gases: Science and Policy Options*, Council on Agricultural Science and Technology (CAST) Report R141 2004, ISBN 1-887383-26-3, CAST: Ames, Iowa, USA

Reeder, J.D., Schuman, G.E., Morgan, J.A. and Lecain, D.R. (2004) 'Response of organic and inorganic carbon and nitrogen to long-term grazing of the shortgrass steppe', *Environmental Management*, vol 33, pp485–495

Rice, C.W. and Owensby, C.E. (2000) 'Effects of fire and grazing on soil carbon in rangelands', in Follet, R., Kimble, J.M. and Lal, R. (eds) *The Potential of U.S. Grazing Lands to Sequester Carbon and Mitigate the Greenhouse Effect*, Lewis, Boca Raton, FL, pp323–342

Robertson, G.P. (2004) 'Abatement of nitrous oxide, methane and other non-CO_2 greenhouse gases: The need for a systems approach', in Field, C.B. and Raupach, M.R. (eds) *The Global Carbon Cycle, Integrating Humans, Climate, and the Natural World*, Scope 62, Island, Washington, DC, pp493–506

Schimel, D.S. (1995) 'Terrestrial ecosystems and the carbon-cycle', *Global Change Biology*, vol 1, pp77–91

Schimel, D.S., House, J.I., Hibbard, K.A., Bousquet, P. Ciais, P., Peylin, P., Braswell, B.H., Apps, M.J., Baker, D., Bondeau, A., Canadell, J., Churkina, G., Cramer, W., Denning, A.S., Field, C.B., Friedlingstein, P., Goodale, C., Heimann, M., Houghton, R.A.,

Melillo, J.M., Moore, B., Murdiyarso, D., Noble, I., Pacala, S.W., Prentice, I.C., Raupach, M.R., Rayner, P.J., Scholes, R.J., Steffen, W.L. and Wirth, C. (2001) 'Recent patterns and mechanisms of carbon exchange by terrestrial ecosystems', *Nature*, vol 414, pp169–172

Schlesinger, W.H. (1999) 'Carbon sequestration in soils', *Science*, vol 284, p2095

Schnabel, R.R., Franzluebbers, A.J., Stout, W.L., Sanderson, M.A. and Stuedemann, J.A. (2001) 'The effects of pasture management practices', in Follet, R., Kimble, J.M. and Lal, R. (eds) *The Potential of U.S. Grazing Lands to Sequester Carbon and Mitigate the Greenhouse Effect*, Boca Raton, FL, Lewis, pp291–322

Scholes, R.J. and Biggs, R. (2004) *Ecosystem Services in Southern Africa: A Regional Assessment*, CSIR, Pretoria

Scholes, R.J. and Merwe, M.R. van der (1996) 'Sequestration of carbon in savannas and woodlands', *The Environmental Professional*, vol 18, pp96–103

Schuman, G.E., Herrick, J.E. and Janzen, H.H. (2001) 'The dynamics of soil carbon in rangelands', in Follet, R., Kimble, J.M. and Lal, R. (eds) The *Potential of U.S. Grazing Lands to Sequester Carbon and Mitigate the Greenhouse Effect*, Boca Raton, FL, Lewis, pp267–290

Sisti, C.P.J., Santos, H.P., Kohhann, R., Alves, B.J.R., Urquiaga, S. and Boddey, R.M. (2004) 'Change in carbon and nitrogen stocks in soil under 13 years of conventional or zero tillage in southern Brazil', *Soil and Tillage Research*, vol 76, pp39–58

Smith, J.U., Smith, P., Wattenbach, M., Zaehle, S., Hiederer, R., Jones, R.J.A., Montanarella, L., Rounsevell, M., Reginster, I. and Ewert, F. (2005) 'Projected changes in mineral soil carbon of European croplands and grasslands, 1990–2080', *Global Change Biology*, vol 11, pp2141–2152

Smith, K.A. and Conen, F. (2004) 'Impacts of land management on fluxes of trace greenhouse gases', *Soil Use and Management*, vol 20, pp255–263

Smith, P. (2004a) 'Soils as carbon sinks – the global context', *Soil Use and Management*, vol 20, pp212–218

Smith, P. (2004b) 'Carbon sequestration in croplands: The potential in Europe and the global context', *European Journal of Agronomy*, vol 20, pp229–236

Smith, P. (2004c) 'Engineered biological sinks on land', in Field, C.B. and Raupach, M.R. (eds) The *Global Carbon Cycle, Integrating Humans, Climate, and the Natural World*, Scope 62, Island, Washington, DC, pp479–491

Smith, P. (2008) 'Land use change and soil organic carbon dynamics', *Nutrient Cycling in Agroecosystems*, vol 81, pp169–178

Smith, P. and Powlson, D.S. (2003) 'Sustainability of soil management practices – a global perspective', in Abbott, L.K. and Murphy, D.V. (eds) *Soil Biological Fertility – A Key To Sustainable Land Use in Agriculture*, Kluwer Academic Publishers, Dordrecht, Netherlands, pp241–254

Smith, P. and Trines, E. (2007) 'Agricultural measures for mitigating climate change: Will the barriers prevent any benefits to developing countries?' *International Journal of Agricultural Sustainability*, vol 4, pp173–175

Smith, P., Powlson, D.S., Glendining, M.J. (1996) 'Establishing a European soil organic matter network (SOMNET)', in Powlson, D.S., Smith, P. and Smith, J.U. (eds) *Evaluation of Soil Organic Matter Models Using Existing, Long-Term Datasets*, NATO ASI Series I, vol 38, Springer-Verlag, Berlin, pp81–98

Smith, P., Powlson, D.S., Glendining, M.J. and Smith, J.U. (1997) 'Potential for carbon sequestration in European soils: Preliminary estimates for five scenarios using results from long-term experiments', *Global Change Biology*, vol 3, pp67–79

Smith, P., Falloon, P., Coleman, K., Smith, J.U., Piccolo, M., Cerri, C.C., Bernoux, M., Jenkinson, D.S., Ingram, J.S.I., Szabó, J. and Pásztor, L. (1999) 'Modelling soil carbon dynamics in tropical ecosystems', in Lal, R., Kimble, J.M., Follett, R.F., Stewart, B.A. (eds) *Global Climate Change and Tropical Soils*, Advances in Soil Science Series, CRC Press, pp341–364, Boca Raton, Fl, USA

Smith, P., Powlson, D.S., Smith, J.U., Falloon, P.D. and Coleman, K. (2000) 'Meeting Europe's climate change commitments: Quantitative estimates of the potential for carbon mitigation by agriculture', *Global Change Biology*, vol 6, pp525–539

Smith, P., Falloon, P., Smith, J.U. and Powlson, D.S. (eds) (2001) *Soil Organic Matter Network (SOMNET): 2001 Model and Experimental Metadata*, GCTE Report 7 (2nd edn), GCTE Focus 3 Office, Wallingford

Smith, P., Falloon, P.D., Körschens, M., Shevtsova, L.K., Franko, U., Romanenkov, V., Coleman, K., Rodionova, V., Smith, J.U. and Schramm, G. (2002) 'EuroSOMNET – a European database of long-term experiments on soil organic matter: The WWW metadatabase', *Journal of Agricultural Science*, vol 138, pp123–134

Smith, P., Andrén, O., Karlsson, T., Perälä, P., Regina, K., Rounsevell, M. and Van Wesemael, B. (2005) 'Carbon sequestration potential in European croplands has been overestimated', *Global Change Biology*, vol 11, pp2153–2163

Smith, P., Martino, D., Cai, Z., Gwary, D., Janzen, H.H., Kumar, P., McCarl, B., Ogle, S., O'Mara, F., Rice, C., Scholes, R.J., Sirotenko, O., Howden, M., McAllister, T., Pan, G., Romanenkov, V., Rose, S., Schneider, U. and Towprayoon, S. (2007a) 'Agriculture', in Metz, B., Davidson, O.R., Bosch, P.R., Dave, R. and Meyer, L.A. (eds) *Climate Change 2007: Mitigation*, Chapter 8, Contribution of Working Group III to the Fourth Assessment Report of the Intergovernmental Panel on Climate Change, Cambridge University Press, New York

Smith, P., Martino, D., Cai, Z., Gwary, D., Janzen, H.H., Kumar, P., McCarl, B., Ogle, S., O'Mara, F., Rice, C., Scholes, R.J., Sirotenko, O., Howden, M., McAllister, T., Pan, G., Romanenkov, V., Schneider, U. and Towprayoon, S. (2007b) 'Policy and technological constraints to implementation of greenhouse gas mitigation options in agriculture', *Agriculture, Ecosystems and Environment*, vol 118, pp6–28

Smith, P., Martino, D., Cai, Z., Gwary, D., Janzen, H.H., Kumar, P., McCarl, B., Ogle, S., O'Mara, F., Rice, C., Scholes, R.J., Sirotenko, O., Howden, M., McAllister, T., Pan, G., Romanenkov, V., Schneider, U., Towprayoon, S., Wattenbach, M. and Smith, J.U. (2008) 'Greenhouse gas mitigation in agriculture', *Philosophical Transactions of the Royal Society London, B*, vol 363, pp789–813

Soussana, J.-F., Loiseau, P., Viuchard, N., Ceschia, E., Balesdent, J., Chevallier, T. and Arrouays, D. (2004) 'Carbon cycling and sequestration opportunities in temperate grasslands', *Soil Use and Management*, vol 20, pp219–230

Stern, N. (2006) 'Stern Review: The economics of climate change', www.sternreview.org.uk accessed 30 March 2008

Tejo, P. (2004) *Public Policies and Agriculture in Latin America During the 2000's*, Comisión Económica para América Latina, Serie Desarrollo Productivo 152 (in Spanish), CEPAL, Santiago, Chile

Trines, E., Höhne, N., Jung, M., Skutsch, M., Petsonk, A., Silva-Chavez, G., Smith, P., Nabuurs, G.J., Verweij, P. and Schlamadinger, B. (2006) *Integrating Agriculture, Forestry, and Other Land Use in Future Climate Regimes: Methodological Issues and Policy Options*, Netherlands Environmental Assessment Agency, Climate Change – Scientific Assessment and Policy Analysis, Report 500102 002, Netherlands Environmental Assessment Agency: Bilthoven, NL

Veldkamp, E. (1994) 'Organic carbon turnover in three tropical soils under pasture after deforestation', *Soil Science Society of America Journal*, vol 58, pp175–180

Vlek, P.L.G., Rodríguez-Kuhl, G. and Sommer, R. (2004) 'Energy use and CO_2 production in tropical agriculture and means and strategies for reduction or mitigation', *Environment, Development and Sustainability*, vol 6, pp213–233

Wang, M.X. and Shangguan, X.J. (1996) 'CH_4 emission from various rice fields in PR China', *Theoretical and Applied Climatology*, vol 55, pp129–138

Wassmann, R., Lantin, R.S., Neue, H.U., Buendia, L.V., Corton, T.M. and Lu, Y. (2000) 'Characterization of methane emissions from rice fields in Asia, III, Mitigation options and future research needs', *Nutrient Cycling Agroecosystems*, vol 58, pp23–36

West, T.O. and Post, W.M. (2002) 'Soil organic carbon sequestration rates by tillage and crop rotation: A global data analysis', *Soil Science Society of America Journal*, vol 66, pp1930–1946

Xu, H., Cai, Z.C. and Tsuruta, H. (2003) 'Soil moisture between rice-growing seasons affects methane emission, production, and oxidation', *Soil Science Society of America Journal*, vol 67, pp1147–1157

5

Anaerobic Digestion and Its Implications for Land Use

Charles Banks, Alan Swinbank and Guy Poppy

A short introduction to anaerobic digestion

Anaerobic digestion (AD) has long been recognized as a means of producing energy in the form of biogas while at the same time stabilizing waste organic matter. Historically the major applications of the process were in the treatment of wastewater biosolids and the stabilization of animal slurries and manures, using simple reactor types and processes. Over the last 30 years the technology has developed rapidly to open up new applications in the treatment of process industry wastewaters, the stabilization of municipal solid wastes and the production of energy from crops grown specifically for this purpose. Along with these practical innovations and technology developments, our understanding of the process and its microbiology has also increased allowing optimization of the process to fit this growing range of roles.

Although AD has a recognized role in energy production, it can also provide environmental benefits such as odour reduction, pathogen control, minimizing sludge production, conservation and management of nutrients, and reduction in emissions of GHG. It can be operated at a number of scales, ranging from individual households, as is increasingly common in rural China and India, to large-scale industrial applications for municipal waste and farm use (Mata-Alvarez, 2003; Hopfner-Sixt et al, 2005)

AD has been widely adopted in Germany and Austria to generate renewable electricity from energy crops, animal manures and other agricultural residues. This development has been supported by progressive national legislation, providing the stimulus to meet European renewable energy obligations. Austria currently produces 70 per cent of its electricity from renewable sources and is on schedule to meet its European target of 78 per cent by 2010. The contribution of biogas towards this target is relatively low, but if fully exploited biogas could contribute 10 per cent of Austria‘s overall electricity demand. In 2005 there were 294 farm-based biogas plants in Austria and 3000 in Germany (Hopfner-Sixt et al, 2005; Weiland, 2006), ranging in capacity from 20 to

1000 kW continuous electrical output, with 20 per cent greater than 500 kW and the majority in the range 100–500 kW.

Currently most biogas plants process energy crops as co-digestates with manures from pigs and cattle, although some new plants now digest energy crops by themselves, as experience and confidence in using these substrates grows. The most widely used crop is maize, but grass silage, green grass, alfalfa, clover, sunflowers, sudan grass and sugar millet are also used. Food wastes are also co-digested in some older plants, but in some cases this may lead to a reduction in the energy tariff received by the farmers. Because of the generally smaller scale of organic farms, which primarily manage grassland and cattle, many of the biogas plants are operated as co-operatives. For environmental protection, all of these plants have the capacity to store digestate for six months, ensuring that bad weather conditions do not prevent spreading of this material.

In both Germany and Austria, on-farm biogas production has been geared towards electricity production, although there are also examples of heat from combined heat and power systems (CHP). In Sweden biogas is widely used as a transport fuel with a zero tax rating, and this has led to an extensive uptake of biogas for public transport (Lantz et al, 2007). The other major user of AD technology on farms in Europe is Denmark, where progressive environmental and renewable energy legislation since the mid-1990s has led to the establishment of co-operatives of between 5 and 80 farmers running centralized digestion facilities, each processing 10,000–200,000 tonnes per year of animal manures co-digested with food waste (Holm-Nielsen, 2000).

In the UK about 45 plants were constructed on farms between 1978 and 1990, mainly with partial financing from the Farm Waste Management Grant Scheme. Twenty-nine of these are reported to be still operating satisfactorily (BIOEXELL, 2005) Since that time there has been little activity in the construction of on-farm digesters, although seven were recently built on Scottish dairy farms in a project funded by the Scottish Executive's Water Environment Division to reduce the risk of diffuse pollution to bathing water from slurry run-off. The first large-scale centralized UK plant for the co-digestion of cattle slurry and food wastes was commissioned in 2002 and was initially run by a co-operative of farmers following the Danish model. This plant benefited from a contract to supply electricity with a subsidy under the Non Fossil Fuel Obligation (NFFO), which required electricity suppliers to purchase a proportion of renewable supplies. The plant later passed into private company ownership and is currently reducing the animal manure input in favour of food wastes, which command a gate fee.

A similar co-digestion concept in which food waste is being treated in conjunction with a smaller proportion of pig slurry is operating on a farm in Bedfordshire. One on-farm biogas plant using energy crops as part of the feed material has recently started operation in Dorset. This is run on the co-digestion model with maize and cattle slurry, although the proportions of cattle slurry are higher than seen in its continental counterparts. There have also been off-farm developments testing the feasibility of source-segregated food waste treatment in a demonstration plant in Ludlow funded by the UK government's 'New Technologies' programme. The end product from this plant is returned to agriculture as a safe and nutrient-rich fertilizer.

Policy and regulatory drivers for farm-based energy production and environmental protection through AD

This section considers the policy and regulatory drivers that relate primarily to the production of energy crops and the management of farm wastes (similar issues relating to the import of waste materials onto the farm are dealt with later). In the context of UK and European rural development, a number of measures are currently being considered or are already in place that may lead to more AD on farms. Adoption is driven by an appreciation of how the process can contribute towards meeting renewable energy targets, providing energy security, reducing GHG emissions and helping to deliver the other environmental improvements noted above.

It is often difficult to separate measures that specifically promote biogas from those aimed at technologies producing other biomass-derived renewable energy. In this context biogas can be regarded as both a renewable energy source (EU Renewables directive 2001/77/EC) and a biofuel for transportation (EU Biofuels directive 2003/30/EC). In the latter case, biofuel materials are often categorized according to whether they are produced by first- or second-generation technologies. Simply put, first-generation technologies are those that take vegetable oils from oilseed crops (or used cooking fats) and turn them into biodiesel; or that produce bioethanol by fermentation of the sugars present in crops such as cereals, sugarcane and wine-grapes. Second-generation technologies are capable of producing biofuel from the whole plant, not just the higher-value extracted parts: for example both biomass liquefaction (via pyrolysis or thermal gasification with further catalytic purification) and biogas production through anaerobic digestion are viewed as second-generation technologies under this definition.

There are four basic ways in which diversification into bioenergy production can be promoted:

1 incentives to use land for the production of bioenergy crops
2 assistance in establishing the infrastructure for bioenergy production
3 incentives to promote the purchase and use of bioenergy
4 rewards for environmental improvement.

Incentives to use land for the production of bioenergy crops

The escalation of cereal prices in the world food crisis of 2007–2008, due to poor harvests and increasing demand outstripping supply, has brought into question the use of land for production of bioenergy crops. Previous estimates indicating an apparent surplus of productive agricultural land in the EU-25 (Höglund-Isaksson et al, 2006) are no longer considered valid. The argument is made that if bioenergy crops are encouraged, less land will be available for growing food; and if these crops were to make a sizeable contribution to the reduction of GHG emissions large areas of land would have to be devoted to this purpose. Previous claims that bioenergy production would be possible using biomass from set-aside land, which under the MacSharry reforms of 1992 could be used for non-food crops (EC Regulation 2461/1999/EC), lost substance with the introduction of the Single Payment Scheme (SPS) in 2005. In September 2007, with soaring world commodity prices, EU farm ministers agreed to reduce the set-aside rate to 0 per cent for the 2008 crop, and in the 'Health Check' reforms of November 2008 set-aside was abolished (Agra Informa, 2007, 2008).

Part of the incentive for energy crop production was being able to grow a cash crop where a food crop could not be grown under set-aside rules, and removing this incentive increases competition for this previously restricted land use. Set-aside land has been widely used in Austria, where 23 per cent of farmers used it to grow silage maize for energy production by AD. In Germany in 2005, 21.4 thousand hectares was also used for this purpose, and a further 43.5 thousand hectares in Germany was grown with aid for energy crop production under the European Community's (EC's) ECS. The ECS subsidy was based on a direct payment of €45 ha^{-1}, in addition to the SPS payment, which is made in full provided that the maximum area under cultivation across the EU-27 is below 2 million hectares. When this area was exceeded the payment was made pro rata: in 2007, for example, claims were submitted for an area of about 2.84 million hectares, leading to a 30 per cent reduction in the payments made (European Commission, 2007). Questions were soon being asked about the effectiveness of the ECS, with the EC suggesting it should be examined in the light both of new incentives for biomass production (e.g. compulsory energy targets and high prices) and of the need to develop second-generation biofuels within Rural Development Measures (CEC, 2007).

The abolition of set-aside and subsidies for energy crop cultivation means there will be no direct land-use incentive to farmers to grow crops for bioenergy production. This means there will soon be open competition for land use between food and energy in the EU-27, driven by the commodity markets and subsidies placed at other points in the production or end use of the bioenergy. This applies equally to crops used for thermal heat and/or power generation and to first- and second-generation biofuels.

These changes underline the fragility of the system currently in place, which gives the farmer a direct payment subsidy for growing dedicated energy crops. For energy crops exported off the farm any subsidy deficiency will have to be covered by the buyer, and for those used on the farm the loss will have to be covered by an increase in the sale value of the energy product. Biogas is potentially less affected by this situation as the feedstocks for energy production can be gained from the non-edible parts of food crops; and the adoption of an integrated approach to sustainable agricultural production, which maximizes the use of these components, could provide additional income above that gained from the sale of the food crop. It is estimated that over 50 per cent of biomass grown in food production never leaves the field, but remains in the form of so-called hidden flows (Gazley and Francis, 2005). The insecurity of the bioenergy market further strengthens the position of AD because of its ability to utilize different feedstocks and its flexibility to adapt and change with changing agricultural patterns.

Assistance in establishing the infrastructure for bioenergy production

Within the three axes of the EC Rural Development Regulation, there are measures that will impact upon the installation and use of anaerobic digesters on farms. Axis 1 provides for grant aid to improve the competitiveness of farm businesses; Axis 2 funds agri-environmental measures; and Axis 3 provides for grant aid to diversify the rural economy and improve the quality of life in rural areas. There is also an overarching provision to aid local initiatives under the so-called Leader programme. Under their Rural Development Programmes, Member States in the EU can give grants to facilitate the establishment of bioenergy projects. Following the Health Check reforms of November 2008, additional funds are being made available for rural development, with four 'new challenges' (climate change, renewable energy, management of water and biodiversity) and one old challenge (the dairy sector) identified for additional support (CEU, 2008). Biogas production (both 'on-farm and local production') is listed as a priority action which Member States must include in their Rural Development Plans from 2010 (CEC, 2008).

The Rural Development Plan for England includes grant aid for building on-farm digesters. These schemes funded through Axes 1 and 3 and Leader will, however, be administered on a regional basis in England, with eight Regional Development Agencies having their own budgets and priorities. The draft plan for the South East England Development Agency (SEEDA), under 'Modernization of agricultural holdings' (Axis 1), notes that funding will be available to support 'Developing renewable energy projects or small scale on-farm renewable energy technologies (such as biogas and anaerobic digestion)' and 'Alternative agriculture (diversification into non-food markets, e.g. raw material energy products, bio energy crops, niche and novel crops and livestock)' (SEEDA, 2007). Another scheme is the Bio-energy Capital Grants Scheme in England originally administered by Defra and now by the Department of Energy and Climate Change (DECC). The grant available in Round 5 is 'a variable rate of up to 40 per cent of the difference in cost of installing the equipment using biomass compared with installing the fossil fuel alternative', to a maximum of £500,000 (DECC, 2009).

Incentives to promote the purchase and use of renewable energy and biofuels

Biogas can be burnt directly to provide heating or heat water or steam in boilers, used as a fuel in combustion engines in vehicles or used for the production of electricity by connecting the engine to a generator. In the latter case greater overall energy efficiency can be achieved by reclaiming the waste heat from the engine in a CHP unit. Like all bioenergy sources, biogas has to compete in price with fossil fuels in the market place and at present a subsidy is required to allow this. In the future this requirement may be reduced due to increasing fossil fuel energy prices and through increased efficiency in renewable energy biomass and biofuel production, but there is no certainty in predictions of this sort. The approach to providing these subsidies varies across the EU-27 and has consequently led to the promotion of different schemes, with the emphasis in some cases on electricity production and in others on biofuels. Because of its flexibility as an energy source, biogas bridges these categories and in most cases can qualify in schemes to promote renewable vehicle fuels, electricity production and/or provision of heat.

Electricity production

Apart from in Scandinavia, biogas production in Europe has mainly been used in electricity generation. The rapid and sustained growth seen in Germany was initiated by the 2000 Renewable Energy Law. The success of the scheme has been attributed to the following key elements:

- It allowed priority connection for producers of renewable electricity to the supply grid.
- Electricity suppliers had an obligation to purchase and transmit the electricity.
- A tariff for purchase of the electricity by the grid operators was set on a national basis and guaranteed generally for a 20-year period.

The actual purchase price depends on a number of factors, which include the size of the installation, date of commissioning, type of biomass used, type of technology and the use made of generated heat. The costs are passed on not by direct subsidy but by apportioning the cost to the consumers. A farmer operating a 'new technology' digester on manure and energy crops, making use of the process heat and who joined the scheme at an early stage would receive between €0.215/kWh for a small plant generating less than 150 kW to €0.169/kWh for a plant generating up to 5 MW. Most plants use conventional technology, however, and not all use the heat generated, leading to a loss of €0.04/kWh on this tariff. An interesting feature of both the Austrian and German schemes is that a portion of the tariff is lost if waste is imported onto the farm for digestion. The underlying logic is that the waste producer should pay a gate fee which compensates for the loss in the energy price, in line with the general principle of the 'polluter pays'. The adopted framework led to a fourfold increase in the number of biogas plants in a period of five years, with around 3000 plants generating 600 MW of electricity by the end of 2005 (Weiland, 2006).

In contrast, in the UK there has been almost no development of energy-crop farming using AD, despite subsidies and an obligation for licensed electricity suppliers in the UK to source some of their supplies from renewables or face a financial penalty. This system has been in place since 2002 and will now extend until 2037, with annual targets that must be met within the government's current aspiration for some 20 per cent of electricity to come from renewables (DECC, 2008). The system works in the following way: the electricity producer sells electricity directly to the supplier and receives the wholesale price, but in addition receives Renewable Obligation Certificates (ROCs); up to 2009 this was one ROC for each megawatt-hour generated. At the end of each year the electricity supplier has to show that obligations to supply renewable electricity are on target either through direct generation of renewable electricity or by buying ROCs from the renewable producers (an online auction is run four times a year by the Non-Fossil Purchasing Agency Ltd). If the supplier has insufficient ROCs a financial penalty is incurred – the 'buy-out price' – for each megawatt-hour of deficiency. This penalty increases each year with inflation and is paid to Ofgen, which then shares out the money received between those suppliers who have redeemed ROCs. While there is insufficient renewable electricity to meet targets, the value of a ROC will therefore be greater than the buy-out price. When the amount of renewable electricity equals the target for that year, there will be sufficient ROCs in circulation for their value to fall to that of the buy-out price. If the number of ROCs exceeds demand then the excess will not be bought, and some producers will therefore not receive a premium price for their electricity.

Where electricity suppliers are also renewable electricity producers, the ROCs they gain are directly redeemable. In future these producer–suppliers (mainly large power companies) may satisfy their requirements for ROCs through their own activities, at which point the producer but non-supplier (usually smaller independent operators) will find ROCs worthless. This has left the potential small-scale producer uncertain of long-term profitability, making it difficult to produce financial projections to secure funding for a project. Contrast this with the market stability created in Germany, where a fixed-rate, feed-in tariff was introduced with guaranteed prices for 20 years, a timeframe over which capital payback could be assured.

The UK Government considers that, by treating all renewables equally in the granting of ROCs, too much support is given to some technologies and too little to others. Thus the Energy White Paper (DTI, 2007) indicated that more ROCs should be given for the newer emerging technologies. This went to consultation, and the Energy Act 2008 gave the UK Government powers to implement a new Renewables Obligation Order in which anaerobic digestion is supported at two ROCs per megawatt-hour from April 2009. In the short term, this makes AD a very attractive proposition as a means of producing electrical power, but the longer-term prospects are still uncertain because of the features of the ROCs' trading system.

These concerns are alleviated to some extent by further provisions within the Energy Act 2008, which gives the Government the power to introduce feed-in tariffs for operators with up to five megawatts electrical capacity: this concept is similar to the system in Germany and Austria and is widely canvassed as superior to the ROC trading scheme (Mitchell et al, 2006). Although the scheme is subject to further consultation, it is hoped that when introduced it will provide both higher rates for newly emerging technologies and a guaranteed outlet for the energy produced. The Energy Act 2008 also makes provision for financial support for renewable heat generation and specifically allows producers of biomethane injected into the gas grid to receive a renewable heat incentive, which is likely to be based on a tariff system, as there are at present no plans for a target-based trading system.

Fuel production

The EU has also set targets for replacing a proportion of fossil fuels in transport use: by 2020, 10 per cent of energy supplied for transport purposes in each Member State will have to come from renewable sources (European Parliament, 2008). In the medium term it is anticipated that these requirements will be met by biofuels. One approach to stimulating the biofuel market in Europe is by direct tax incentives, as transport fuels are in most cases subject to taxation; but the fuels to which this applies and the application method varies between states. The system for taxing gases used for road fuel in the UK, including natural gas, biogas and liquefied petroleum gas (LPG), differs from that for liquid fuels. In 2008 biogas and natural gas were taxed at the same rate of £0.166/kg, compared with £0.2077/kg for LPG. As one kilogram of biogas is approximately equivalent to the energy content of 1.4 l of diesel, the current tax on biogas is equivalent to 11.86 pence/l compared with 50.35 pence/l on mineral diesel (a differential of 38.49 pence/l). The equivalence between biogas and natural gas will end in 2009–2010, with the tax on natural gas increasing in line with other road fuels, while the 2007 differential between biogas and other fuels will be maintained at least until 2012. There is thus currently quite a large incentive to use biogas and even natural gas, as a vehicle fuel, except where there is a concessionary rate of fuel tax for certain activities. Examples of such rates include diesel for agricultural use or for public service vehicles. There may thus be little incentive for farmers to use biogas to power farm equipment or for bus companies to use biogas vehicles. In Sweden biogas is exempt from fuel duty for all classes of vehicle, whereas diesel is subject to tax even in public service vehicles. This makes biogas buses, refuse collection vehicles, taxis and even trains an attractive economic proposition and has led to their widespread introduction.

Tax incentives are not the only means of providing a subsidy to biofuels, however, and in April 2008 the UK put in place the Renewable Transport Fuels Obligation (RTFO) policy as an alternative mechanism. The accounting system uses RTFO certificates that are issued 'according to the quantity of renewable fuel on which duty has been paid'. The Government's intention is that the RTFO will last until 'at least 2020', but target figures past the 5 per cent of aggregate fuel fixed for 2010 have not been agreed. The buy-out price until 2010 has been set at £0.15/l for biodiesel and bioethanol, but with a guarantee that 'the total package of support (buy-out + duty incentive) would be £0.35 per litre in 2009/10 and £0.30 per litre in 2010/11' (Department for Transport, 2009).

Although biogas is included in the RTFO scheme, it is unclear what the future position might be, as the Department of Transport has stressed that the emphasis in supporting biofuels will shift away from fuel tax rebates to buy-out and RTFO certificates. The 2008 budget announced that the rebate would cease in 2010 and that the RTFO buy-out price would be set at £0.30/l (HM Treasury, 2008). As this is greater than the fuel tax currently payable on biogas, sale of the RTFO certificates would nominally result in a direct revenue income for biogas. As part of its biofuels policy, the UK Government is developing Environmental Assurance Schemes in which RTFO suppliers will need to demonstrate the carbon savings claimed. From 2010 the Government aims to reward biofuels under the RTFO according to the amount of carbon saved and from 2011 only if they meet appropriate 'sustainability standards'.

GHG emissions and the EU trading scheme

EU efforts to meet the Kyoto Protocol targets for reduction in GHG emissions are brought together in the European Climate Change Programme (ECCP), which provides measures to reduce emissions and enhance carbon sinks. The promotion of sustainable agriculture is explicitly mentioned among these, and guidance is given on measures to mitigate or offset the contribution of agriculture to GHG emissions. AD can contribute by providing a carbon-neutral renewable energy source and helping reduce methane emissions by improved slurry management.

The EU's GHG trading scheme began in January 2005 and currently only covers CO_2 emitted from power generation and heavy industry, although there are plans to widen this remit. In addition, in the UK a new scheme – the Carbon Reduction Commitment (CRC) – was announced in the Energy White Paper (DTI, 2007). This will extend emissions trading to large private and public sector organizations not yet covered by the EU scheme. Currently carbon trading is not possible within the agricultural sector of the EU but it is beginning to occur worldwide, with AD potentially playing a major role. When electricity is produced from renewable sources, the displaced CO_2 emissions can be traded, for example on the European Climate Exchange (ECX). This provides another financial incentive for AD by trading both avoided CO_2 emissions and CO_2 savings achieved by displacement of fossil fuels. This system is already providing income for pig farmers in Latin America from industrial giants in the US (Telegraph, 2006). Farm-based AD can gain carbon credits for displacement of fossil fuels but currently cannot gain credit for offsetting its own emissions.

Benefits and risks of AD in a farming environment

The major benefit from AD in farming is in the improved management of animal manures and slurries. Although these are rarely a valuable energy source from an economic perspective, there are good reasons to consider them for digestion from an environmental point of view. The management of all farm wastes will, however, have some environmental risks associated with it that also need to be considered.

Mitigation of GHG emissions

Agriculture is responsible for about 10 per cent of the total CO_2 equivalent of GHG in the EU and notably contributes 58.1 per cent of methane (CH_4) and 64.5 per cent of nitrous oxide (N_2O) emissions. These gases have a greater GHG potential than CO_2 and arise as a result of manure and soil management practice (CH_4 and N_2O) and enteric fermentations (CH_4) (ECCP, 2006).

Methane

Although the greatest share of CH_4 emissions is from the rumen of livestock, it has been estimated that manure storage systems could account for 60–70 million tonnes of CO_2-equivalent in the EU (Höglund-Isaksson et al, 2006). If the potential energy in manure was captured by AD this could displace CO_2 that would otherwise be generated by fossil fuel usage. Of the options considered for methane emissions, reduction in the EU-only AD is thought to be within the current economic boundary permit price of around €20 per tonne CO_2-equivalent, although this boundary price could rise to as much as €65 by 2030 (EEA, 2005). Methane emissions are considered an important element in the strategy for GHG reduction as the gas has a short atmospheric lifetime of about 12 years. Because of this, reducing global methane emissions could have a significant and rapid positive effect on atmospheric warming and yield important economic and energy benefits. AD processes designed to capture the energy potential of manures, although efficient, are optimized in such a way that there is usually a residual methane potential in the digestate. This methane is best captured in a covered storage tank and mixed with the biogas from the main digestion tank. By combining AD with closed digestate storage, significant reductions in GHG emissions from slurries produced in livestock farming can be achieved (Wulf et al, 2003; Amon et al, 2005). The application to land of both untreated and digested manure will lead to some methane emissions but the magnitude of these is small compared with those occurring from open or atmospherically vented storage tanks (Clemens and Huschka, 2001).

Nitrous oxide

Very little N_2O emission occurs in digestate storage tanks (ECCP, 2006) and the gas is mainly formed in soils by microbial transformation of nitrogen. This occurs in the application of both mineral and organic fertilizers but to a greater extent with the latter (Velthof and Oenema, 1995; Boeckx and Van Cleemput, 2001). The form of nitrogen in fertilizers can also influence N_2O emission rates (Clayton et al, 1997). Very few studies have been carried out to compare N_2O emissions between untreated and digested organic fertilizers. In laboratory experiments Petersen et al (1996) found a significant reduction of N_2O emissions due to anaerobic digestion of slurry. This result could not be confirmed in field experiments using surface application, but incorporation of the fertilizer into the soil showed lower N_2O emissions from the digested slurry (Petersen, 1999). Rubaek et al (1996) have also reported a significant reduction in N_2O emissions due to anaerobic digestion of slurry. Clemens and Huschka (2001) suggest that the reduced oxygen demand of digested slurry leads to a reduction in N_2O generation immediately after application. In a field experiment on arable and grasslands in Germany (Wulf et al, 2003), slightly less N_2O was emitted after application of digested slurry compared with untreated slurry. These experiments suggest that slurry digestion leads to some improvement, but there is insufficient data to provide a measure of the CO_2 equivalents that could be mitigated as a result. Further work is needed to quantify any improvement if it is to be used as the basis for a future subsidy reflecting environmental benefit through reduced N_2O emissions.

Other gaseous emissions

These include NH_3, NO_x and N_2 and not only represent a direct loss in nutrients (NH_3 and N_2) but also contribute to acid deposition (NO_x). These losses are a result of transformations in the soil and can arise from the application of both mineral and organic fertilizers. Reductions can be achieved by optimizing application of fertilizer products in both space and time. The role of AD in minimizing these losses is considered here only as the difference between using untreated and digested organic nitrogen sources and not in comparison with mineral fertilizer applications, although these also enter into soil-based transformations with subsequent losses.

Ammonia emissions

As complex organic matter breaks down in the digestion process it produces ammonia. This makes the digestate more useful as a fertilizer product, but at the same time can lead to a higher emission potential if the material is held in open tanks. In the UK there is currently no obligation to cover a storage tank of either slurry or digestate, although this is good farming practice and a requirement in many parts of Europe (Edelman et al, 2001). During and immediately following application, ammonia emissions from digested slurry tend to be higher (Amon et al, 2005) but also to decline more rapidly, probably as a result of better soil infiltration (Wulf et al, 2003; Gericke et al, 2007). These emissions can be minimized by using soil-injection or trailing-shoe application rather than splash plates or broad spreading (Edelman et al, 2001).

Nutrient management

Livestock farming and the growth of crops for food and industrial use leads to a net export of nutrients from the agricultural system, as most of the processed products are consumed in the urban environment and eventually end up in solid and liquid waste streams. After a relatively brief period in the human economy, the nutrients contained in these materials are returned to the land, the atmosphere or the aquatic environment through degradation or thermal processing. Apart from the case of wastewater treatment biosolids, there are currently few mechanisms, processes or management strategies that return organic matter and its nutrients back into the agricultural system. To compensate for this continuous loss of nutrients and to maintain productivity, farming relies heavily on the use of mineral fertilizers. There is potential for greater substitution of mineral fertilizers by joining together the urban and agricultural cycles as a means of returning plant nutrients to the land. Such a strategy presents many challenges and some associated risks: these include maintaining public, animal and plant health; protecting soil structure; balancing plant nutrient requirements; and providing organic-based nutrient products that fit with modern agricultural practice. Anaerobic digestion cannot provide a universal solution to bring about 'circular agriculture' (SAIN, 2008), but it may form part of a wider management strategy that will reduce dependence on chemically fixed nitrogen and on mined minerals such as phosphate and potash.

Any management strategy for nutrients has to take into account transformations, interactions and losses both in the soil and in waste processing (Figure 5.1). In the case of nitrogen, for example, losses from soil occur as a result of

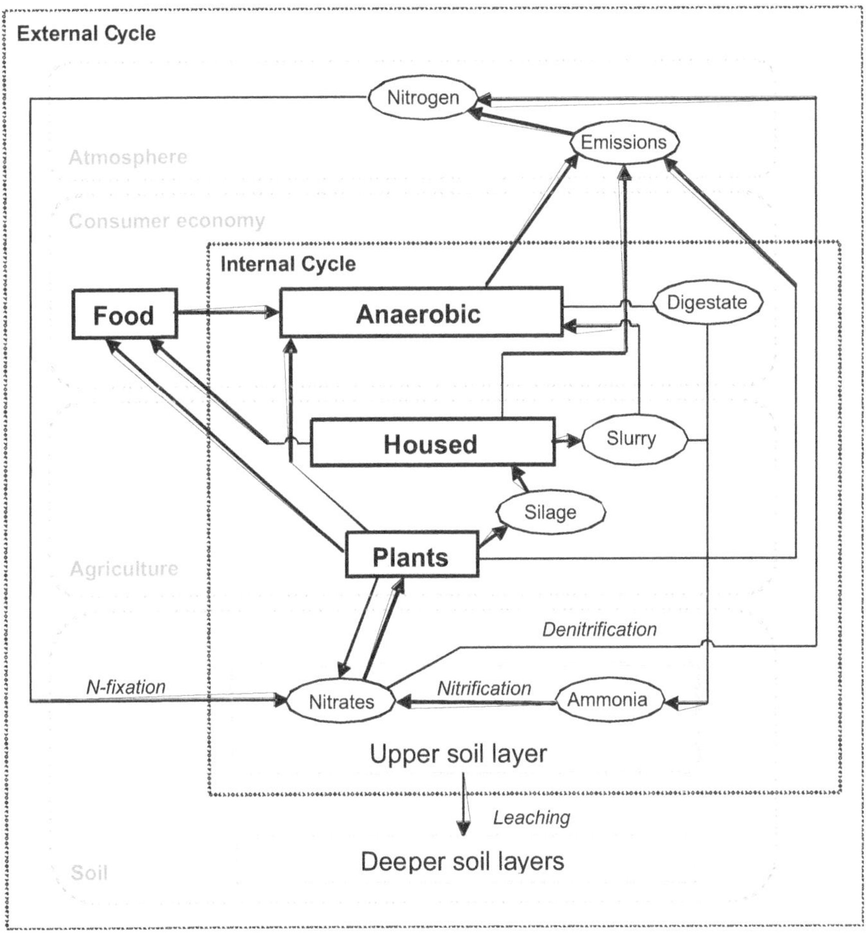

Note: N is potentially available inside internal cycle

Figure 5.1 *Simplified scheme of nitrogen pathways and plant availability for an anaerobic digester using farm input material*

permeation, surface run-off and transformation into gaseous forms. Maintaining nitrogen in a plant-available form such as ammonia and minimizing losses should be considered in any nitrogen management strategy.

Phosphorus and potassium are not lost to the atmosphere but can be leached out of soils and enter the aquatic environment, where phosphorus can impact upon the ecosystem by leading to unwanted changes in species composition and growth. Conservation of phosphorus within the nutrient cycle is particularly important as currently known, economically exploitable reserves have an estimated lifespan of 60–130 years (Steen, 1998). Potassium has no known negative environmental impacts but its efficient use is important as, like phosphorus, it is a limited resource. Anaerobic digestion can play a major role in better nutrient management both within the farming environment and in the wider context of combining urban and farm waste management, in a number of ways.

Energy farming

Crops grown specifically for on-farm biogas production are harvested green and contain all the plant nutrients taken up for growth, with little loss occurring before they reach the storage place or the digester. The AD process conserves nutrients, with only a small amount of nitrogen being lost in the biogas as ammonia (Yadava and Hesse, 1981). During the process nitrogen, which in plants mainly exists in an organically bound form, is partly converted into a mineral form (ammonium). In the soil further conversion to nitrate may take place, but both forms are directly plant-available, although some losses from the internal cycle (plant-available) into the external cycle (not plant-available) may occur through emissions and leaching (Figure 5.1). Nutrient conservation with on-farm biogas production therefore depends on a relatively simple cycle, with the major issues being the timing and means of digestate application and any implications this might have within a rotational crop cycle.

Dairy and livestock

Animal slurries are rich in nutrients, and on a farm that grows its own animal feed the application of these slurries back to land should not lead to an excess nutrient load. Where there is intensification with extensive import of feed materials, for example in housed pig and poultry operations or where nutrient-rich dietary supplements are used, slurry management poses a greater problem as the nutrient load may exceed the land capacity of the farm. In some countries this has led to restrictions on the number of housed animals, together with the adoption of AD as part of a nutrient management programme (Kjær and Madsen, 1998). The surplus of phosphorus in slurry from dairy and beef production where phosphorus-rich feed supplements are used can also lead to run-off endangering sensitive aquatic environments. AD will not solve this problem, and it has been suggested the solution lies in changing the diet or by increasing the land area used for spreading (Aarts et al, 2000). Overall phosphorus budgets for UK agriculture indicate an annual surplus of 12–16 kg/ha and this is most likely to occur on dairy farms.

On both intensive and non-intensive livestock farms AD is useful since it can improve the availability of nutrients to crops when digestate is applied compared with untreated slurry. This allows an increased nutrient uptake during the active growth phase of the crop and reduces the likelihood of run-off or permeation into water over a longer time frame. The additional benefit is higher crop yields when compared with the use of untreated material, and in some cases digested slurry can produce yields comparable with those from the application of mineral fertilizer (Tafdrup, 1994; Asmus and Linke, 1987; Bergström and Kirchmann, 2006; Möller et al, 2006). In dairy and livestock farming, nutrients are exported from the farm in milk and through animals for slaughter, leading to a requirement for nutrient import. In the case of nitrogen this can be offset by growing legumes.

Arable and grassland

Because cross-boundary losses of nutrients are greatest from crop cultivation, arable farms have the greatest potential for closing the loop between urban sinks and agricultural nutrient needs. They also pose the greatest practical difficulties, as modern arable farming is a precision business in which nutrients are applied to meet crop growth requirements using sophisticated mechanized equipment designed for 'clean and homogeneous' chemical fertilizer products. Is there a place on an arable farm for a digester that imports urban organic wastes to transform into products that safely satisfy agronomic needs? Alternatively, can centralized anaerobic digestion facilities produce a marketable and economic product, and would this be acceptable to the farmers, commodity dealers, supermarkets and the public? The answer to both of these questions is not simple, and as noted earlier there are a number of regulatory issues and policy drivers that may strongly influence the outcome. These are discussed in a later section.

Grass plays a large part in the UK's agricultural economy with 37 per cent of agricultural land as grass leys or bare fallow and 35 per cent as rough pasture (Defra, 2005). The widespread adoption of energy farming as an alternative to livestock could potentially lead to a different management approach to land currently turned over to grass production. If grasslands are lost to arable crop production this could lead to a corresponding loss in biodiversity and release of soil carbon. It is important therefore to consider whether grasslands could be used for energy production in place of livestock grazing, by cutting and harvesting to maintain biodiversity and providing the feedstock for energy production through AD. On a European scale the contribution of grasslands to bioenergy production is relatively small, but in the UK this could be a major source as our climate favours high yields compared with dryer continental conditions.

Soil enhancement

Although the use of organic-based fertilizers has some drawbacks in present-day farming, there are a number of positive effects. These include improvements to soil structure, increasing the water-holding capacity of sandy soils, improving drainage in clay soils and increasing soil biological activity. The magnitude of these effects is difficult to assess and in most cases continued use is required to benefit a field in the longer term.

Indirect benefits of AD

Biodiversity and farming patterns

The changing pattern of land use in Europe has seen an increasing proportion of agricultural area devoted to organic production, with 6.1 million ha in EU-25 in 2005 (Eurostat, 2007). There is also a focus on the protection of biodiversity, with around 12–13 per cent of the agricultural and forestry area designated under Natura 2000. High natural value farming systems play an important role in preserving biodiversity and habitats as well as in landscape protection and soil quality. In most Member States, these farming systems account for between 10 per cent and 30 per cent of the agricultural area (2006/144/EC). Equally it is recognized that extensively cultivated farmland usually has a high biodiversity value and that abandonment of productive land can have an adverse effect, although crop diversity is essential to provide 'ecological stepping stones' for conservation.

With changes in land use, particularly with the end of set-aside, farm-based bioenergy using AD could contribute towards environmentally orientated farming. In areas of intensive arable farming there may be a longer-term requirement to develop new bioenergy cropping systems and varieties to reduce environmental pressure, control nutrient input, give a higher energy yield and require less use of heavy machinery. This may mean a move towards the use of multi-year perennial species such as coppice willow and energy grasses such as miscanthus or switch grass, which provide erosion protection and reduce nitrogen inputs. Alternatively, by encouraging the continuation of local agricultural practice through an alternative output for the biomass, the introduction of on-farm AD could be managed to have a low impact on the countryside and utilize traditional high-productivity fodder crops, catch crops and ley crops, with inter-cropping or improved rotations to minimize fallow and bare ground conditions. These options are sometimes overlooked, for example in a recent report by Grover et al (2005), which examined large energy grasses, SRC and oil crops for energy production and crops for textiles, pharmaceuticals, health and beauty products By making use of such practices AD can contribute to the non-competitive production of energy and food in the UK agricultural system.

Rural economic development, employment and social issues

The productive use of land itself leads to employment within the rural community, and diversification within agriculture for the production of non-food crops may lead to further opportunities if the products are used close to the farm itself. It is not possible to give firm estimates of the number of jobs that could be created as a result of widespread adoption of on-farm AD, but as it is a farm- or community-based technology the running of the plant and its maintenance will provide additional employment in addition to that created as a direct result of growing energy crops or from managing farm residues. Estimates made in Austria (Walla and Schneeberger, 2003) indicate that biogas plants using only slurries and manures require on average 1.1 h labour per day and those fermenting energy crops 1.25 h. Plants digesting organic waste have a labour requirement of 1.7 h/day which could rise to 5.3 h/day depending on the degree of impurity of waste and the equipment used for pretreatment. The maintenance of AD plant requires engineering skills that are similar to the agricultural engineering requirements of servicing mechanized farm equipment.

If biogas is used for electricity generation around half of the net energy produced is in the form of heat from cooling the generator and its exhaust. This heat presents many opportunities for creating new enterprises or for contributing towards community needs. Surplus heat can be used to heat glasshouses, and carbon dioxide recovered from biogas scrubbing can be used to enrich the glasshouse atmosphere to give enhanced plant growth. This opens up opportunities for the development of market gardening and for satisfying the need for year-round production of a variety of salad and vegetable products to meet market demand and reduce imports. Digester excess heat has also been used in Europe to feed district central heating schemes, with examples ranging from small towns supplied by the larger, centralized digestion facilities typical in Denmark, to rural village applications for the smaller, farm-scale operations typical in Austria and Germany.

Where biogas is used for transport vehicle fuel or for supply into gas distribution grids, there is a need for gas cleaning to remove CO_2 and trace gases. This can most effectively be done at a larger scale, which again introduces employment opportunities and further develops a skill base in this type of technology. Biogas as a transport fuel will not satisfy the volume requirements of mass private transport, but could make a significant contribution to fleet use for public transport, haulage and farm vehicles. This can be stimulated by favourable taxation under the transport fuel obligations and by supporting the infrastructure to supply public service vehicles.

The UK Government is currently supporting a demonstration project in Surrey to investigate the use of biogas in heavy vehicles; the biogas used having been exempted from fuel duty altogether. An extension of this approach to rural public transport systems would also help reduce air emissions associated with heavy vehicles. Biogas is also used extensively in taxis, with both Volvo and Mercedes producing a production-line biogas car for fleet taxi use, and there are small vans available from most major vehicle manufacturers. Vehicle use for biogas opens up many possibilities for economic development, with farmers and co-operatives able to make direct contracts for fuel supply to fleet users.

Jobs in the bioenergy industries are therefore likely to be located either in rural areas or at existing refinery sites. Transport costs for the product (biomass) to the consumer are significant in both economic and energy terms. Decentralized structures for energy supply can contribute to minimizing this cost or, in certain cases, ensure supply through the use of local fuels in rural areas, which are often the most vulnerable parts of traditional centralized energy-distribution systems.

Biomass sources and on-farm digestion scenarios

As noted above, a number of different types of biomass are potential substrates for energy production through AD. Currently by far the largest source is animal manure, approximately 1578 million tonnes annually in the EU-27 (Ho, 2008). This is equivalent to 66 per cent of potential digestion substrate on a tonnage basis, with the UK contributing between 10 and 15 per cent of this load. As an energy source, animal manures are far from ideal, as most of the 'goodness' has already been extracted by the animal itself: consequently the potential yield in energy terms is currently only 33 per cent of the total potential from all biomass types suitable for digestion. Fodder crops such as maize and grass have up to eight times the biogas potential on a wet-tonne basis, and although these currently account for only about 8 per cent of the potentially available substrate for energy production through digestion, their contribution could be as high as 27 per cent. A further 22 per cent of biomass is available as plant residues, including straw, which could provide another 30 per cent of the potentially available energy. The remainder of available biomass is from municipal, commercial and industrial sources, includes waste and residues from parks and gardens and could contribute 10 per cent of the potentially available energy.

Animal manures and slurries

Until December 2008 an AD plant in England or Wales that produced biogas from animal slurry or manure required a permit to operate under the Waste

Management (England and Wales) Regulations (2006) (SI 2006 937). It also required an exemption from the regulations for spreading digestate back to land (involving a £565 fee for each 50-ha parcel of land spread, with a 25-day notification procedure), as this material was classified as a waste. The annual cost of the permit and of obtaining the exemptions is possibly one of the major reasons why farm-based digestion has not been attractive to farmers in these countries, especially considering that there is no licence or fee requirement for spreading raw slurry or manure. From December 2008, the Environment Agency revised its policy on how regulatory controls are applied to the digestion of agricultural manure and slurry and to use of the resulting digestate as a fertilizer on agricultural land in England and Wales. These changes take into account the recently published Quality Protocol for anaerobic digestate (WRAP, 2009a), which defines the point at which a waste material may become a non-waste, not subject to waste regulatory controls. Where agricultural manure and slurry or a mixture of both with crops grown specifically for AD is the only feedstock to an AD plant, the digestate output will now not be regarded as waste if it is used in the same way that undigested manure and slurry would normally be used, that is, spread on agricultural land as a fertilizer. The change in policy reflects the revised view of the Environment Agency that the digestate is a fully recovered material and is likely to have improved fertilizer properties and a reduced environmental impact compared with undigested manure and slurry. This also eliminates a further potential barrier to the adoption of AD on farms, as farmers working under Quality Assurance Schemes to supply major supermarket chains did not wish to jeopardize their position by applying a classified exempt waste governed by Waste Management Regulations onto their land. Equally, supermarkets and other retailers were sceptical about purchasing food from farms subject to these regulations, despite the strict standards applied to processing and the reduced health risks compared with the application of untreated manure.

The biogas produced from AD plants will, however, continue to be subject to waste regulatory controls, except where it has been treated to an agreed standard that would allow it to be used as a transport fuel or for injection into the national gas grid. The farm must therefore be an authorized site which holds an Environmental Permit or have its waste operations registered as exempt. The type of authorization required will depend on the scale and nature of the processing carried out and will typically cover both the digestion process and the burning of the resultant biogas.

The biogas yield from animal slurries depends on the efficiency of the animal's digestive system. For example the expected biogas yields are around 0.25, 0.48 and 0.37 m^3/kg ODM (organic dry matter) added to the digester from a dairy cow, sow and litter, and laying hen respectively. The actual volume of slurry will depend on the amount of washwater and bedding material used, but typical ODM values would be 8 per cent, 5 per cent and 24 per cent. Examples of the required digester volumes for different animal types and the potential energy yields are shown in Table 5.1. In the case of poultry litter the digestion process is limited by the amount of ammoniacal nitrogen released during the degradation and a loading rate of 4 kg ODM/m^3 per day may not be achievable in practice.

Based on the Environment Agency's guidance on exemptions from waste management licence requirements, indicative numbers of livestock whose slurry and manure output would not exceed the capacity of a 1000-m^3 digester are also shown in Table 5.1. The digester volume is likely to be the limiting factor. Digestion of slurry alone is often considered not economically viable due to the dilute nature of the material and the resulting low volumetric methane yields. For example, a 1000-m^3 digester treating cattle slurry would produce about 750 m^3 biogas per day and generate only 50 kW continuous electrical output if this biogas is used in a CHP unit.

The possibility of capital grants and financial incentives for the production of renewable energy will increase the profitability of animal waste digestion. Even with subsidies to promote AD, however, such as those offered in Denmark in the 1990s (Raven and Gregersen, 2007) and more recently in Germany and Austria, most successful schemes have improved the gas productivity of slurry digesters by supplementing the base load with high energy yielding co-substrates such as food-processing wastes or energy crops.

Energy crop digestion

The production of renewable energy from biomass and in particular the production of biofuels has attracted considerable debate concerning the energy balance around the process. In some cases analysis has shown this to be negative, with more fossil fuel energy being applied to production than is gained from the biofuel itself. For a renewable energy source to be viable, its overall

Table 5.1 *Digestion parameters for animal manures and slurries*

	Units	Dairy cows	Sow and litter	Laying hens
Digester loading rate	kg ODM m^{-3} day^{-1}	3	2	4
Retention time	days	27	24	60
Digester vol per head livestock	m^3 $unit^{-1}$	1.950	0.310	0.007
Daily energy production per head	kWh $unit^{-1}$ day^{-1}	7.90	1.50	0.05
Indicative no of livestock to meet waste exemption criteria	No.	500	3,200	140,000

net energy output must be positive and the qualification for biofuel subsidy in the future will be dependent on demonstrating carbon savings.

To calculate energy balance and carbon savings, an analysis of the whole system is required covering crop (biomass) production, conversion of biomass into fuel and processing of the fuel into usable energy. The final energy efficiency is best considered on a per hectare basis and should take into account both the direct energy inputs, which include fuel and power usage, and indirect energy inputs such as the full life-cycle embodied energy in plant and machinery, and the energy used in the manufacture of fertilizer and agrochemicals. The amount of energy required for crop production is directly related to the type of crop grown and reflects the number of operations carried out in the field and the application of fertilizer, agrochemicals and irrigation water. Examples of energy requirements for various crop operations have been reported in the literature and are summarized for some potential energy crops that might be grown under UK conditions in Table 5.2 (Salter and Banks, 2009).

The digester itself has both an embodied energy deficit and an operational energy requirement for mixing and for maintaining an optimum temperature. It is possible to calculate these based on the equipment design, construction materials, differential temperatures, type and efficiency of insulation and so on, to give the continuing energy needs and the inputs that must be written off over the lifetime of the plant. As already mentioned, biogas as an energy carrier has a range of possible uses, some of which require purification (gas grid injection and vehicle fuel) and others which have a conversion energy loss (electricity and heat production).

Table 5.2 *Energy requirements for crops*

Crop	Sowing period	Indirect GJ $ha^{-1}yr^{-1}$	Fuel GJ $ha^{-1}yr^{-1}$ ($l\ ha^{-1}yr^{-1}$)	Fertilizer & sprays GJ $ha^{-1}yr^{-1}$	Others (e.g. labour) GJ $ha^{-1}yr^{-1}$	Total GJ $ha^{-1}yr^{-1}$	Crop yield tFM $ha^{-1}yr^{-1}$	Energy demand GJ $ha^{-1}yr^{-1}$
Annual								
Maize	Spring	1.92	2.48	10.5	0.2	15.1	40	0.38
Fodder beet	Spring	3.76	2.76	10.2	0.3	17.0	80	0.21
Lupin	Spring	1.81	1.9	4.3	0.19	8.2	30	0.27
Perennial								
Perennial	1 year	2.4	2.4 (59)	12.1	0.5	17.4	33	0.53
Ryegrass	2, 3	4.2	3.7 (94)	12.1	0.1	20.1	42	0.48

Note: FM fresh matter

If the biogas is used to provide heat then little further processing is required. The gas is consumed in a boiler and converted into heat at approximately 85 per cent efficiency. If the biogas is consumed in a CHP unit, a minimal amount of gas scrubbing may be required to remove hydrogen sulphide and other impurities. Approximately 85 per cent of the energy value of the biogas will be converted to energy in the form of electricity (at approximately 30–40 per cent efficiency) and of heat (at approximately 50–60 per cent efficiency). If the biogas is used as a biofuel then CO_2 needs to be removed to reduce CH_4 concentration, after which the gas is compressed to reduce the storage volume. Both of these processes consume electricity. For example upgrading may require up to 0.75 kWh m^{-3} upgraded biogas (Murphy et al, 2004). After digestion the energy implications of using digestate must also be considered. These include the direct energy costs of land application (Berglund and Börjesson, 2006) and the potential energy savings from displacement of some of the mineral fertilizer requirement. Various scenarios for energy production and use can be considered, and the energy boundaries associated with these are illustrated in Figure 5.2, which shows crop production with digestate used as fertilizer, biogas used in CHP to generate the heat and electricity necessary for the AD plant and surplus biogas exported as an upgraded vehicle fuel.

An energy balance for the production of farm-based electricity and/or biofuel is shown in Table 5.3, based on a single-stage concrete digester with a working capacity of 3800 m^3 and a life of 30 years. The digester is fed

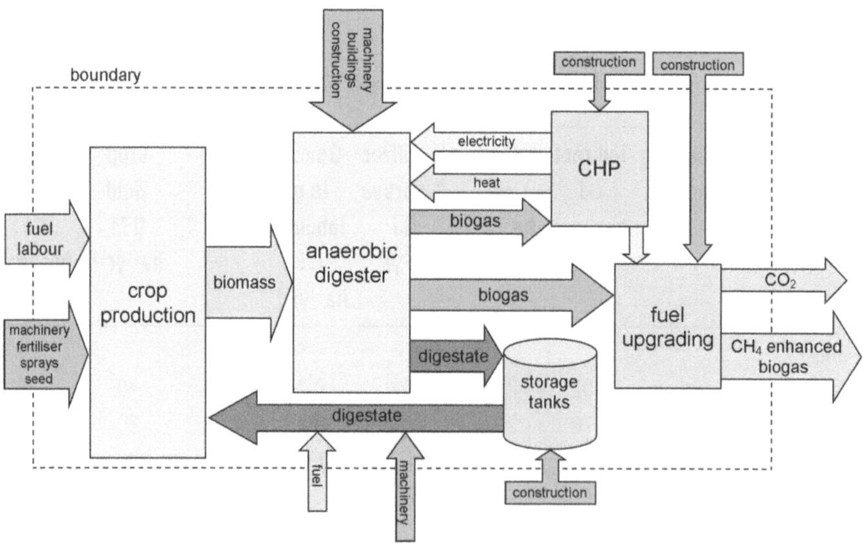

Figure 5.2 *Energy boundaries for the AD process as part of an energy-crop production system*

Table 5.3 *Energy balance for electricity and biofuel production for four crops*

			Maize	Fodder beet	Lupin	Perennial ryegrass
Requirements for crop production	Area of crop required	ha	369	392	1063	620
	Nitrogen fertilizer	kg N ha^{-1}	160	180	0	170
	Energy for crop production (excluding fertilizer)	GJ $year^{-1}$	–2589	–3622	–6726	–5085
	Energy for crop transport	GJ $year^{-1}$	–184	–348	–637	–370
	Energy for digestate transport & spreading	GJ $year^{-1}$	–316	–742	–820	–575
	Energy for mineral fertilizer	GJ $year^{-1}$	–2893	–3042	–1991	–6819
Embodied energy	Digestion plant – CHP	GJ $year^{-1}$	–931	–931	–931	–931
	Digestion plant – biofuel	GJ $year^{-1}$	–1006	–1006	–1006	–1006
Primary energy reduction	CH_4 in biogas	$10^6 m^3 year^{-1}$	1.46	1.33	1.33	1.47
CHP energy output	Digester parasitic energy	TJ $year^{-1}$	–4	–6.4	–6.4	–5.3
	Surplus electricity	TJ $year^{-1}$	15.1	13.2	13.2	14.9
	Surplus heat	TJ $year^{-1}$	22.5	18.4	18.4	21.7
	Electricity yield	kWh ha^{-1} $year^{-1}$	11,367	9354	3449	6676
Energy balance for CHP – mineral fertilizer		TJ year–1	26.6	16.5	14.1	17.5
Biofuel energy output	CH_4 required for CHP and gas upgrading	$10^6 m^3 year^{-1}$	–0.28	-0.3	–0.31	–0.31
	Surplus CH_4 in upgraded biofuel	$10^6 m^3 year^{-1}$	1.18	1.03	1.02	1.16
	Energy value of biofuel	TJ $year^{-1}$	42.0	36.6	36.6	41.4
	Energy in diesel equivalent	10^6 l $year^{-1}$	1.17	1.02	1.02	1.16
	Biofuel yield	l ha^{-1} $year^{-1}$	3180	2610	961	1867
Energy balance for biofuel – with mineral fertilizer		TJ $year^{-1}$	20.8	10.6	8.1	11.3

Note: TJ = terajoules

three kg ODM m^{-3} day^{-1} and the land area required to maintain this reflects the different yields, ODM content and CH_4 potentials of four crops. The embodied energy includes the digester, CHP and upgrading facilities, silage clamps for crop storage and tanks for storage of digestate during periods when it cannot be applied to the fields.

It is clear that crops with low yields (such as lupins) or high-energy inputs (such as perennial ryegrass) are less productive sources of energy than those such as maize which have been bred for high biomass production. The scenarios have been simplified to illustrate the concept of energy and material movement within crop-based AD and in reality may be more complex depending on the design, mode of operation and location of the digester. For example using digestate as biofertilizer to replace mineral fertilizer could improve the energy balance by around 10 per cent for maize and fodder beet and as much as 27 per cent for perennial ryegrass. It is clear, however, that AD can provide a very favourable energy yield as either electrical output or litres equivalent of transport fuel. If the heat from CHP can be utilized the energy benefits are still greater. The net energy output per hectare of land is high in comparison with those first-generation biofuels that can be grown in UK climatic conditions.

The approximate costs at 2008 prices of installing a 3800-m^3 digester with CHP unit would be around £1 million. From April 2009, with double ROCs and the levy exemption certification (LEC), the value of one kilowatt-hour of electrical energy could be approximately £0.15, giving annual electricity sales worth £0.635 million if using maize. This revenue would have to cover the capital costs of the digester and all costs associated with crop production. The current subsidy is now similar to that offered in Germany and Austria, where farmers have reported rates of return of around £600–700 net profit per hectare of land turned over to energy production. As a result of increases in cereal prices in 2007–2008 and abolition of the ECS and of set-aside, many farmers have switched to selling their crop in place of digesting it, with the digester being 'mothballed' while the selling price for animal feed outstrips the return from energy production.

Where large numbers of farm digesters have been built, this has been on the basis of having a guaranteed price for the energy output for a period at least as long as the repayment period for the capital investment, as in Germany, Austria and Denmark. The planned introduction in the UK of renewable energy tariffs for the sale of electricity and possibly of heat may provide greater financial security for small-scale operators such as farmers by taking them outside of the ROC scheme, but the level of these tariffs and the period over which they will be guaranteed is still unknown. Systems offering the potential for other sources of income such as gate fees are therefore an attractive alternative.

Utilization of non-farm organic wastes

In England and Wales if manure and slurry is mixed with other waste feedstocks the resultant digestate is currently classified as a waste and is subject to waste regulatory controls. This is likely to change in the future in line with the recommendations made in the Quality Protocol for digestate (WRAP, 2009a), which was recently published as a final draft after public consultation but still has to be notified before the European Commission's Technical Standards Committee (ECTSC). If the protocol is adopted it will allow digestate produced from a number of clearly defined certified feedstocks to be used without the need for waste controls. The protocol specifically excludes wastes such as residual municipal solid waste that has not been source segregated as this is known to have a high level of contamination. As a further safeguard the digestate will also have to show compliance to a standard approved by the ECTSC. The protocol is likely to make it easier to import wastes for digestion both on-farm and in centralized digestion facilities in urban areas, the digestate from which will be available for on-farm use.

Recent history has seen the establishment of bovine spongiform encephalopathy (BSE) and the catastrophic impact of foot and mouth disease, swine fever and other animal diseases. As well as promoting the return of organic matter and nutrients to land, AD can play a role in ensuring biosecurity. For this purpose, the primary regulation to control the end use and disposal of animal by-products (ABPs) was introduced by the EC in 2002 (Animal By-products Regulations, ABPR; EC Regulation 1774/2002). This covered all animals or parts of animals not intended for human consumption and categorized these according to the perceived risk and the requirements for treatment prior to disposal. Category 1 materials are very high risk and require special handling and disposal; category 2 includes animal wastes and animal parts; and category 3 includes material from healthy animals destined for the human food chain which at some stage has been rejected for that purpose and passed into the waste stream. Category 3 is by far the greatest volume of material as it includes wastes from the manufacture of food products (which contain meat or have been in contact with meat) and food wastes from the retail sector, catering establishments and the home. Both category 2 and 3 wastes can be processed through anaerobic digestion provided additional measures are introduced in accordance with the ABPR (Table 5.4).

With the exception of manures and low-risk materials from slaughterhouses and dairies, category 2 and 3 ABP material requires treatment to achieve an end product which must be free from *Salmonellae* (absence in 25 g) and *Enterobacteriaceae* (less than 1000 colony-forming units per gram). In general this has to be achieved in a two-stage process which involves particle size reduction, heat treatment and biological stabilization, for which anaerobic digestion

Table 5.4 *Summary of the process treatment requirements for ABP material (EC 1774/2002)*

ABPR classification	Type of material	Treatments required in conjunction with anaerobic digestion
2	Manure, stomach and gut contents, milk colostrum	No requirement for pre- or post-treatment provided animals are disease free
	All other category 2 material	Sterilization (133°C for 20 min), marking
3	Food processing wastes not covered in categories 1 & 2	Particle size reduction to 12 mm and heat treatment for 60 min at 70°C
	Catering wastes (commercial and domestic kitchens, supermarket waste	Either: Particle size reduction to 12 mm and heat treatment for 60 min at 70°C Or: In accordance with national legislation based on the requirements of the ABPR

is a suitable option. For catering waste, national governments can derogate from this requirement providing they can show the same or better performance in destruction of the pathogen indicator organisms. For example in the UK it is possible to treat catering waste in either a composting or an anaerobic digestion plant provided there is a minimum guaranteed retention time and an approved temperature profile in the biological reactor.

In addition to these process requirements there is also a requirement to implement strict hygiene controls to prevent bypass between incoming ABP material and the final product; any access to vermin, rodents, birds and insects; or cross-contamination between containers, receptacles and vehicles. The operators must also: ensure that ABP material is properly stored and treated as soon as possible after arrival; have access to a laboratory equipped to carry out the required analysis; adhere to strict documented cleaning regimes within the premises; establish and implement methods of monitoring and checking the critical control points; and ensure the digestion residues comply with the microbial standards and are stored in such a way as to prevent recontamination. Finally, the facility must be authorized by a competent authority, which in England and Wales is the Animal Health Office (Defra) and enforced by the relevant local authority.

These requirements are onerous and add considerably to both the capital and operating costs of running a digester. These costs can be offset by charging a gate fee for waste containing ABP material and also, to some extent, by the enhanced volumetric biogas production achieved in the digester resulting from a feedstock with a high bio-convertible calorific content. Food wastes could be treated alone in centralized facilities or used for on-farm co-digestion with animal slurry, with the final bio-fertilizer product returned to agricultural land use. In either case, the major driver will be the reception and treatment of the waste material, as the gate fee is likely to exceed the revenue from energy sales. For example one tonne of domestic food waste has a biogas potential of around 150 m^3 which is sufficient to generate 320 kWh of electricity. If the gate fee for the waste is £50/tonne this is equivalent to an additional income of £0.15/kWh, which together with the wholesale electricity price, double ROCS and climate levy gives a total revenue of £0.30 for every kilowatt-hour generated. This is being driven by the EU Landfill directive (1999/31/EC), which introduced demanding targets for reduction in the amount of biodegradable municipal waste sent to landfill. The incentive for diversion of wastes from landfill is fiscal, and from April 2008 had reached £32/tonne in the UK with further annual increases of £8/tonne up to £48/tonne in 2010. The policy has led to businesses that can offer an alternative to landfill being able to meet at least part of their operating and capitalization costs by levying gate fees, which could be as high as £70/tonne and still compete with landfill gate fees and taxes in 2010. Even if local authorities in the UK were prepared to pay landfill gate fees and taxes, this option is further restricted by the requirement to comply with the landfill allowances imposed in order to meet the targets of the EU directive (1999/31/EC). Failure to meet these will incur fines that would further increase costs to the local authority. This situation leads to opportunities for the creation of merchant waste facilities that can divert organic wastes from landfill, and AD has been promoted as one such method in the Defra New Technologies Waste Implementation programme.

Digestate from the anaerobic treatment of food waste is now being supplied back to agricultural land from large-scale centralized facilities operated by Biocycle, BiogenGreenfinch and AnDigestion, with further plants in the planning and construction stage. There are estimated to be 6.7 million tonnes of household kitchen food wastes and 12–13 million tonnes of commercial and industrial food waste in the UK (WRAP, 2009b). It is likely that in the next few years a growing proportion of this food waste will be captured in source-segregated waste streams and be available to AD for treatment and energy production.

Closing the nutrient cycle

From a technical perspective the release of nutrients during degradation and the transformation of these into plant-available forms will happen in most waste mixes. The remaining issue is the quality and suitability of the digestate product, and this frequently reflects both misplaced priorities and the manner in which the waste market is economically driven in the UK and most of the developed world. If the priority was the production of a quality-assured product meeting a market specification in terms of nutrient balance, and the economic reward came for achieving this, greater attention would be paid to the blend of waste materials required to meet this specification. If the first economic reward is a gate fee for the reception of waste, and an anaerobic digester is primarily regarded as a means of producing energy and rendering the material less noxious, then less emphasis is placed on obtaining a consistent and tailored digestate nutrient balance to meet the needs of different crop types. As a result the nutrient composition of digestates can vary greatly depending on the input material, and this generally makes them less competitive than other organic fertilizers.

In practice, the application rate for a digestate is mainly calculated on the basis of its nitrogen content, and this may lead to over- or undersupply of other nutrients which in turn may pose environmental risks. Figure 5.3 (Muskolus et al, 2009) shows the distribution of nutrients in digestates produced from a number of different input sources including municipal waste, agricultural slurries and energy crops, with the nutrient requirements of some major crops and their actual application rates to common UK crops and grassland in 2006 (MAFF, 2000; Defra, 2007). It is clear that in energy farming, where digestate is taken from a single crop and applied back to that crop, the nutrient requirements are well aligned, but application to a different crop may lead to imbalance. The digestate from cattle slurry is very variable, probably reflecting seasonal changes in the diet of the animals; whereas that from pigs is more consistent and is more closely aligned to the requirements of cereal crops. Food waste digestate and slaughterhouse wastes, for example blood, are rich in nitrogen but low in phosphorus and could potentially be blended or co-digested with animal slurries to improve the nutrient mix with the aim of matching it more closely to crop requirements, particularly those of cereal crops. Alternatively digestate could be blended with single-mineral fertilizers to improve the nutrient balance.

Nutrient balance is not the only consideration in the use of digestate as a fertilizer substitute, however, and factors such as the window for application and associated on-farm storage capacities, the effect on soil structure due to compaction resulting from spreading machinery, and the different chemical forms in which the nutrients are available need to be taken into account (Muskolus et al, 2009).

The future

AD has the potential to work in harmony with traditional land use and agricultural practices and to bring about significant changes in energy production. For a sustainable outcome it is essential that the drivers and incentives are developed to ensure a balanced programme of rural diversification and land use.

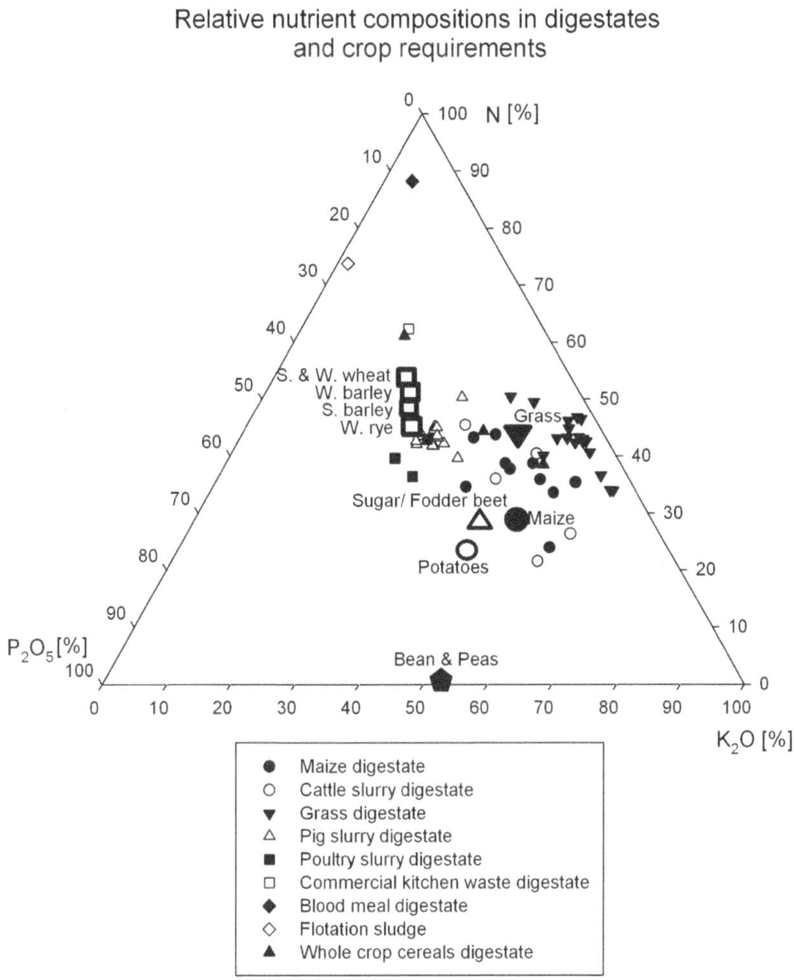

Figure 5.3 *Simplified scheme of N, P_2O_5 and K_2O contents in relation to each other as available in digestate and as required by crops*

There is, however, an urgent need to evaluate fully the role that biogas can play in rural development and in contributing to the cross-cutting issues of energy production, climate change mitigation, nutrient management, reduction in farm chemical usage, and direct and indirect farm energy usage. The potential contribution has been recognized in some countries by the implementation of progressive policies but has not always resulted in a well-balanced approach maximizing all of the benefits of AD. There is still a need for research and support for the development and introduction of low-cost technical solutions; the development of a coherent policy on the promotion and use of digestate and its status as a certified organic fertilizer; and further work to establish the environmental benefits of the technology.

Although the UK has made significant progress in beginning to remove the regulatory barriers to AD and providing the financial incentives for investment in this technology, the future is still unclear. There are still doubts about the longer-term pricing structure for renewable energy, the priorities of the Regional Development Agencies in allocation of pump-priming funds to farms and the volatility of food and feed markets. There are large incentives in the form of renewable energy subsidies and gate fees for the digestion of segregated food waste from domestic and commercial sources, and it is likely that merchant waste processors will take advantage of this and make use of the land bank for digestate spreading. Without food wastes or other higher-energy waste substrates to blend with animal manures for co-digestion it is unlikely that farm-based AD plant will be commercially viable. At the moment there appear to be no drivers, incentives or regulatory controls to encourage this, and in current market conditions slurries will continue to be spread to land untreated, representing a missed opportunity to realize the substantial environmental improvements that can be brought about by AD.

Acknowledgements

The authors wish to acknowledge the input and support of Dr Andrew Salter and Dr Andreas Muskolus, whose original research is reported in this chapter. Thanks are also due to the Relu Programme for its support of the work on the role of anaerobic digestion in the rural economy.

References

Aarts, H.F.M., Habekotte, B. and Keulen, H. van (2000) 'Phosphorus (P) management in the "De Marke" dairy farming system', *Nutrient Cycling in Agroecosystems*, vol 56, no 3, pp219–229

Agra Informa (2007) 'EU confirms zero set-aside rate for 2008 harvest', *Agra Europe*, no 2278, 28 September, pEP/5

Agra Informa (2008) 'CAP Health Check – A detailed summary', *Agra Europe*, no 2337, 21 November, pEP/6

Amon, B., Kryvoruchko, V., Amon, T. and Zechmeister-Boltenstern, S. (2005) Methane, nitrous oxide and ammonia emissions during storage and after application of dairy cattle slurry and influence of slurry treatment', *Agriculture, Ecosystem and Environment*, vol 112, pp153–162

Asmus, F. and Linke, B. (1987) 'Zur pflanzenbaulichen Verwertung von Gülle-Faulschlamm aus der Biogasgewinnung' *Feldwirtschaft*, vol 28, pp354–355

Berglund, M. and Börjesson, P. (2006) 'Assessment of energy performance in the life-cycle of biogas production' *Biomass and Bioenergy*, vol 30, no 3, pp254–266

Bergström, L. and Kirchmann, H. (2006) 'Leaching and crop uptake of nitrogen and phosphorus from pig slurry as affected by different application rates', *Journal of Environmental Quality*, vol 35, no 5, pp1803–1811

BIOEXELL (2005) *Biogas from AD – BIOEXELL Training Manual*, BIOEXELL – European Biogas Centre of Excellence, University of Southern Denmark, Esbjerg, Denmark

Boeckx, P. and Van Cleemput, O. (2001) 'Estimates of N_2O and CH_4 fluxes from agricultural lands in various regions in Europe', *Nutrient Cycling in Agroecosystems*, vol 60, nos 1–3, pp35–47

Clayton, H., McTaggert, I.P., Parker, J., Swan, L. and Smith, K.A. (1997) 'Nitrous oxide emissions from fertilised grassland: A 2-year study of the effects of N fertiliser form and environmental conditions', *Biology and Fertility of Soils*, vol 25, no 3, pp252–260

Clemens, J. and Huschka, A. (2001) 'The effect of biological oxygen demand of cattle slurry and soil moisture on nitrous oxide emissions' *Nutrient Cycling in Agroecosystems*, vol 59, no 2, pp193–198

Commission of the European Communities (2007) 'Communication from the Commission to the Council and the European Parliament preparing for the "Health Check" of the CAP reform', COM(2007)722, CEC, Brussels

Commission of the European Communities (2008) 'Proposal for a Council Regulation establishing common rules for direct support schemes for farmers under the common agricultural policy and establishing certain support schemes for farmers', COM(2008)306/4, CEC, Brussels

Council of the European Union (2008) Press Release 2904th meeting of the Council – Agriculture and Fisheries, Brussels, 18–20 November 2008, 15940/08 (Presse 335) Provisional Version, CEU, Brussels

Defra (2005) 'e-Digest of Environmental Statistics', Department for Environment, Food and Rural Affairs, http://www.defra.gov.uk/environment/statistics/index.htm accessed 1 February 2009

Defra (2007) *The British Survey of Fertiliser Practice, Fertiliser Use on Farm Crops for Crop Year 2006*, Farming Statistics and Register Policy Branch SSFE Division, Defra Analysis and CAP Strategy Group, Defra, York, UK

Department for Transport (2009) 'Renewable Transport Fuel Obligation: frequently asked questions', http://dft.gov.uk/pgr/roads/environment/rtfo/faq accessed 13 August 2009

Department of Energy and Climate Change (2008) *Reform of the Renewables Obligation*, Government Response to the Statutory Consultation on the Renewables Obligation Order 2009, DECC, HMSO, London

Department of Energy and Climate Change (2009) 'Bio-energy Capital Grants Scheme: Round 5', DECC, http://www.decc.gov.uk/bioenergy-grants/ accessed 26 January 2009

Department of Trade and Industry (2007) *Meeting the Energy Challenge. A White Paper on Energy*, CM 7124, DTI, HMSO, London

Edelmann, W., Baier, U., Engeli, H. and Schleiss, K. (2001) *Oekobilanz der Stromproduktion aus landwirtschaftlichem Biogas*, ARBI, Switzerland

EEA (2005) *Briefing 02/2005*, European Environment Agency, Copenhagen

European Climate Change Programme (2006) 'ECCP I review: Agriculture. Contribution of the CAP to climate change mitigation', Presentation by A. Moreale to Meeting 1,

31 January 2006, ECCP, http://circa.europa.eu/Public/irc/env/eccp_2/library?l=/eccp_agriculture/presentations_meeting/31-01-06_finalpdf/_EN_1.0_&a=d accessed 1 February 2009

European Commission (2007) 'Biofuels: Aid per hectare of energy crops reduced as the area exceeds 2 million hectares', Press Release IP/07/1528, 17 October, EC, Brussels

European Parliament (2008) 'EP seals climate change package', Background Paper from the European Parliament's Press Office, 17 December, Brussels

Eurostat (2007) 'Different organic farming patterns within EU-25: An overview of the current situation', *Statistics in Focus*, 69/2007, European Commission, Luxemburg

Gazley, I. and Francis, P. (2005) *UK Material Flow Review*, Office for National Statistics, London

Gericke, D., Pacholski, A. and Kage, H. (2007) 'NH_3-Emissionen bei der ackerbaulichen Nutzung von Gärresten aus Biogasanlagen', *Mitteilungen der Gesellschaft für Pflanzenbauwissenschaften*, vol 19, pp280–281

Grover, D., Aubrey, R. and Webster, R. (2005) *Research to Ascertain the Environmental and Economic Implications of Alternative and Non-food Crops: Impacts on the Countryside*, A study for the Countryside Agency, Institute for Sustainable Development in Business, Nottingham Trent University, Nottingham

HM Treasury (2008) *Budget 2008, Stability and Opportunity: Building a Strong, Sustainable Future*, HC 388, HMSO, London

Ho, M.W. (2008) *The Biogas Economy Arrives,* Special Report, The Institute of Science in Society, London

Höglund-Isaksson, L., Winiwarter, W., Klimont, Z. and Bertok, I. (2006) 'Emission scenarios for methane and nitrous oxides from the agricultural sector in the EU-25', Background document prepared for the Topic Group Agriculture and Forestry under the ECCP-I review, International Institute for Applied Systems Analysis (IIASA)

Holm-Nielsen, J.-B. (2000) *Danish Centralised Biogas Plants – Plant Descriptions*, Bioenergy Department, University of Southern Denmark, Esbjerg

Hopfner-Sixt, K., Amon, T. and Amon, B. (2005) 'Anaerobic digestion of energy crops – State of the art of biogas technology', 14th European Biomass Conference, 17–21 October 2005, Paris. ETA Renewable Energies, Florence, Italy

Kjær, S. and Madsen, J. (1998) 'Environmental policy for intensive livestock production in Denmark', in Ho Y. and Chan Y. (eds) *Proceedings of the Regional Workshop on Area-Wide Integration of Crop–Livestock Activities*, FAO Document Repository, http://www.fao.org/WAIRDOCS/LEAD/X6105E/x6105e11.htm#P0_0 accessed 1 February 2009

Lantz, M., Svensson, M., Bjornsson, L. and Borjesson, P. (2007) 'The prospects for an expansion of biogas systems in Sweden – Incentives, barriers and potentials', *Energy Policy*, vol 35, pp1830–1843

MAFF (2000) *Fertiliser Recommendations for Agricultural and Horticultural Crops (RB209)*, 7th edn, Ministry of Agriculture, Fisheries and Food, London

Mata-Alvarez, J. (2003) *Biomethanization of the Organic Fraction of Municipal Solid Wastes*, IWA Publishing, London

Mitchell, C., Bauknecht, D. and Connor, P.M. (2006) 'Effectiveness through risk reduction: A comparison of the renewable obligation in England and Wales and the feed-in system in Germany', *Energy Policy*, vol 34, no 3, pp297–305

Möller, K., Stinner, W. and Leithold, G. (2006) 'Biogas in organic agriculture: Effects on yields, nutrient uptake and environmental parameters of the cropping system', Paper presented at Joint Organic Congress, Odense, Denmark

Murphy, J.D., McKeogh, E. and Kiely, G. (2004) 'Technical/economic/environmental analysis of biogas utilisation', *Applied Energy*, vol 77, no 4, pp407–427

Petersen, S.O. (1999) 'Nitrous oxide emissions from manure and inorganic fertilizers applied to spring barley', *Journal of Environmental Quality*, vol 28, no 5, pp1610–1618

Petersen, S.O., Nielsen, T.H., Frostegard, A. and Olesen, T. (1996) 'O_2 uptake, C metabolism and denitrification associated with manure hot-spots', *Soil Biology and Biochemistry*, vol 28, no 3, pp341–349

Raven, R.P.J.M. and Gregersen, K.H. (2007) 'Biogas plants in Denmark: Successes and setbacks', *Renewable & Sustainable Energy Reviews*, vol 11, no 1, pp116–132

Rubaek, G.H., Henriksen, K., Petersen, J., Rasmussen, B. and Sommer, S.G. (1996) 'Effects of application technique and anaerobic digestion on gaseous nitrogen loss from animal slurry applied to ryegrass (*Lolium perenne*)', *Journal of Agricultural Science*, vol 126, no 4, pp481–429

SAIN (2008) Sustainable Agriculture Innovation Network Proceedings of the meeting on 'Circular Agriculture – Science, Technology and Policy' held in conjunction with Yangling Agro-Science Forum, November 2008, http://www.sainonline.org/SAIN-website(English)/pages/News/presentation.html accessed 1 February 2009

Salter, A.M. and Banks, C.J. (2009) 'Establishing an energy balance for crop based anaerobic digestion', *Water Science and Technology*, vol 59, no 6, pp1053–1060

SEEDA (2007) *South East England and London Regional Development Plan, Rural Development Programme 2007–2013*, South East England Development Agency, http://www.seeda.co.uk/Work_in_the_Region/Rural_Issues/Rural_Development_Programme_for_England/ accessed 1 February 2009

Steen, I. (1998) 'Phosphorus availability in the 21st century: Management of a non-renewable resource', *Phosphorus & Potassium*, vol 217, pp25–31

Tafdrup, S. (1994) 'Environmental impact of biogas production from Danish centralized plants', in Mitchell, C.P. and Bridgwater, A.V. (eds) *Environmental Impacts of Bioenergy, IEA Seminar*, Snekkersten, Denmark, 20–21 September 1993, CPL Press, Newbury, UK, pp138–143

Telegraph (2006) 'Business profile: The plutocrat of poo', *The Daily Telegraph*, http://www.telegraph.co.uk/finance/2919099/Business-profile-The-plutocrat-of-poo.html accessed 1 February 2009

Velthof, G.L. and Oenema, O. (1995) 'Nitrous oxide fluxes from grassland in the Netherlands: II, Effects of soil type, nitrogen fertilizer application and grazing', *European Journal of Soil Science*, vol 46, no 4, pp541–549

Walla, C. and Schneeberger, W. (2003) 'Farm biogas plants in Austria – An economic analysis', Paper presented at the Conference 'EU-Enlargement – Chances and Risks for the Rural Area', 13th Annual Meeting of the Austrian Society of Agricultural Economists, Ljubljana/Domzale, Slovenia

Weiland, P. (2006) 'Biomass digestion in agriculture: A successful pathway for the energy production and waste treatment in Germany, *Engineering in Life Sciences*, vol 6, pp302–309

WRAP (2009a) 'Anaerobic digestate', http://www.environment-agency.gov.uk/static/documents/Business/090210_AnaerobicDigestate_QP_v28.pdf accessed 15 August 2009

WRAP (2009b) 'Non-household food waste', http://www.wrap.org.uk/retail/food_waste/nonhousehold_food.html, accessed 1 February 2009

Wulf, S., Brenner, A., Clemens, J., Döhler, H., Jäger, P., Krohmer, K.H., Maeting, M., Rieger, C., Schumacher, I., Tschepe, M., Vandré, R. and Weiland, P. (2003) *Untersuchung der Emission direkt klimawirksamer Spurengase (NH_3, N_2O und CH_4) während der Lagerung und nach der Ausbringung von Kofermentationsrückständen sowie Entwicklung von Verminderungsstrategien (DBU-AZ 08912)*, Bonner Agrikulturchemische Series 16, Bonn, Germany

Yadava, L.S. and Hesse, P.R. (1981) *The Development and Use of Biogas Technology in Rural Areas of Asia (A Status Report 1981)*, Improving soil fertility through organic recycling, FAO/UNDP Regional Project RAS/75/004, Project Field Document No. 10, Rome, Italy

6

Watery Land: The Management of Lowland Floodplains in England

Joe Morris, Helena Posthumus, Tim Hess, David Gowing, Jim Rouquette

Introduction

Agricultural flood defence and land drainage improvement schemes in floodplain areas were a major component of Government support for agriculture and farmers in Britain in the post-World War II period. More recently, however, changing priorities in rural and environmental policy (evident for example in the reform of the European Common Agricultural Policy, the Water Framework Directive (WFD) and strategies such as 'Making Space for Water'; Defra, 2005a) have encouraged a re-appraisal of land management options for floodplain areas. Options such as the restoration of wetland areas for nature conservation or the storage of floodwater to alleviate flooding in lowland urban areas are being considered (Defra, 2009). However, in the face of strengthened prices for agricultural commodities and calls for greater national food security, there is renewed interest in the agricultural potential of Britain's lowland floodplains and fens. Many of the challenges for land use management and policy in Britain are currently being played out in these areas.

This chapter reviews the policy arena affecting floodplain management and explores the scope for floodplains to provide multiple benefits (including agriculture, nature conservation and floodwater storage) using an ecosystem-functions framework to balance the diverse interests of various stakeholders.

Context

Lowland floodplains provide a range of ecological and socio-economic services to society (Hale and Adams, 2007). In England, 1.02 million ha of agricultural land are within the indicative floodplain, that is, they have an annual risk of flooding of 1 per cent or greater from rivers or 0.5 per cent or greater from coastal flooding (Table 6.1). Although the indicative floodplain accounts for only 9 per cent

Table 6.1 *Classification of agricultural land in the indicative floodplain, England*

Agricultural land classification	Typical land use	Total (thousand ha)	Indicative floodplain (thousand ha)	Proportion of total area in indicative floodplain (%)
Grade 1	Intensive arable	355	204	57%
Grade 2	Intensive arable	1849	239	13%
Grade 3	Extensive arable	6292	379	6%
Grade 4	Dairy and grazing livestock	1840	186	10%
Grade 5	Grazing livestock	1101	11	1%
Total potential agricultural land		11,437	1019	9%
Non-agricultural land		657	31	5%
Urban areas		952	72	8%
Total		13,046	1122	9%

Source: NE, 2002 (Crown Copyright, All rights reserved)

of the total agricultural area of England, it includes some of the most fertile and productive areas, having been 'reclaimed' and 'improved' for agricultural purposes over hundreds of years. Fifty-seven per cent of Grade 1 agricultural land falls within the indicative floodplain, and the capital value of agricultural land at risk of flooding has been estimated at £7.1 billion (Defra, 2001). The management of hydrological regimes, in the form of flood alleviation and land drainage, has been key to maintaining the agricultural productivity of this land.

Lowland floodplains can support unique habitats such as floodplain grazing marsh, floodplain meadow, lowland fen, reedbed and wet woodland (Table 6.2) that perform key ecological functions and provide a valuable repository of biodiversity. However, the extent of these habitats has declined dramatically over the past 50 years as a result of changing land use (Fuller, 1987; Wilson et al, 2005), inappropriate grazing management, water level and ditch management, drainage, water abstraction and diffuse pollution (Turner et al, 2000; Tockner and Stanford, 2002; Hale and Adams, 2007; NE, 2008).

Ecosystems approach for integrated floodplain management

The ecosystems approach is a useful method to capture the variety of benefits and services provided by floodplains, as these satisfy the interests of a range of

Table 6.2 *Extent of wetland habitats and SSSI notification*

Habitat	Total resource (thousand ha)	Resource within SSSI (%)	SSSI area in favourable/ recovering condition (%)
Blanket bog	255	69	70
Coastal & floodplain grazing marsh	235	16	69
Fen	22	89	60
Lowland raised bogs	10	88	63
Reedbed	6	84	81
Total	259	47	69

Note: Areas given for fens are overestimates; the total includes other habitats existing within the mosaic

Source: NE, 2008

stakeholders, including: land managers, flood management agencies, nature conservation organizations, local communities, government, and society as a whole. The concept of 'ecosystem functions' represents the capacity of natural processes (methods of continuous operation) to provide goods and services (items that confer benefit and advantage) to meet human needs, directly or indirectly (Turner et al, 2000; Groot et al, 2002; MEA, 2003; Groot, 2006; Brauman et al, 2007; Zhang et al, 2007).

Ecosystem functions and services

In the context of lowland floodplains, the following five ecosystem functions can be identified (Groot et al, 2002):

1 The *production* function (linked to consumption functions) is the capacity of the ecosystem to provide resources, for example food from agriculture and raw materials.
2 The *regulation* function is the capacity of the ecosystem to regulate essential ecological processes and life support systems, for example regulating climatic, water, soil, nutrients, ecological and genetic conditions. Floodwater storage is a significant regulation function of flood plains.
3 The *carrier* function is the capacity of the ecosystem to provide space and location for activities and processes, for example habitation, energy generation and recreation.
4 The *habitat* function is the capacity of the ecosystem to provide unique refuges and nurseries for plants and animals, helping with the conservation

of genetic, species and ecosystem diversity (habitats are sometimes treated as part of carrier function).

5 The *information* (or *cultural*) function is the capacity of the ecosystem to contribute to human well-being through knowledge and experience, and sense of relationship with context, for example spiritual experiences, aesthetic pleasure, cognition and recreation.

These functions and related processes provide a range of services, such as agricultural production, outdoor recreation and landscapes, which are valued by the people, groups and organizations that make up society. These services for the most part involve some aspect of 'use', usefulness and hence utility. Some services, such as the value associated with the preservation of endangered habitats and species, involve 'non-use' benefits. Services and underlying functions perceived by society as most useful have generally been given greatest priority, such as crop production on farm land. Furthermore, as discussed later, societal preferences commonly find expression in the definition and distribution of property rights that give entitlement to flows of benefits for the use of resources. The concept of agricultural land tenure is a case in point, giving primacy to agricultural use and the production function.

The central role of water management

Land use in floodplains is largely determined by hydrological conditions described in terms of flooding and groundwater levels, such as:

- how frequently land is inundated by floodwater
- how long land remains inundated
- at what time of year flooding occurs
- how quickly flood water drains away
- the seasonal variation in the depth to the water table.

High levels of protection from flooding and the control of soil-water tables to avoid waterlogging are required to support intensive arable farming. Less intensive land uses, such as wet grassland, can tolerate lower standards of flood protection and land drainage and tend to be associated more with provision of non-market goods such as nature conservation and amenity.

Plant growth rates are reduced in waterlogged (i.e. saturated or inundated) soil due to anoxic conditions in the root zone, restricted nutrient uptake and lower soil temperatures. If waterlogged conditions persist, this can lead eventually to plant death. The impact depends on the timing in relation to growth stage and the tolerance of individual species to wet conditions. High water tables, for example (within 0.3 m of the surface), cannot support arable cropping, and agricultural land use is confined to grassland (Morris et al, 2004). Under extreme wet conditions, only plants that are adapted to waterlogging will prosper. They may have a high biodiversity but low agricultural value. Wet soils also have a reduced bearing capacity, such that machinery and animals can cause soil compaction and structural damage. For this reason, wet floodplains often cannot be farmed in the winter and can only support agriculture in the summer.

Inundation by floodwater not only raises the water table and submerges plants, but can cause physical damage to plants due to the erosive effects of flowing water. Physical assets can be damaged or destroyed and operations interrupted. The more valuable (in economic terms) are the crops in the ground or the assets at risk, the greater is the potential for damage. Table 6.3 shows the flooding and drainage regimes required for various floodplain uses. Typically cereal crops would not be grown on land that floods for short periods in winter more frequently than 1 in about 8 years, vegetable crops more than 1 in 20

Table 6.3 *Classification of flood and drainage regimes and related land use and habitat types*

Drainage	Rapid	Moderate	Slow
Winter flooding only			
Short-duration flooding	Arable, pasture, hay meadow, woodland	Pasture, hay meadow, woodland	Pasture, woodland
Medium-duration flooding	Hay meadow, pasture, woodland	Pasture, woodland	Pasture, swamp, woodland
Long-duration flooding	Pasture, woodland	Pasture, woodland	Swamp, pasture, woodland
Flooding at any time of the year			
Short-duration flooding	Hay meadow, pasture, woodland woodland	Woodland, pasture	Swamp, pasture, woodland
Medium-duration flooding	Pasture, woodland	Pasture, woodland, swamp	Swamp, pasture
Long-duration flooding	Swamp, woodland	Swamp	Swamp

Notes: Soil drainage is a function both of soil conductivity and drainage infrastructure. Rapid soil drainage: following inundation, water table typically falls by more than 30 cm in less than ten days in winter. Moderate soil drainage: following inundation, water table typically falls by more than 30 cm in less than 30 days in winter. Slow soil drainage: water table does not fall below 30 cm following an inundation event in winter until late April. Short-duration flooding: typically three days of surface water per event; Medium-duration flooding: typically less than two weeks of surface water per event. Long-duration flooding: typically more than two weeks of surface water per event

Source: Morris et al, 2005)

years and intensive grassland 1 in about 4 years. Tolerance to summer flooding is considerably lower.

Similarly, the position of the water table at different times of year controls the viability of different land uses. Figure 6.1 shows the seasonal water table requirements for conservation objectives in the Parrett catchment, in Somerset. The line represents the ideal water table regime that would provide a mixture of habitat and agricultural opportunities (Morris et al, 2008). If the water table should rise or fall to the extent that it is outside the white band, various ecosystem services would be affected. For example, a water table within 20 cm of the surface between mid-March and June would be unfavourable for the flood-meadow plant community, whereas a water table more than 35 mm from the surface will be unfavourable for habitats suited to breeding waders.

Land-use decisions will therefore determine the way in which the hydrological regime of the floodplain is managed. Embankments can be built or channels can be modified to regulate the frequency of inundation; under-drainage can be installed, in some cases with pumping of ditches, to regulate field water tables; and arterial drainage can be improved to evacuate excess water more rapidly. The resulting hydrological regime will be compatible with some uses, but may be in conflict with others. For example, it is evident that the development of floodplains to promote biodiversity associated with wetlands is likely to preclude intensive agriculture (Morris et al, 2004).

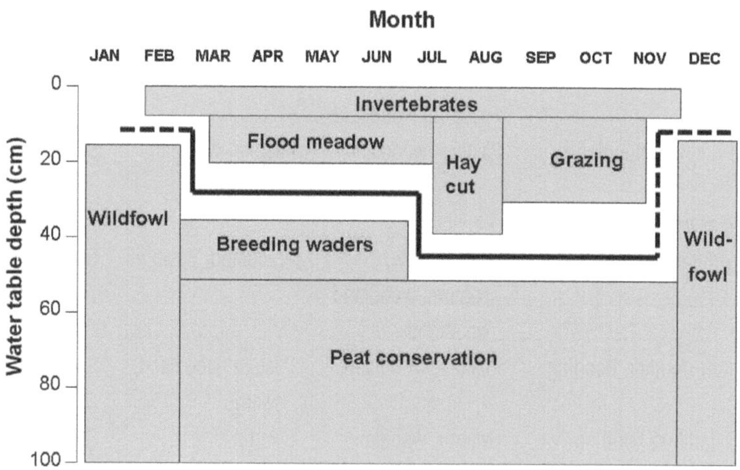

Note: The shaded zones represent ranges of water-table depth that would be potentially detrimental to the named interest. Thus, the unshaded zone represents a soil-water regime that has potential to meet the needs of a range of objectives at a site

Source: Morris et al, 2008

Figure 6.1 *Water table regime requirements for environmental characteristics in the Parrett catchment*

Similarly, floodplains can form a mechanism for flood management, that is, washlands (EA, 2002), and simultaneously enhance biodiversity by providing wetland habitat such as grazing marshes, fens and reedbeds (Morris et al, 2008). However, in order to maximize the flood storage function, it may be necessary to control the timing and frequency of flooding in such a way as is incompatible with habitat requirements, but concordant with the needs of agricultural production.

Hydrological regimes also determine the suitability for infrastructure and space for residential use. If land is flooded frequently, causing severe damage to existing infrastructure and residential properties, flood risk might reach unacceptable levels, compromising these ecosystem carrier functions. Other ecosystem services, such as GHG production, species, landscape value and recreation, are also indirectly affected by the hydrological regime as they are strongly dependent on land use (or habitats and agricultural production).

The desired functions and associated levels of hydraulic control depend on the local context and the position of the floodplain within a catchment. In some cases, floodplains will be assigned mainly for flood management, in other cases mainly for nature conservation, or in some cases a compromise may be possible by integrating these objectives (Morris et al, 2005). However, the sum of ecosystem goods and services delivered by the floodplain will depend critically on the hydrological regime.

Historical development of floodplain management, 1930–2000

Historically, floodplains have been used for various purposes, strongly influenced by changing needs, circumstances and a range of Government policy interventions. In the context of general economic depression, the 1930 Land Drainage Act, supported by grant aid, facilitated large-scale public works to improve agricultural productivity by alleviating flooding and establishing arterial drainage networks to evacuate excess water and control field water levels. Many of these projects were carried out by regional River Boards and local Internal Drainage Boards (IDBs), in some cases involving large-scale pumping schemes. Capital costs were shared among tax payers and local ratepayers, with the latter (including farmers) paying for operations and maintenance. Thus, improved drainage for agriculture was regarded as a 'public good', worthy of public funding, delivering benefits beyond those enjoyed by farmers themselves. Many agricultural flood defence and land drainage schemes were built in the inter-war period, but the potential benefits were not always taken up, partly due to the low prices of agricultural produce and lack of incentive for farmer investment in follow-up field drainage (Bowers, 1998; Scrase and

Sheate, 2005). Given the strategic importance of securing national food supplies, the war years (1939–1945) witnessed further public support for improved drainage, with large areas converted from pasture to arable land as part of a programme of 'digging for victory'.

In the post-1945 period, as food rationing continued, there were clear priorities for domestic food production. The 1947 Agriculture Act promoted 'efficient' agriculture to maximize production, minimize imports and ensure cheap food. Guaranteed commodity prices, ploughing grants and capital grants for improvement of farm assets, including field drainage, were introduced to ensure food supply at reasonable prices and provide fair rewards to farmers (Morris, 1992; Hodge, 2001; Werritty, 2006). The National Agricultural Advisory Service, established in 1948, also included technical assistance to farmers on field drainage.

At a time of general world food shortages and with the UK facing a foreign exchange crisis, severe floods in the spring of 1947 inundated 279,000 ha of agricultural land, almost half of which was arable. A concerted effort and substantial national investments were made to repair damaged embankments and replant crops and the impact on the nation was much less than it could have been (MAF, 1948). However, the event served as a reminder of the importance of land drainage and flood defence to national food security. The major floods in eastern England in 1953, inundating much of the previously 'reclaimed' and intensively farmed lowland fenland areas, reinforced the importance of sea defences in lowland coastal areas. During the following decades, large investments were made to raise defences in both urban and rural areas against river and coastal flooding (Johnson et al, 2005; Werritty, 2006).

From the initiation of the CAP by the European Community in 1958 (Dobbs and Pretty, 2004) and through to the 1980s, European agricultural policy continued to promote increased production of and self-sufficiency in food and fibre, simultaneously supporting farming income through commodity price support. Following UK membership of the European Community in 1973, the UK Government's 1975 White Paper 'Food from Our Own Resources' (HMSO, 1975) continued to emphasize agriculture's import-saving role.

In England and Wales, capital grants were provided for farm improvement measures and as a consequence drainage activity increased throughout the period from the 1940s to the early 1970s. Field drainage was eligible for grant aid of up to 60 per cent of capital expenditure (Bingham, 1983) and, by 1972, around 100,000 ha $year^{-1}$ were being drained (Armstrong, 1978), most of which was on arable land with existing, but deteriorated drainage systems (Trafford, 1978).

At the local scale, the intensification of agriculture arising from the CAP resulted in the removal of hedgerows and woodland, land drainage and conversion of pastures into arable land (Ogaji, 2005), subsequently increasing water pollution, land degradation and biodiversity loss (Mayrand et al, 2003). The drive for increased production also led to commodity surpluses and, of particular concern at the time, a realization that the EU CAP regime was financially unsustainable.

In recognition of the threats posed to important natural habitats, international agreements (e.g. the Ramsar Convention on Wetlands, 1971) and legislation (e.g. the EU Birds Directive, 1979) were made to protect and restore endangered habitats and species (Hodge and McNally, 2000). The introduction of the UK Wildlife and Countryside Act (1981) gave public agencies a duty to consider nature conservation, to designate some wetlands as SSSIs and to allow floodplain and river restoration to emerge (Adams et al, 2004). It was also apparent that (as in the pre-war period) the full benefits of land drainage improvement schemes were not being realized as farmers were not following up with on-farm investments, especially on later schemes (Morris and Hess, 1986).

A substantive change occurred in British agricultural policy in the mid-1980s. Milk quotas were introduced in 1984 and, following the 1986 Agriculture Act, agri-environmental schemes in the form of Environmentally Sensitive Areas (ESAs). The period of agricultural enhancement was over. Grants for field drainage were gradually withdrawn – from around £30 million per year in 1984–1985 to £2 million per year in 1988–1989. As a result the installation of new field drainage fell dramatically.

Reflecting a major change in policy, the programme of ESAs was designed to provide financial incentives to encourage farmers to adopt or maintain practices that protect and enhance the farmland environment and wildlife. Beginning in 1987, 22 areas in England and Wales were designated as ESAs, within which farmers were automatically eligible for membership. Several ESAs contained options for floodplain areas where the aim was to maintain or enhance the wetland resource by paying farmers to pursue appropriate agricultural practices (Hodge and McNally, 2000).

The commitment to agri-environmental objectives was further strengthened in 1991 with introduction of the Countryside Stewardship Scheme (CSS). This invited landowners outside designated ESAs to competitively bid for funding to support actions to sustain the beauty and diversity of the landscape and to improve and extend wildlife habitats, including wetland sites. Further rounds of CAP reform extended the scope of agri-environmental schemes and essentially made set-aside compulsory for a majority of arable farmers in exchange for new arable area support payments. Lands liable to flooding or poor drainage became obvious candidates for extensification, either through set-aside or agri-environment options.

Current situation: reform

The current period is characterized by the Agenda 2000 CAP Reforms, the 2003 Mid Term Review (implemented in 2005) and attempts to wean agriculture off a high level of support, either because of the expense or because of contravention of the principles of market liberalization according to the World Trade Organization (Latacz-Lohman and Hodge, 2003; Herzog, 2005). In 2005, financial support to farmers was further decoupled from commodity prices and production with the introduction of the Single Payment Scheme (SPS) (Defra, 2008a). Direct production subsidies were replaced with a new, single support payment, based on a combination of historical entitlements and farm size, linked to compliance with standards which aimed to protect the environment, animal health and welfare (Cross Compliance rules; see Defra, 2008a). ESA and CCS schemes were reformulated under a new Environment Stewardship Scheme, with a range of options of differing environmental intensity from 'entry' to 'high' levels, including an option for 'inundation grassland' suited to floodplains.

Following several decades without major flood events, flooding largely receded from public view. Indeed, a succession of dry years, especially in the mid-1970s and the early 1990s focused attention on the risks of droughts rather than floods. During Easter 1998, however, a widespread fluvial flood occurred after exceptional rainfall. Extensive fluvial floods occurred in autumn 2000, suggesting that the exceptional floods in 1998 were not so exceptional after all. Increasing concern about climate change further highlighted the need to plan for more frequent extreme flood events and a realization that structural flood defences cannot provide protection for all potential flood risks (ICE, 2001; Johnson et al, 2005).

The lessons learned in 1998 and 2000, and the drive for environmental enhancement (e.g. WFD) promoted a policy switch from flood defence to flood risk management (Werritty, 2006). In many respects this was a return to the focus on integrated land drainage that had always been a characteristic of the approach to water management in the countryside. Reinforced by concern about impending climate change and increased flood risk, the Government-sponsored Foresight Flood and Coastal Defence Project drew on available science to explore possible long-term flooding futures for the UK under alternative development scenarios (Evans et al, 2004). The project identified very large increases in potential economic damage, predominantly in urban areas, and called for a reappraisal of policy responses. It also confirmed the potential role of rural and floodplain land in flood risk management. These points were reinforced by the Pitt Review (Pitt, 2008), 'Learning the lessons from the 2007 floods'.

Supported by Foresight and other evidence (O'Connell et al, 2004; FRMRC, 2007), a strategic approach to flood risk entitled 'Making Space for Water' (Defra, 2005a) sets out 'to manage the risks of flooding and coastal erosions in an integrated and holistic way', taking account of social and environmental factors as well as economic costs. This means prioritizing investment and resources in areas where flood risk is greatest and alleviation gives greatest net benefit. This necessarily leads to a focus on urban areas where flood damage costs are highest. With respect to the rural space, Defra (2005a) suggests that greater use can be made of rural floodplains to reduce flood risk. This can be done, it is argued, by creating wetlands and washlands, river-corridor widening and river restoration, provided that impacts are beneficial to the environment and do not adversely impact people (EA, 2003; Defra, 2005a). As mentioned above, however, it cannot always be assumed that wetland and flood storage options are compatible. Neither might making space for water in floodplains necessarily provide the degree of flood control that will reliably prevent flooding in urban areas downstream.

In the English case, the Government continues to fund the maintenance of existing defences on rural floodplains where the costs are justified by the full range of benefits provided by the defences. Cost-benefit analysis of flood alleviation expenditure is now supplemented by appraisal of target outcomes, which include creation of wetland areas (Defra, 2007), and landowners may qualify for financial payments for the retention and/or storage of floodwater on their land (Defra, 2005b; JNCC, 2008).

Given some of these changing drivers, rural land use in some floodplain areas has shifted from predominantly agricultural production to land uses that need less protection against flooding and can deliver multiple benefits. There is a growing commitment to deliver sustainable flood risk management solutions through land use and land management change, among others through the Catchment Flood Management Plans (EA, 2008). These include reducing standards of protection for non-residential, rural parts of the catchment and increasing standards elsewhere. For its part, the Catchment Sensitive Farming programme aims to join up flood risk management with other objectives such as the control of diffuse pollution from farmland (Defra, 2008b).

It is apparent that an array of policies shapes the management of lowland floodplains, addressing issues such as agriculture, biodiversity, pollution and flooding. Figure 6.2 (and Appendix 1) show the current policies concerning land use (top left grey circle), water management (bottom grey circle) and biodiversity (top right grey circle). It also shows the scale at which these interventions operate, from international treaties in the outer circle, through EU directives, legislation and policies in England to the individual policy mechanisms affecting floodplain management in the inner circle.

Since the early 1930s, Acts of Parliament, which have been amended over time, have allocated the responsibilities and permissive powers for land drainage and flood defence. Most of the policies concerning biodiversity and environmental protection are set at European level (e.g. Birds Directive,

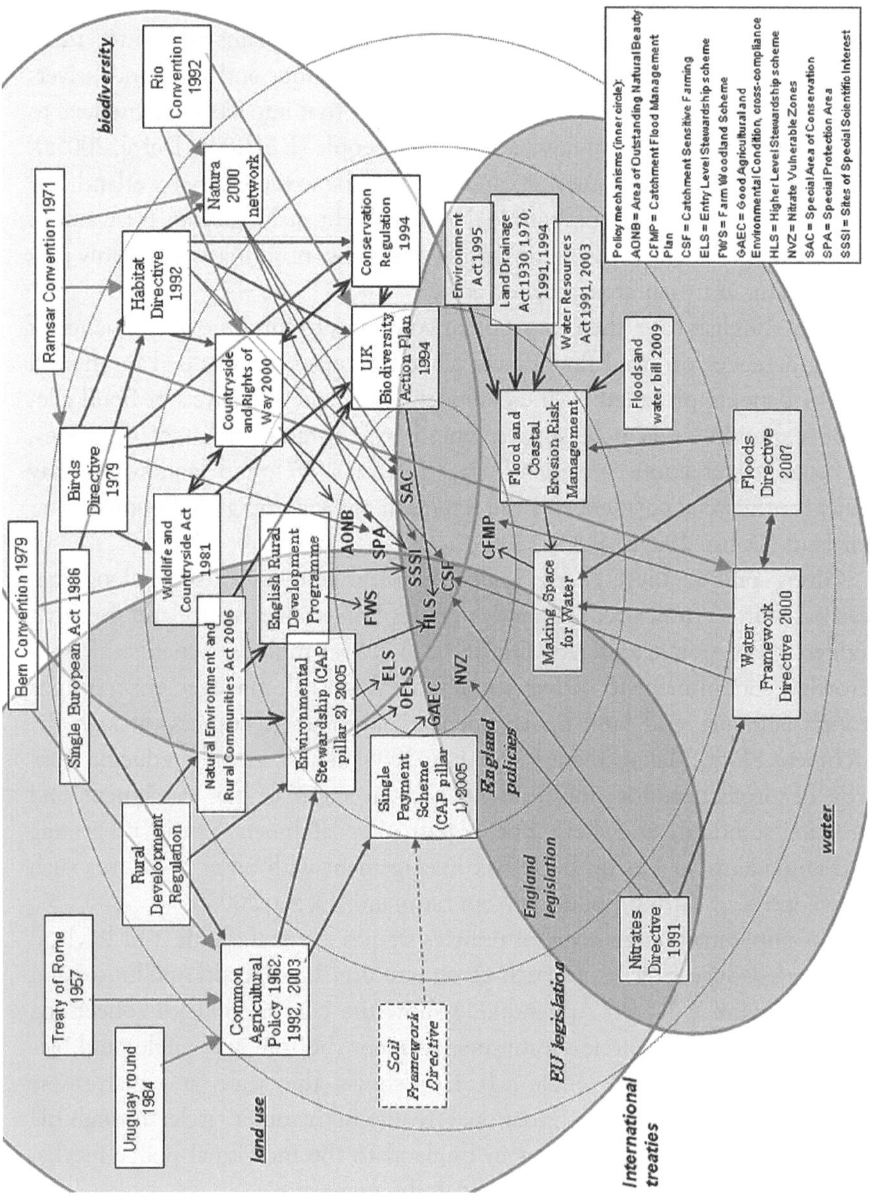

Figure 6.2 *Current policies influencing the management of lowland floodplains in England*

Habitat Directive), and translated into legislation (e.g. Wildlife and Countryside Act) and policies (e.g. UK Biodiversity Action Plan) at national level. It is only more recently that agreements on water management have been reached at European level (e.g. WFD, Floods Directive). The objectives of protection and enhancement of natural resources and biodiversity are typically delivered by a mix of policy instruments. These combine regulatory measures such as habitat protection or limits on nitrogen fertilizer, economic mechanisms such as agri-environment payments for wetland management and/or voluntary arrangements undertaken by land owners in collaboration with NGOs such as the Farming and Wildlife Advisory Group (FWAG) and Linking Environment and Farming (LEAF).

Although priorities appear to have switched in favour of environmental enhancement, the requirement for farmers to maintain land in 'good agricultural and environmental condition' in order to qualify for income support under the SPS confirms the strategic importance of maintaining the productive capacity of the land resource. The food shortages and price spikes of 2007/2008, exacerbated by international demand for bioenergy, confirm the relevance of such capacity in reserve. Indeed, much of this capacity exists in areas of lowland Britain that are almost universally dependent on land drainage and in many cases on some degree of flood protection.

In many respects, the new Floods and Water Bill, (Defra, 2009) reveals the emerging multi-purpose agenda for floodplains in England. Here, the Government seeks to manage flood risk while achieving 'wider policy objectives such as maintaining good soil quality, landscape and healthy resilient natural environments'. It specifically refers to working 'with natural processes of flooding and erosion at a local scale'. Thus, as far as rural floodplains are concerned, from an initial focus on flood defence and drainage for agriculture, there is now clear recognition of the important 'links and dependencies between different policy areas and activities' – an attempt to join things up in floodplains.

Property rights and stakeholders

The concept of ecosystem services, referred to earlier, reflects an anthropogenic perspective whereby 'stakeholders' have an interest in and derive value from particular services. Stakeholders usually lay claim to the benefits bestowed by ecosystem services and their underlying functions through the operation of property rights (Bromely, 1991). Property rights determine whether an individual or organization has entitlement to own, possess or use a particular resource and/or derive benefit from the flow of services that it provides. Property rights are not absolute, but rather conditional on, and derived from, social preferences that change over time (Tawney, 1948; Bromley and Hodge, 1990).

Historically, property rights have been associated with the direct use of ecosystems to provide goods and services, such as the products of agriculture and industry. Hence, uses and property rights have typically focused on the production (and related consumption) functions of ecosystems, with values reflected in the price of goods and services traded in the market. For the most part, property rights for natural resources have been given to resource-using industries. In principle, the award of entitlement for agricultural land tenure, for example, reflects a preference by society to allocate land to farmers for the production of food and fibre. Indeed, in post-World War II Britain, food security was clearly a 'public good' and government support to farming a clear indication of this. However, a shift towards so-called multi-functionality, whereby the management of land is required to deliver a range of services, means that conventional property regimes based on use for a single purpose become inappropriate or unworkable (Sandberg, 2007).

Thus, changing priorities for land and water use, for example for flood-water storage or enhancement of biodiversity, may call for a redefinition of property rights and entitlements, as well as changes in policy to effect this. Failure to adequately recognize the value of ecosystems services is commonly associated with ill-defined property rights. The ecosystem functions of regulation, habitat and information, which are particularly associated with public rather than private benefits, are often excluded from formal entitlements. Furthermore, they may be lost or compromised when primacy is given to formal rights that serve private, mainly production and consumption interests. Indeed, externalities due to the actions of individuals that are borne or enjoyed by third parties without compensation or payment are indicative of a failure of property rights. Current holders of property rights often have little incentive to conserve, produce and enhance the provision of such public services without some form of outside inducement (Goldman et al, 2007).

Property rights are held by stakeholders, who can as a result exert influence or control over a particular phenomenon such as land resources. Stakeholders can be classified according to their degree of influence and also by degree of interest, as shown in Figure 6.3 (Johnson and Scholes, 2002; Posthumus et al, 2008a). 'Key players' have a high influence and interest. 'Context setters' are highly influential but have little interest, whereas 'subjects' have a high interest but limited influence. The 'crowd' are stakeholders with relatively little interest and influence. Interest and influence can, however, change over time as circumstances change and stakeholders move from one category to another.

It is apparent that society is reclaiming entitlements to ecosystem services through the evolution of environmental policies across UK landscapes (Hodge, 2001). In the case of lowland floodplain, there is evidence of a range of interventions, such as limits on nitrogen fertilizer or the timing of some field

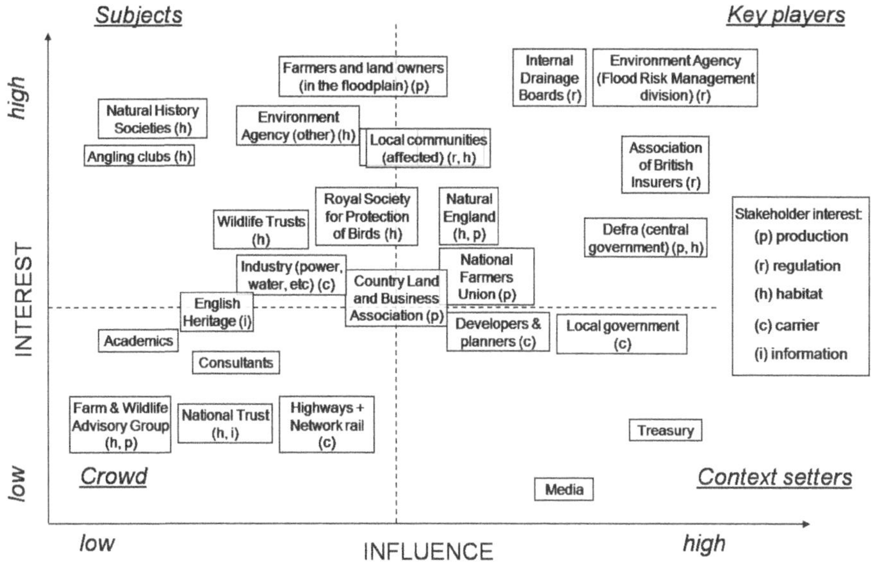

Figure 6.3 *Interest and influence of stakeholders regarding ecosystem functions of lowland floodplains*

operations, voluntary agreements such as seasonal limits on water abstraction for irrigation, and negotiated financial agreements such as Environmental Stewardship.

The aforementioned classification of stakeholders varies across locations and scale of influence, whether national or local. The dominant interests of the key players in floodplain management have traditionally been with production, regulation and, where urban development is concerned, with carrier functions. Key players typically hold formal land ownership or occupancy rights, such as farmer–producers, or statutory powers associated with the regulation of water, such as IDBs and the Environment Agency. Organizations such as Local Government and the Association of British Insurers (ABI) also have particularly high interest in the carrier function as this relates to human settlements exposed to flood risk. Stakeholders with an interest in habitat and information functions have tended to have relatively limited influence. An NGO conservation organization might, however, purchase farm land and with it the entitlement to convert land use to wildlife habitats. Alternatively, it might lobby key institutional land owners who are sensitive to public scrutiny. It might form strategic alliances with local land owners to promote publicly funded programmes to enhance biodiversity on farmland (Reed et al, 2009). Some of these initiatives can be observed in practice, as discussed below.

Integrated floodplain management: Case study of the Beckingham Marshes

The challenge of managing land and water resources to meet diverse interests is being considered in a Relu research project entitled 'Integrated floodplain management' (Relu, 2008a). Options of land and water management that can deliver a range of ecosystem services, and thereby serve a range of stakeholder interests, have been explored in the benefit areas of eight agricultural flood defence schemes throughout England.

By way of example, Beckingham Marshes, one of the study areas, occupies 900 ha of floodplain of the River Trent in Nottinghamshire, opposite the town of Gainsborough. Prior to 1945, the area was covered by a mix of grassland, marsh and willows, and provided uncontrolled flood storage for Gainsborough. Flood defences (designed for 1 in 20-year flood events) were built in the 1960s to provide flood storage (2,000,000 m^3) to reduce the probability of inundation of adjacent urban areas as well as protection for arable farming, supported by land drainage and a pumping station. By 1983, 82 per cent of the land area was used for arable farming.

Before the scheme, the land was purchased from institutional land owners by the River Trent Water Board, the predecessor of the Environment Agency (EA), who awarded life-long tenancies to farmers. Thus, property rights held by the EA support the provision of emergency flood storage (regulation function) to protect adjacent urban areas in the floodplain (carrier) and enable arable farming (production).

A recent collaborative venture between the EA and the Royal Society for the Protection of Birds (RSPB) intends to promote the habitat function in the Marshes. This involves the RSPB taking over some agricultural tenancies and a planned reversion (initially) of about 10 per cent of the area to wet grassland for nature conservation. Modifications to hydrological regimes are required to achieve the desired changes in ecosystem functions and services: from production to habitat provision. The RSPB intend to raise water levels in the floodplain in order to create habitats for lapwings (*Vanellus vanellus*). However, this could reduce floodwater storage capacity. It will reduce agricultural output on wetland sites and also possibly on adjacent land. As a consequence, there is potential conflict between the different interests of farming, flood management and habitat provision. For their part, farmers have reported that the adoption of wet grassland options would require financial 'compensation' in the form of lower land rents or environmental payments.

Scenarios of land and water management

Different scenarios of land and water management have been developed in order to analyse the impact of different management regimes on ecosystem goods and services delivered by floodplains:

- *Current situation*: based on information collected during farmer interviews in 2006.
- *Maximum agricultural production*: comprises intensive agricultural land use (as originally envisaged when the scheme was enacted in the 1960s). Land use is defined by soil, climate and current and past land use patterns. The hydrological regime is characterized by controlled drainage and low flood frequency.
- *Maximum biodiversity within an agricultural system*: seeks to enhance biodiversity with the imposed constraint that the predominant land use remains agriculture. Land use options are those promoted by current agri-environment schemes, in particular the Higher Level Stewardship (HLS) scheme (Defra, 2005b). The hydrological regime attempts to balance the requirements for agriculture and wet habitats, but typically consists of medium-flood frequency and just surface rather than deep drainage. Local soil conditions, topography and historical context, together with local and regional conservation and land-use priorities are used to determine specific habitat types.
- *Maximum biodiversity outside an agricultural system*: seeks to enhance biodiversity, guided by local and national Biodiversity Action Plan targets. The hydrological regime is characterized by frequent flooding and sometimes slow surface drainage. The same criteria are used for determining habitat types as for the previous scenario.
- *Maximum floodwater storage*: seeks to maximize the attenuation of the flood hydrograph by 'flood peak capping'. A flood protection standard is designed so that the timing of the start of filling the storage provided by floodplains is delayed to accommodate peak flows. This requires a hydrological regime with controlled low-flood frequency (one in ten years), rapid evacuation of surface water and low in-field water tables.
- *Maximum income*: seeks to maximize the income derived from the land. The land use for this scenario is determined by one of the previous scenarios with the highest estimated profitability (net margin) for land management.

Indicators were defined to assess the delivery of ecosystem goods and services under the different scenarios. Table 6.4 gives selected indicators for ecosystem goods and services provided within the floodplain.

Table 6.4 *Indicators for ecosystem goods and services provided by lowland floodplains and fens*

Function	Uses, goods and services	Indicator
Production	Agricultural production	Gross output (£ ha^{-1} $year^{-1}$)
	Financial return	Net margin (£ ha^{-1} $year^{-1}$)
	Employment	Labour (mandays ha^{-1} $year^{-1}$)
	Soil quality	Soil carbon flow (kg C ha^{-1} $year^{-1}$)
Regulation	Floodwater storage	Time to fill capacity (days)
	Water quality	Nutrient leaching (kg NO_3 ha^{-1} $year^{-1}$)
	Greenhouse gas balance	Global Warming Potential (kg CO_2-equiv. ha^{-1} $year^{-1}$)
Habitat	Habitat provision	Habitat conservation value (score)
	Wildlife	Species conservation value (score)
Carrier	Transport	Risk exposure road infrastructure (£ ha^{-1} $year^{-1}$)
	Settlement	Risk exposure residential properties (£ ha^{-1} $year^{-1}$)
	Space for water	Area inundated through fluvial flood (ha $year^{-1}$)
Information	Recreation	Potential recreational use (score)
	Landscape	Landscape value (score)

Table 6.5 summarizes key characteristics of the different scenarios (see also Posthumus et al, 2008b). The 2006 scenario features current land use: winter cereals in rotation with oilseed rape, field beans and peas, are the predominant crop type as they are most suitable to the heavy clay soils. Grassland managed by the RSPB supports an extensive beef system, and it is assumed that under other, 'wetter' scenarios this system will prevail. The *maximum production* scenario features similar land use to the 2006 scenario, but farming systems are more intensive. The recently established wet grassland is converted back into arable land under this scenario.

The most appropriate land use to increase biodiversity in an agricultural context (*maximum biodiversity within an agricultural system*) is a combination of wet grassland for breeding waders and species-rich hay meadow, which supports an extensive beef system. Maximizing biodiversity outside of an agricultural context would enable the creation of a large area of reed bed and wet woodland, along with some wet grassland.

For the scenario *maximum floodwater storage*, a land-cover type (either improved grass or winter cereals) is chosen with high transpiration rates and relatively low sensitivity to flooding. Since Beckingham Marshes is predominantly arable at present, cereals are grown under this scenario.

Table 6.5 *Scenario characteristics for Beckingham Marshes*

	2006	Max production	Max biodiversity & agriculture	Max biodiversity	Max floodwater storage	Max income
Mean water table depth (m)	0.5	0.6	0.4	0.1	1.0	0.3
Annual flood probability (%)	10%	10%	50%	50%	10%	60%
Land use	Cereals, grazing marsh	Cereals, root crops	Grazing marsh, floodplain meadow	Reed beds, wet woodland, grazing marsh	Cereals	Grazing marsh
Livestock	Extensive beef	n/a	Extensive beef	Extensive beef	n/a	Extensive beef

Source: Posthumus et al, 2008b

The scenario *maximum farm income* features grazing marsh with extensive beef as this land use type provides the highest income when 2006 prices for agricultural produce are used (wheat price is taken as £68/tonne). For this scenario it is assumed that farmers receive £335/ha annually under the HLS scheme for maintenance of wet grazing marshland suitable for breeding waders.

Table 6.6 presents the results for a sub-set of indicators representing ecosystem goods and services for the different scenarios. The scenario outcomes highlight synergies and conflicts between ecosystem goods and services delivered by lowland floodplains. There is clear synergy between agricultural production and floodwater storage, as shown in the scenario *maximum floodwater storage*. There also exists a clear synergy between wet habitats and low environmental impact (GHG emissions and nitrate leaching). However, intensive agricultural production generates relatively high environmental impacts such as high GHG emissions, high risk of nitrate leaching and low biodiversity.

The financial performance of different land uses under each scenario is sensitive to farm output and input prices and agri-environment payments. This sensitivity has implications for the design and implementation: (i) of hybrid or composite land and water management scenarios that will be beneficial and robust under a range of future possible conditions, and (ii) of policy and

Table 6.6 *Scenario outcomes for selected indicators, Beckingham Marshes*

Indicator	Scenario					
	2006	Max production	Max biodiversity & agriculture	Max biodiversity	Max floodwater storage	Max income
Agricultural production	****	*****	***	*	*****	***
Financial return	**	**	*****	****	**	*****
Employment	**	***	***	*****	***	***
Soil quality	**	*	****	*****	**	*****
Floodwater storage	*****	*****	*****	**	*****	*****
Water quality	**	*	***	*****	*	***
GHG balance	***	*	*****	****	*	*****
Habitat provision	**	*	*****	*****	*	****
Species	***	***	*****	****	***	*****
Transport	*****	*****	*****	*	*****	*****
Settlement	*****	*****	*****	*	*****	*****
Space for water	*	**	*****	*****	*	*****
Recreation	**	**	*****	*****	**	*****
Landscape	****	*	*****	****	*	*****

Note: * = worst, ***** = best

support regimes that will make such scenarios appealing to the main stakeholders, especially land managers, conservationists, flood managers and local communities (Posthumus et al, 2008b).

In 2006, extensive arable (cereals) and livestock (beef) farming systems were not profitable as they generated negative net margins (excluding single farm payments decoupled from production). Indeed, farmers interviewed in the first quarter of 2007 commented that it was more profitable to create habitats for breeding waders and receive agri-environment payments than to undertake conventional farming. In 2007, prices for feed wheat doubled to over £150/tonne as a result of shortages of cereals on the world market. A wheat price of about £140/tonne would make wheat production more profitable than opting for the HLS payments, and the maximum farm income and maximum production scenarios would converge at this price. Farmers interviewed in early 2008 suggested that if cereal prices stayed high they would consider abandoning agri-environment schemes in favour of a return to cereal production. However, by late 2008, feed wheat prices had fallen back to about £85/tonne as a result of increased global and national supplies induced by high prices.

Ecosystem services and stakeholder interests

It is apparent that dominant interests, influences and property rights at Beckingham have varied over time. As explained earlier, the main stakeholders for the past three decades have been the EA, the IDBs and the farmers, serving regulatory, production and (downstream) carrier functions. The RSPB have been able to enhance habitat functions through acquisition of land and adoption of Government-sponsored agri-environment schemes. Though initially the RSPB intended to collaborate with local tenant farmers, it appears that differences of opinion have arisen over how best to manage wet grassland in ways that combine farming and wildlife interests. Table 6.7 summarizes the ecosystem goods and services and stakeholder interests for the Beckingham Marshes (see Reed et al, 2009).

Table 6.7 *Ecosystem functions, goods and services and stakeholder interests in Beckingham Marshes*

Function	Uses, goods and services	Values	Stakeholders
Production	Agricultural production	Economic gains from crop and livestock production	Farmers, Defra
Regulation	Floodwater storage, drainage, carbon cycle	Avoided damage due to flooding, tradeable services	EA (flood risk management), IDB, farmers, local industry, RSPB
Habitat	Maintenance and enhancement of biodiversity	Contribution to Biodiversity Action Plan (BAP) targets	RPSB, EA (conservation), Wildlife Trust, On Trent Initiative, local residents
Carrier	Transport and settlements, industrial site	Location for housing, roads, local industry; property and service values, costs of alternatives	Local residents, local industry, farmers, local government
Information	Amenity, landscape, recreation, history	Enjoyment of countryside and related benefits: willingness to pay	RPSB, local residents, local authority

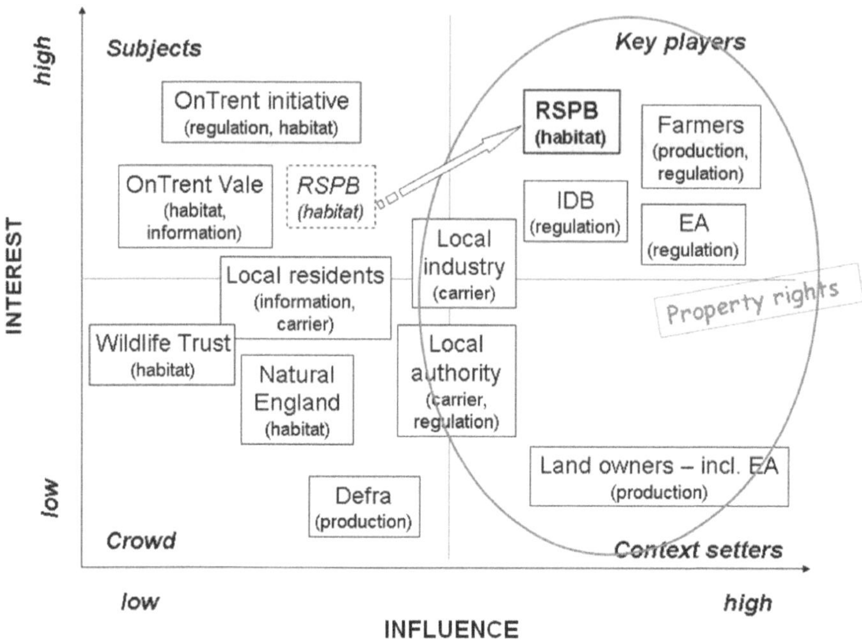

Source: Reed et al, 2009

Figure 6.4 *Interest–influence matrix for stakeholders in Beckingham Marshes*

Mapping the Beckingham stakeholders on a matrix of interest and influence confirms that stakeholders concerned with regulation and production functions are the key players (Figure 6.4). However, more recently other stakeholders are able to promote their interests in habitat and information functions by encouraging the adoption of multi-functional agri-environment schemes. It is possible, however, that conflicts may arise as more stakeholders with different interests find they can exert greater influence, and some form of bargaining and reconciliation may be required (see Reed et al, 2009).

Conclusions and policy implications

Lowland floodplains deliver a range of benefits, both market and non-market goods and services, to society. Though some of these benefits can be delivered simultaneously in synergy, other benefits tend to conflict with each other and are either exclusive or have to be compromised. The management of lowland floodplains is clearly a product of policy interventions that have promoted particular objectives at different times. Once the recipient of large-scale public investments in agricultural flood defence and drainage, lowland floodplains are now recognized as providing a range of valued ecosystems goods and services, including water regulation, carbon sequestration, landscapes and wildlife, and recreation and amenity. As we argue here, these latter interests are finding expression in new policy initiatives which attempt to integrate land and water management in flood-prone areas such as, for example, Defra's 'Making Space for Water' and 'Catchment Sensitive Farming'.

The example of the Beckingham Marshes illustrates the potential advantage of an ecosystem framework for assessing land and water management options. It is clear that the type of goods and services rendered by floodplains reflect dominant stakeholder interests and influences, as shaped by prevailing incentives and property rights and entitlements. An ecosystems approach could help to reinforce the value and importance of non-market goods and services provided by floodplains, with policies that target particular outcomes independently of agriculture.

However, using an ecosystems framework requires a much more integrated, joined-up approach to natural resource management. This may imply a restructuring of the current institutional framework as policies and entitlements tend to focus on one or two ecosystem functions only. New, disaggregated property regimes may be required whose various elements explicitly refer to particular uses and services, giving separate entitlements, possibly at different time of year, to different services such as agricultural use, flood storage, conservation or public access. In this respect, clear insights are required of both the demand for and supply of ecosystem services, at present and into the future. This begs answers to the questions raised in the Great Land Use Debate (Relu, 2008b), namely: What is land for? Who should decide? and What are the best ways of ensuring that the future stock of land and water resources, including those in floodplains, provides the diverse flows of services required to meet future needs?

Appendix 1 Overview of current policies affecting the management of lowland floodplains in England

Policy or legislation	Description	Policy instruments
International treaties		
Treaty of Rome 1958	Establishment of European Economic Community	Regulatory: common market for agriculture and trade
Bern Convention 1979	Intergovernmental treaty to conserve and protect wild plant and animal species and their natural habitats	Regulatory: Birds Directive, Habitat Directive, Wildlife and Countryside Act
Rio Convention 1992: Convention on Biological Diversity	Intergovernmental treaty to conserve biodiversity, use components of biodiversity sustainably, and share benefits arising from commercial and other utilization of genetic resources in a fair and equitable way	Regulatory: UKBAP
Ramsar Convention 1971	Intergovernmental treaty to enhance national action and international co-operation for the conservation and wise use of wetlands and their resources	Regulatory: designation of suitable wetlands for the List of Wetlands of International Importance
EU policies		
CAP	EU policy to support agriculture (pillar 1) and rural development (pillar 2) and to control EU agricultural markets. The objectives of pillar 2 are to create a productive and sustainable rural economy, and to conserve and enhance the rural environment	Regulatory and economic instruments (CAP pillars 1 & 2)
Rural Development Regulation	European regulation for support for rural development under the second pillar of the CAP	Regulatory, with economic instruments
Habitat Directive (92/43/EEC)	EU directive to conserve natural habitats for selected species of flora and fauna. Linked with Birds Directive (79/409/EEC)	Regulatory: UKBAP, Natura 2000

Policy or legislation	Description	Policy instruments
Birds Directive (79/409/EEC)	EU directive to conserve and manage wild birds	Regulatory: SPAs
Natura 2000 network	The Natura 2000 network is a European wide network of protected areas recognized as sites of community importance under the Habitat Directive	Regulatory: designation of protected areas and species; SACs, SPAs, SSSIs
Water Framework Directive (2000/60/EC)	EU directive to improve and protect the ecological status of the water environment and encourage sustainable water resource management	Regulatory: Making Space for Water, Catchment Flood Management Plans, Catchment-Sensitive Farming
Nitrates Directive (1991/676/EEC)	EU directive to reduce diffuse water pollution by nitrate from agriculture	Regulatory: Restrictions on timing and quantity of nitrogen fertilizers in designated nitrate vulnerable zones (NVZ).
Flood Directive (2007/60/EC)	EU directive to prevent and limit floods and their damaging effect on human health, environment, infrastructure and property	Regulatory: flood risk maps, flood risk management plans
National legislation		
Land Drainage Act 1991, 1994	An Act of Parliament that includes the current legal basis for permissive powers of local authorities and drainage boards to carry out flood management and land drainage works. The 1994 Act added environmental and recreational obligations	Regulatory: imposed duties on authorities responsible for flood defence to further conservation, enhance beauty and take account of the effect of their work on the beauty and amenity of affected area
Environment Act 1995	Act of Parliament that established the permissive powers and responsibilities of the EA regarding flood management	Regulatory: outlines duties and powers for flood defence and other functions
Water Resources Act 1991	Act of Parliament that includes the current legal basis for permissive powers and responsibilities of the National Rivers Authority (now EA)	Regulatory: duties and powers of NRA (now EA) regarding, among others, flood defence, abstraction, fisheries, pollution

Policy or legislation	Description	Policy instruments
Water Act 2003	Act of Parliament that further enhances the EA's supervisory duties and powers to carry out flood defence and drainage works. Provides new powers to allow establishment and abolitiont of Regional Flood Defence Committees and to revoke local flood defence schemes and drainage works	Regulatory: enhances supervisory duties and powers EA
Wildlife and Countryside Act 1981	Act of Parliament that makes it an offence to harm listed wild birds, animals and plants, and notifies designation, protection and management of SSSIs	Regulatory: designated areas as SSSI
Conservation Regulations 1994	Act of Parliament that transposes the Birds and Habitat Directives into national law	Regulatory: designation and protection of sites and species
Countryside and Rights of Way 2000	An Act of Parliament that creates statutory rights access on foot to certain types of land, to strengthen nature conservation legislation and to facilitate better management of Areas of Outstanding Natural Beauty	Regulatory: increases protection of SSSIs and strengthens wildlife enforcement regulation
Natural Environment and Rural Communities Act 2006	An Act of Parliament that established the permissive powers and responsibilities of Natural England regarding nature conservation	Regulatory: outlines duties and powers for nature conservation
National policies		
Flood and Coastal Erosion Risk Management	Government strategy to reduce risks to people and the developed and natural environment from flooding and coastal erosion by encouraging the provision of technically, environmentally and economically sound and sustainable defence measures	Regulatory: guidelines for flood defence and drainage, Making Space for Water

Policy or legislation	Description	Policy instruments
Making Space for Water	FCERM strategy to manage the risks from flooding and coastal erosion by employing an integrated portfolio of approaches to reduce the threat to people and their property and to deliver the greatest environmental, social and economic benefit	Regulatory: e.g. planning, Catchment Flood Management Plans, managed realignment
England Rural Development Programme	Government strategy to support environmental improvement and rural development	
Biodiversity Action Plan (UKBAP) 1994	Government strategy for biodiversity conservation (in response to Rio Convention)	Regulatory: list of priority species and habitats
Regulatory instruments		
Single Payment Scheme, CAP pillar 1, 2005	Governmental scheme to support agriculture; support is decoupled from production but linked with Good Agricultural and Environmental Condition guidelines addressing environment, public and plant health, animal health and welfare, and livestock identification and tracing	Regulatory: farmers receive financial support if they comply with standards set out under the Good Agricultural and Environmental Condition guidelines
Environmentally Sensitive Area 1987	Agri-environment scheme to protect the landscape, wildlife and historic interest of designated areas in England	Regulatory: management guidelines for designated areas, supported with economic incentives
Special Area of Conservation	Strictly protected sites designated under the Habitat Directive	Regulatory: protection of designated areas
Special Protection Area	Strictly protected sites designated under the Birds Directive	Regulatory: protection of designated areas
SSSIs	Designated areas to conserve and protect wildlife and geological and physiographical heritage	Regulatory: management guidelines for designated areas, supported by economic incentives

Policy or legislation	Description	Policy instruments
Catchment Flood Management Plans	Large-scale planning document which identifies long-term sustainable policies for the holistic management of flood risks in a defined river catchment	Regulatory: planning guidelines for catchment management
Nitrate Vulnerable Zones	Government action programme to limit and reduce water pollution caused by losses of nitrate from agriculture	Regulatory: compliance with guidelines in designated vulnerable areas
Economic instruments		
Countryside Stewardship Scheme 1991	Agri-environment scheme to sustain the beauty and diversity of landscapes, improve and extend wildlife habitats, conserve archaeological sites and historic features, improve opportunities for countryside enjoyment, restore neglected land or features, create new habitats and landscapes	Economic instruments: competitive scheme, 10-year agreements
Environmental Stewardship, CAP pillar 2, 2005	Agri-environment schemes to encourage farmers to conserve wildlife (biodiversity), maintain and enhance landscape quality and character, protect the historic environment and natural resources, and promote public access and understanding of the countryside. Replaces ESA and CSS	Economic instruments; various tiers available: entry level scheme, organic entry level scheme, higher level scheme
Woodland Grant Scheme	Aim: to create and manage woodlands to enhance the environment	Economic instruments
Voluntary instruments		
Catchment-Sensitive Farming	Government initiative to tackle diffuse water pollution from agriculture	Voluntary, supported by advice and economic instruments

References

Adams, W.M., Perrow, M.R. and Carpenter, A. (2004) 'Conservatives and champions: River managers and the river restoration discourse in the United Kingdom', *Environment and Planning A*, vol 36, pp1929–1942

Armstrong, A. (1978) *A Digest of Drainage Statistics 1971–78*, Field Drainage Experimental Unit technical Report 78/7, HMSO, London

Bingham, S.P. (1983) *A Guide to the Development of Grants for Agriculture and Horticulture in England and Wales – 1940–82*, Economics Division III, Ministry of Agriculture, Fisheries and Food, London

Bowers, J. (1998) 'Inter-war land drainage and policy in England and Wales', *Agricultural History Review*, vol 46, pp64–80

Brauman, K.A., Daily, G.C., Duarte, T.K. and Mooney, H.A. (2007) 'The nature and value of ecosystem services: An overview highlighting hydrological services', *Annual Review of Environment and Resources*, vol 32, pp67–98

Bromley, D.W. (1991) *Environment and Economy: Property Rights and Public Policy*, Basil Blackwell, New York

Bromley, D.W. and Hodge, I.D. (1990) 'Private property rights and presumptive policy entitlements', *European Review of Agricultural Economics*, 17, pp197–214

Defra (2001) *National Appraisal of Assets at Risk from Flooding and Coastal Erosion, Including the Potential Impact of Climate Change*, Department for Environment, Food and Rural Affairs, London

Defra (2005a) *Making Space for Water: Taking Forward a New Government Strategy for Flood And Coastal Risk Management in England*, Department for Environment, Food and Rural Affairs, London

Defra (2005b) *Higher Level Scheme Handbook*, Department for Environment, Food and Rural Affairs, London

Defra (2007) 'Making Space for Water: Outcome Measures', Department for Environment, Food and Rural Affairs, London, http://www.defra.gov.uk/Environ/fcd/policy/strategy/sd4/default.htm accessed 17 August 2009

Defra (2008a) 'Farming: Single Payment Scheme', Department for Environment, Food and Rural Affairs, London, http://www.defra.gov.uk/farm/singlepay/index.htm accessed 17 August 2009

Defra (2008b) 'Catchment Sensitive Farming', Department for Environment, Food and Rural Affairs, London, http://www.defra.gov.uk/farm/environment/water/csf/index.htm accessed 17 August 2009

Defra (2009) 'Draft Flood and Water Management Bill', April 2009, Department for Environment, Food and Rural Affairs, London and the Welsh Assembly, Cardiff, Cm 7582, ISBN 978-0-10-175822-2

Dobbs, T.L. and Pretty, J.N. (2004) 'Agri-environment stewardship schemes and "multi-functionality"', *Review of Agricultural Economics*, vol 26, no 2, pp220–237

Environment Agency (2002) *Agriculture and Natural Resources: Benefits, Costs and Potential Solutions*, EA, Bristol

Environment Agency (2003) *Strategy for Flood Risk Management (2003/4–2007/8)*, EA, Bristol

Environment Agency (2008) 'Position statement: The Environment Agency's position on land management and flood risk management', EA, Bristol, http://publications.environment-agency.gov.uk/pdf/GEHO0609BQDS-E-E.pdf accessed 17 August 2009

Evans, E.P., Ashley, R., Hall, J.W., Penning-Rowsell, E.C., Saul, A., Sayers, P.B., Thorne, C.R. and Watkinson, A.R. (2004) 'Foresight Future Flooding, Scientific Summary', vol 1, *Future Risks And Their Drivers*, Office of Science and Technology, London

FRMRC (2007) Flood Risk Management Research Consortium. http://www.floodrisk.org.uk/ accessed 17 August 2009

Fuller, R.M. (1987) 'The changing extent and conservation interest of lowland grasslands in England and Wales: A review of grassland surveys 1930–1984', *Biological Conservation*, vol 40, pp281–300

Goldman, R.L., Thompson, B.H. and Daily, G.C. (2007) 'Institutional incentives for managing the landscape: Inducing cooperation for the production of ecosystem services', *Ecological Economics*, vol 64, no 2, pp333–343

Groot, R. de. (2006) 'Function analysis and valuation as a tool to assess land use conflicts in planning for sustainable multifunctional landscapes', *Journal of Landscape and Urban Planning*, vol 75, pp175–186

Groot, R. de, Wilson, M. and Boumans, R. (2002) 'A topology for classification, description and valuation of ecosystem goods and services', *Ecological Economics*, vol 41, pp393–408

Hale, B.W. and Adams, M.S. (2007) 'Ecosystem management and the conservation of river-floodplain systems', *Landscape and Urban Planning*, vol 80, nos 1–2, pp23–33

Herzog, F. (2005) 'Agri-environment schemes as landscape experiments', *Agriculture, Ecosystems and Environment*, vol 108, pp175–177

Hodge, I. (2001) 'Beyond agri-environmental policy: Towards an alternative model of rural environment governance', *Land Use Policy*, vol 18, pp99–111

Hodge, I. and McNally, S. (2000) 'Wetland restoration, collective action and the role of water management institutions', *Ecological Economics*, vol 35, no 1, pp107–118

ICE (2001) *Learning to Live with Rivers*, Institution of Civil Engineers, London

Johnson, C.L., Tunstall, S.M. and Penning-Rowsell, E.C. (2005) 'Floods as catalysts for policy change: Historical lessons from England and Wales', *Water Resources Development*, vol 21, no 4, pp561–575

Johnson, G. and Scholes, K. (2002) *Exploring Corporate Strategy*, 6th edn, Financial Times/Prentice Hall, Harlow, UK

Joint Nature Conservation Committee (2008) *The Conservation (Natural Habitats, etc) Regulations 1994*, JNCC, http://www.jnnc.gov.uk accessed 17 August 2009

Latacz-Lohman, U. and Hodge, I. (2003) 'European agri-environmental policy for the 21st century', *The Australian Journal of Agricultural and Resource Economics*, vol 47, no 1, pp123–139

Mayrand, K., Dionne, S., Paquin, M. and Pageot-LeBel, I. (2003) *The economic and environmental impacts of agricultural subsidies: An assessment of the 2002 US Farm Bill and Doha Round*, Unisfera International Centre, Montreal

MAF (1948) *Harvest Home, the Official Story of the Great Floods of 1947 and Their Sequel*, Ministry of Agriculture and Fisheries, HMSO, London

Millennium Ecosystem Assessment (2003) *Ecosystems and Human Well-Being: A Framework for Assessment*, Millennium Ecosystem Assessment Series, Island Press, Washington DC, http://www.millenniumassessment.org accessed 17 August 2009

Ministry of Agriculture, Fisheries and Food (1975) *Food from Our Own Resources*, Agriculture White Paper, 6020cmd. HMSO, London

Morris, J. (1992) 'Agricultural land drainage, land use change and economic performance: experience in the UK', *Land Use Policy*, vol 3, no 9, pp185–198

Morris, J. and Hess, T.M. (1986) 'Farmer uptake of agricultural land drainage benefits', *Environment and Planning A*, vol 18, pp1649–64

Morris, J., Hess, T.M., Gowing, D.J.G., Leeds-Harrison, P.B., Bannister, N., Vivash, R.M.N. and Wade, M. (2004) 'Integrated washland management for flood defence and biodiversity', Research report 598, English Nature, Peterborough

Morris, J., Hess, T.M., Gowing, D.J.G., Leeds-Harrison, P.B., Bannister, N., Vivash, R.M.N. and Wade, M. (2005) 'A framework for integrating flood defence and biodiversity in washlands in England', *International Journal of River Basin Management*, vol 3, no 2, pp1–11

Morris, J., Bailey, A.P., Lawson, C.S., Leeds-Harrison, P.B., Alsop, D. and Vivash, R.M. (2008) 'The economic dimensions of integrating flood management and agri-environment through washland creation: A case from Somerset, England', *Journal of Environmental Management*, vol 88, pp372–381

Natural England (2002) *Agricultural Land Classification*, NE, http://www.magic.gov.uk/staticmaps/maps/alc_col.pdf accessed 17 August 2009

Natural England (2008) *State of the Natural Environment 2008*, NE, Peterborough

O'Connell, P.E., Beven, K.J., Carney, J.N., Clements, R.O., Ewen, J., Fowler, H., Harris, G.L., Hollis, J., Morris, J., O'Donnell, G.M., Packman, J.C., Parkin, A., Quinn, P.F., Rose, S.C., Shepherd, M. and Tellier, S. (2004) *Review of impacts of rural land use and management on flood generation*, R&D Technical Report FD2114, Defra, London

Ogaji, J. (2005) 'Sustainable agriculture in the UK', *Environment, Development and Sustainability*, vol 7, pp253–270

Pitt, M. (2008) *The Pitt Review: Learning the Lessons from the 2007 Floods*, Cabinet Office, London

Posthumus, H., Rouquette, J.R., Gowing, D.J.G., Hess, T.M., Morris, J. and Trawick, P. (2008a) Workshop report: 'Integrated floodplain management', Stakeholder workshop, 28 April 2008, London, http://www.relu.ac.uk/research/projects/Report_Int FloodpManag_28Apr2008.pdf accessed 17 August 2009

Posthumus, H., Rouquette, J.R., Morris, J., Hess, T.M., Gowing, D.J.G. and Dawson, Q.L. (2008b) 'Integrated land and water management in floodplains in England', in Samuels et al (eds) *Flood Risk Management: Research and Practice*, Taylor & Francis, London

Reed, M.S, Graves, A., Dandy, N., Posthumus, H., Hubacek, H., Morris, J., Prell, C., Quinn, C.H. and Stringer, L. (2009) 'Who's in and why? A typology of stakeholder analysis methods for natural resource management', *Journal of Environmental Management*, vol 90, pp1933–1947

Relu (2008a) *Integrated Land and Water Management in Flood Plains*, Rural Economic and Land Use Programme, http://www.relu.ac.uk/ accessed 17 August 2009

Relu (2008b) *The Great Land Use Debate*, Festival of Social Science/National Science and Engineering Week 7–16 March 2008, Rural Economic and Land Use Programme, http://www.relu.ac.uk/ accessed 17 August 2009

Sandberg, A. (2007) 'Property rights and ecosystem services', *Land Use Policy*, vol 24, pp613–623

Scrase, J.I. and Sheate, W.R. (2005) 'Re-framing flood control in England and Wales', *Environmental Values*, vol 14, no 1, pp113–137

Tawney, R.H. (1948) *The Acquisitive Society*, Harcourt Brace and World, New York

Tockner, K. and Stanford, J.A. (2002) 'Riverine floodplains: Present state and future trends', *Environmental Conservation*, vol 29, pp308–330

Trafford, B.D. (1978) 'Recent progress in field drainage: Part 2', *Journal of the Royal Agricultural Society of England*, vol 139, pp31–42

Turner, R.K., Bergh, J.C.J.M. van der, Söderqvist, T., Barendregt, A., Straaten, J. van, Maltby, E. and Ierland, E. van (2000) 'Ecological-economic analysis of wetlands: scientific integration for management and policy', *Ecological Economics*, vol 35, pp7–23

Werritty, A. (2006) 'Sustainable flood management: Oxymoron or new paradigm? *Area*, vol 38, no 10, pp16–23

Wilson, A.M., Vickery, J.A., Brown, A., Langston, R.H.W., Smallshire, D., Wotton, S. and Vanhinsbergh, D. (2005) 'Changes in the numbers of breeding waders on lowland wet grasslands in England and Wales between 1982 and 2002', *Bird Study*, vol 52, pp55–69

Zhang, W., Ricketts, T.H., Kremen, C., Carney, K. and Swinton, S.M. (2007) 'Ecosystem services and dis-services to agriculture', *Ecological Economics*, vol 64, no 2, pp253–260

7

Ecosystem Services in Dynamic and Contested Landscapes: The Case of UK Uplands

Klaus Hubacek, Nesha Beharry, Alan Bonn, Tim Burt, Joseph Holden, Federica Ravera, Mark Reed, Lindsay Stringer and David Tarrasón

Introduction

In future, society may increasingly question the ways in which uplands are used. Growing populations will need to feed themselves under very different climatic conditions, on a shrinking land base (as sea levels rise), and compete for food with the rapidly growing middle classes of the developing world. In addition, they will need to balance the economic incentives and impacts of policies that favour land use changes towards, for example, the production of biofuels. At the same time, and under a changing climate, upland areas must continue to provide the many other services we have come to expect from them, for example, supplying surrounding cities with drinking water, without compromising biodiversity and important landscape features. This dilemma is captured well in the concept of 'ecosystem services', which in its simplest form refers to those services provided by nature for human well-being.

Ecosystem services can be grouped as: provisioning services (ecosystem products, for example food and fibre); regulating services (including process such as climate stabilization, erosion regulation and pollination); cultural services (non-material benefits from ecosystems, for example, spiritual fulfilment, cognitive development and recreation) and supporting services (necessary for the production of other ecosystem services, for example, soil formation, photosynthesis and nutrient cycling). Ecosystem services are essential to human existence and well-being (MA, 2005). They operate at different spatio-temporal scales, ranging from global processes to impacts at a catchment scale, to intricate processes on small patches, which are little explored to date. Ecosystems services such as the purification of water, mitigation of floods or pollination of plants 'are pervasive' and often unnoticed by human beings (Daily, 1997). The importance of ecosystem services is often only appreciated once they are lost or when they result in negative effects such as flooding (Gowdy, 1997), and so the depreciation of ecosystem capital is typically undervalued (Daily, 2000).

The ecosystem services concept promoted by the Millennium Ecosystem Assessment and similar publications (Daily, 1997; MA, 2005) has been very stimulating and led to significant debates. While there are still many open questions, the ecosystem services concept has been widely adopted by research and policy. It aims to conceptualize the 'complex links between ecosystems and human wellbeing' (MA, 2005) and recognizes that different stakeholders are likely to value ecosystem services differently. As such, it emphasizes the need for the decentralization of control over ecosystem service management and has become *en vogue* as an important integrating concept for disparate interest groups and research disciplines. For example, ecologists have long recognized the service flows coming from functioning ecosystems and can now use a framework to communicate this better to a wider community. At the same time, the ecosystem services concept allows resource economists to see their framework of economic valuations adopted and applied to the natural world. It may also present an opportunity to public land managers and private land owners for additional income streams from the provision of ecosystem services, especially in times of falling resource prices and increasing restrictions on unfettered production (Brown et al, 2007). There is also evidence that the concept can attract new funding and wider support for nature conservation (Goldman et al, 2008).

While there is a strong support following the MA's popularization of the concept, and the Ecosystem Approach has already been incorporated in UK legislation with the cross-government public service agreement (PSA) target (Defra, 2007b), there are still considerable scientific challenges to incorporate it into policy tools and agendas (Kremen et al, 2008): For example, how can we maximize synergies and resolve trade-offs and resulting conflicts between ecosystem services at different scales, as we increasingly make new and diverse demands on our land? Who has the right and the power to make decisions about how land is used in future? Should future land use be shaped by a relatively small number of property-rights owners (usually based on land ownership or withdrawal rights), if this marginalizes less powerful groups who are significantly affected by the decisions that are made? Can new markets, fiscal schemes and other incentives promote the delivery of ecosystem services?

British uplands offer a particularly interesting context in which to explore these questions (Bonn et al, 2008). Upland areas are often recognized for their outstanding beauty and for their supply of ecosystem services in addition to food and fibre, such as flood prevention, carbon sequestration and water provision (Bonn et al, 2008; Wilby et al, 2006). Often the beneficiaries of these services are in distant urban areas, leading to a mismatch of costs incurred by those who manage ecosystems services (usually land managers or farmers) and those who enjoy their benefits (e.g. tourists, local residents, consumers of drinking water). The costs are often borne by the land owners, who can range from well-endowed private estate owners or industrial water companies owning large areas of land to hill-farming communities facing falling incomes and reductions in the agricultural labour force, an ageing demographic structure and farmland abandonment (Burton et al, 2005). Costs to support land management are to some degree borne by tax payers through agri-environment schemes and costs of environmental degradation are, for example, borne by water customers paying for water treatment costs through their water bills.

Any given parcel of upland may have multiple uses and users at any single point in time. For example, a given upland area may concurrently play a role in: sheep or timber production; maintaining water supplies; provision of recreation opportunities through, for example, walking, climbing or game shooting; biodiversity conservation; and delivery of important amenity values through relatively attractive landscapes and scenery. Only with careful management and negotiation between different stakeholder groups with different priorities is it possible to deal with any trade-offs and maximize synergies. To complicate the situation further, these synergies and trade-offs must be negotiated under conditions of uncertainty, an ever-incomplete scientific knowledge base and changing environments, and in the context of evolving preferences and multiple (sometimes conflicting) policies.

This chapter uses the UK uplands as a case study to explore how synergies and trade-offs between ecosystem services and the different priorities accorded to them by different actors may be managed, highlighting some of the key considerations that need to be taken into account. The next section discusses how upland ecosystem services have changed and are likely to change in the future, in response to various environmental, economic and policy drivers. In doing so, it explores some of the conflicts that may emerge as a result. This then leads to an examination of the ecosystem services concept in relation to uplands.

A land of many uses: dynamic and contested upland ecosystem services

An environment shaped by human influence

The UK's uplands, cherished by residents and visitors, have been shaped by millennia of past human activity (Reed et al, 2009). Indeed, 'the major lineaments of the landscape, its openness, its cover of peat and the confinement of improved land to the values are a human creation' (Simmons, 1989, p258). Mesolithic hunter-gatherers opened the predominantly deciduous forest cover by burning the forest edge and openings, thus creating patches and a more varied resource base for grazing animals and hunting. This resulted in a mosaic of different land covers and, through changing hydrology and favourable climatic conditions, led to peat formation (Simmons, 1989). With the onset of agriculture in the Neolithic period, a slash-and-burn type deforestation was practised. This initially shifting process of forest clearing was later continued on a more permanent basis by medieval farmers and monasteries to allow for grazing of large cattle and sheep flocks. By then the forest cover had almost completely been replaced by grass and shrub vegetation (Simmons, 1989) and created a patchwork associated with a diverse range of animal species, becoming an important resource for livestock and game (Sotherton et al, 2008) which is now cherished by locals and tourists for both aesthetic reasons and the recreation opportunities such a diverse landscape brings.

The process of conversion of the upland ecosystem in response to human needs has continued dramatically in the more recent past. Prior to the 19th century, much hill-farming was extensive (Davies, 2006). Land use for most of the 20th century, however, was influenced by a focus on intensified production. Post-war agricultural policy was designed to promote food self-sufficiency and this goal was supported by both financial and infrastructural measures. These included grants to finance the improvement of hill farming land through the application of lime and fertilizers, and landscape drainage to increase the area and condition of heather (*Calluna vulgaris*) for the benefit of increased sheep and red grouse (*Lagopus lagopus scoticus*) production. Economic and social incentives led in the 19th and 20th centuries to growing flock sizes and, in some places, overgrazing by sheep (Anderson and Yalden, 1981; Condliffe, 2008). This was subsequently combined with conflicting demands placed on the landscape by grouse-moor owners, ramblers, and foresters. While in the second half of the 20th century sheep numbers vastly increased, rising input costs and a lack of skilled labour meant that some more extensive forms of management such as shepherding became less widespread.

The tenor of agricultural policy in the UK over the last 60 years until the 1990s has been firmly set on an aim to meet the nation's need for domestically produced food at reasonable prices with fair rewards to farmers and agricultural workers. Agricultural support schemes such as guaranteed prices and farm capital grants provided incentives for upland farmers to increase output through improved grassland management and increased stocking rates. As a result there were dramatic increases in sheep numbers between the 1950s and 1980s. This was followed by attempts to address market imbalances through 'headage' quotas (i.e. a limit on production-based subsidy) and payments for 'extensification' in the mid-1980s, although major reform did not begin until the early 1990s (Anderson and Yalden, 1981; Condliffe, 2008; Gardner et al, 2008). The most recent CAP reform sought to remove production-based subsidies and replace them with decoupled direct payments attached to cross-compliance with environmental and health standards and 'Good Environmental and Agricultural Conditions' (Gardner et al, 2008). It is still unclear whether the CAP reforms and other schemes will encourage management that can deliver the desired ecosystem and economic goods and services.

Other than those areas where there has been an emphasis on coniferous afforestation (e.g. Galloway and mid-Wales), livestock production and grouse-moor management has been the main productive land use in the uplands (Hubacek et al, 2008). Hill farming has been important in sustaining habitats and landscapes, for example, through maintaining traditional landscape features such as hedges and drystone walls and traditional farm buildings. On the other hand, hill farming can also have negative effects through the creation of farm tracks, habitat deterioration and soil erosion arising from heavy grazing pressure (Gardner et al, 2008). On a wider scale, grazing and burning, particularly of heather moorland, can have major effects on the species composition of the uplands (Crowle and McCormack, 2008; English Nature, 2001).

A number of specific policies (e.g. Less Favoured Area Scheme) were designed to address the structural disadvantages of upland farming due to factors such as climate, topography, altitude and remoteness. This is based on the assumption that if farming ceased in these areas there would be further out-migration and land abandonment. There is government commitment to continue the support for upland communities due to their perceived contribution to: (i) the environment (wildlife and landscapes); (ii) the social fabric in relatively remote rural areas; and (iii) the economy through livestock production and maintaining the assets on which other economic activities such as tourism depend (Burton et al, 2005; Defra, 2003).

Land management and ecosystems services

Since 2001, agri-environment schemes have shifted from supporting solely provisioning services such as livestock production through headage payments to include payments linked to good farming practice aiming also at a range of other ecosystem services. ESA schemes aim to establish sustainable stocking rates in sensitive uplands. The scheme is voluntary and farmers are paid to reduce their stock. While existing ESA agreements are beginning to reach the end of their term, they have now been replaced by a two-tier 'Environmental Stewardship Scheme', whereby farmers receive subsidies for developing and maintaining agro-environmental plans.

These policies to protect and enhance upland ecosystems are important, as uplands support a range of internationally rare species, including birds such as dunlin (*Calidris alpina*) and peregrine (*Falco peregrinus*). Many upland areas are protected under national and international conservation law due to their biodiversity value under the European Birds and Habitats Directive (79/409/EEC, 92/43/EEC), leading to the UK Biodiversity Action Plan in 1997 and the 2010 SSSI PSA target. In England, uplands cover 12 per cent of total land, of which 53 per cent are designated as SSSIs (Crowle and McCormack, 2008). In 2007, Natural England estimated, based on the Common Standards Monitoring protocol, that only 17.3 per cent of upland SSSIs in England were in favourable condition. Unfavourable conditions were mainly due to overgrazing, inappropriate burning and drainage (Crowle and McCormack, 2008), and management agreements have therefore been put in place to improve current conditions in order to achieve 2010 conservation targets.

A critical aspect of this debate is the role of rotational burning in sustaining cultural services such as opportunities for recreational game shooting as well as provision for upland biodiversity. Burning is used to maintain mosaics of heather at different stages of maturity to provide habitat for red grouse. Regulations and codes regarding burning extend back to the medieval period (Dodgshon and Olsson, 2006) and are regularly revised (for example, the Scottish Muirburn Code, and the England and Wales Heather and Grass Burning Code and Regulations; Defra, 2007a).

Appropriate burning of heather moorland is claimed to protect against wildfire risk by reducing the quantity of combustible material while creating a mixture of habitats that improves grouse densities. However, in some areas, long-term grouse management has converted blanket bogs into dry heather moorland and so reduced the diversity of shrubs and the moss and lichen layer (Chambers et al, 2007). The impact of grouse-moor management on breeding moorland birds is debated. While heather burning and predator control benefit some species, other species with different habitat preferences are disadvantaged (Tharme et al, 2001, Sotherton et al, 2008, Pearce-Higgins et al, 2008).

Little is known about the effects of burning on regulating ecosystem services such as peat erosion control or provision of water quantity and quality (Holden et al, 2007). However, among other drivers of environmental change, recent data suggest that while burning drives changes in vegetation patterns, it is the vegetation which can then have a strong influence on water quality, in particular water discolouration (Armstrong et al, in review; Neff and Hooper, 2002). Water discolouration associated with DOC is a growing problem in the UK, with some studies showing a 65 per cent increase in DOC over the last 12 years. Discolouration is a problem because potable chlorination of DOC-rich waters can be associated with trihalomethanes, which are strictly regulated as they are potential carcinogens.

The relationship of moorland management and run-off attenuation or flood mitigation remains unclear (Holden et al, 2007). The EU's Water Framework Directive, through the use of integrated River Basin Management Plans, aims to protect and improve the environmental status of all river catchments in the EU, promote sustainable use, and reduce the effects of floods and droughts so that all catchments achieve 'good ecological status' by 2015 (Wilby et al, 2006). Challenges to achieving this include effects of changing agricultural subsidies on land use, the uncertain impacts of climate change and scientific uncertainty around the controls of water discolouration in the uplands and how it can be effectively managed.

Some upland areas saw a massive expansion in coniferous woodland production especially after World War I, as efforts sought to create a strategic reserve of timber as a matter of national security (Condliffe, 2008). While traditional practices would probably have been relatively benign, the practice of deep ribbon ploughing meant that there was increased vulnerability to soil erosion and a number of pollution events occurred (Burt et al, 1983). New guidelines for forestry in upland catchments have largely removed this threat of pollution but other concerns remain about upland afforestation in some regions, notably stream water acidification in south west Scotland (Puhr et al, 2000) and, more generally, concerns over decreased water yield.

Recently, in the context of climate-regulating ecosystem services, wooded areas and especially SRC have become of interest for providing CO_2-neutral biofuels. But even more important is the role of upland peatlands as carbon stores, as they represent the largest terrestrial carbon store in the UK. In addition, upland peatlands represent one of the few long-term stores of carbon that can accumulate on the land surface through good management. Models suggest that across the UK as much as 400,000 tonnes of carbon a year could be stored in this way (Worrall, personal communication) or the equivalent of the carbon emissions of two per cent of car traffic in England and Wales per year. However, currently many degraded upland peatlands lose more carbon than they absorb through gaseous and fluvial pathways (Holden, 2005), so the identification and restoration of damaged peatlands to retain their carbon stores could have significant beneficial impacts for climate regulation.

Finally, the uplands provide significant cultural services by, for example, offering important opportunities for recreation. Much of the upland economy is based on the tourism industry and in 2005 £9.4 billion was spent on tourism and leisure services in England (Curry, 2008). During the foot-and-mouth crisis in 2001, when much of the uplands were closed for visitor access, it is estimated that the tourism industry in the UK as a whole lost as much as £8 billion and many businesses either closed or were severely scaled back as a result (Curry, 2008). The range of tourism activities includes both active physical exercise (e.g. mountain biking, walking) and more passive visits where the tourists stray only a short distance from their cars but nevertheless enjoy the spectacular landscapes (Davies, 2006).

Beyond ecosystem services: facilitating sustainable uplands

This section discusses a number of challenges that emerge from the application of the ecosystem services framework to dynamic UK upland environments. These will then be used to inform research and policy frameworks that could help enable future land use changes to support and enhance the provision of multiple ecosystem services.

Safeguarding dynamic and complex ecosystem services: the challenge

Ecosystem services are dynamic and complex. It is challenging to predict, value and safeguard future ecosystem services that will emerge in response to changing human needs and priorities. Ecosystem services are an anthropocentric concept. As such, they change in response to demands that people put on their environment. It is challenging to value aspects of the natural environment that do not (currently) have human use or are not perceived as valuable. There is a growing literature about synergies and trade-offs between existing ecosystem services, but ecosystem services may transform over time as our needs and perceptions change and our knowledge about them increases. The use of current ecosystem services might compromise our ability to realize future uses. For example, peatlands can provide fuel to heat people's homes (Parish et al, 2007), but historic removal of peat has compromised the ability of some former peatland ecosystems to sequester carbon, provide for wildlife habitats, water purification and run-off attenuation.

As our needs and priorities change, new ecosystem services are realized and/or prioritized over others. However, at best this is done through a complicated process of negotiation between different land users and stakeholders; at worst it is driven by optimizing a certain output (usually food, fibre and water) at the expense of other services. When these multiple land uses occur together in space, this can sometimes lead to trade-offs between ecosystem services that are impacting on each other or are even mutually exclusive. For example, re-wilding conservation strategies may be compatible with continued recreational use, but are not compatible with continued sheep grazing or grouse management.

As people adapt their livelihood strategies and governments alter their policies in response to future drivers of change (or simply as a response to lobbying or priorities spilling over from other policy arenas), we are likely to see shifts in the priorities given to different ecosystem services. Indeed, we have witnessed the support of farming interests in uplands over the last 60 years until the 1990s, and more recently, there has been growing interest in carbon management and proactive water-quality management in upland catchments alongside more traditional uses such as recreation and conservation. Given high fuel and food prices, we might see a renewed interest in food security and potential reversal of these trends.

Trade-offs between ecosystem services at different scales require coordination at the landscape scale

Land use and land management practices may create trade-offs between ecosystem services at different scales (Goldman et al, 2007). One example of the trade-offs at the local level can be seen through the lens of heather burning, as management for recreational game shooting interests may favour some species and be a dis-benefit to other species of conservation concern. At a regional scale, large-scale burning can have potential implications for water-quality erosion or wildfire control as well as for habitat provision for wildlife.

At the regional level, management for flood prevention as well as for drinking water might provide a useful lens, as such hydrological services are most effectively influenced at the catchment scale. Land management, such as creation of short-sward vegetation through sheep grazing, drainage through gripping or increasing surface roughness and infiltration rates through re-vegetation of bare peat areas, can increase or attenuate run-off from land parcels. But it is the spatial arrangement of flow paths and their connectivity that determine flooding downstream, that is, whether the spatial arrangement of flow paths cause flood peaks from sub-catchments to coincide to form a flood event downstream (Lane et al, 2004). In addition, local saturation levels of land parcels and their connectivity within the stream networks may determine water quality and disease control across the catchment, for example concerning transport of bacteria in sheep dung (such as *Cryptosporidium*) in the stream system.

Some bacteria pose a significant health risk and can be very difficult or expensive for water companies to treat. Thus, when planning for water-quality regulation, the location of fields and their stocking rates are important, determining both the quantity of sheep dung as well as run-off properties through sward height. This may lead to recommendations for some fields or farmers to decrease sheep stocking densities, whereas for other fields there is less concern. So some farmers, due to the location of their land, can operate in a certain way which can generate a certain income but not result in downstream problems, whereas if others operated in the same way this may lead to significant downstream problems.

This approach highlights the need for land parcels to be managed in a coordinated way across landscapes rather than as independent units, especially if ecosystem services that integrate entire drainage basins are to be sustained.

Similar arguments have been made with regards to nitrates (Burt and Haycock, 1992). Water protection zones would be designated only locally, whereas farming in other parts of the catchment would, by implication, go on unhindered.

Despite the benefits of more integrated management, this level of coordination is typically not encouraged and not automatically in the interest of the property-right holder (Goldman et al, 2007). Institutional or financial incentives for landowners need to be designed in a way that takes account of the landscape nature of many ecosystem services. This may involve spatial differentiation that aligns incentives to landowners' heterogeneity in participation costs, involving opportunity, transaction and direct costs of protection and thus avoiding efficiency losses (Wünscher et al, 2008). Policies would need to pay attention to these emerging properties across scales and at the same time be much more fine-grained, allowing intensive farming practices to be directed away from vulnerable areas (e.g. from sub-catchments where impacts on water quality would be disproportionate). Within a catchment, for example, differential payments/incentives could be given based on priorities similar to the points system for landscape features for the Entry Level Stewardship (ELS) scheme (it requires a basic level of environmental management from a wide range of over 50 options, including hedgerow management, stone wall maintenance, low-input grassland, buffer strips and arable options). On the other hand, this means that some farmers fare better than others, depending where they are. This is the inevitable result of managing heterogeneous units of land providing different services as well as posing different risks depending on distance to waterways, soil types and topography, and other environmental properties.

Mapping and modelling ecosystem services in collaboration with stakeholders

Mapping and modelling ecosystem services offers a valuable way to assess the sustainability of current land management and to evaluate proposed adaptations, but can only be developed and applied in relation to the objectives of those using the land. Until now, land has often been zoned based on 'single-purpose policies' and they have tended to focus on land's suitability for development, food and timber production, and wildlife, landscape and recreation (Swales and Woods, 2008). A precondition for evaluation and negotiation is a systematic characterization of ecosystem services in biophysical terms. This may include, for example, a classification of ecosystem services and their quantification and mapping for each location, taking into account current and potential land use as well as local and global consumers (Daily, 2000; Naidoo et al, 2008). This would allow the visualization of ecosystem service flows between different regions, allowing the identification of 'free riders' (those obtaining the benefits provided by ecosystem services without necessarily incurring the costs of their maintenance) as well as the identification of areas accruing 'ecological debts' (an imbalance between one's 'fair share' of natural resources and one's actual usage of those resources).

To assess the impacts of alternative land use and management options, scenario modelling by providing information on *potential* ecosystem services for each location and identifying synergies and trade-offs will be necessary. Currently we have not reached this stage yet, and projects mapping ecosystems services have only just started (but see Chan et al, 2006; Naidoo and Ricketts, 2006; Troy and Wilson, 2006; van Jaarsveld et al, 2005). One way forward is to assess ecosystem services in biophysical terms. This means, for example, for provisioning services quantifying the flows of goods harvested in an ecosystem in physical units; or for most regulation services this requires a spatially explicit analysis of the biophysical impact of the service on the environment in or surrounding the ecosystem such as the reduction of peak flows downstream through changes of land cover upstream (Hein et al, 2006). As cultural services such as education, a sense of belonging or spiritual values are dependent on human interpretation, they cannot be easily expressed in biophysical terms (or in monetary terms), while visitor numbers or number of shooting days may provide a proxy for provision of recreational services.

Ecosystem service indicators offer a valuable way to assess the sustainability of current land management and to evaluate proposed future management options and its interactions with other services. However, such indicators can only be developed and applied in relation to the objectives of those using the land. Since a single land parcel can have multiple uses and values accorded to it by different actors, this results in trade-offs among different groups.

Sustainable management across all dimensions (environmental, social and economic) also often relies on the substitution of different capital assets (e.g. liquidating natural capital such as forests to generate financial capital). If, as the ecosystem service framework suggests, sustainability is ultimately dependent on the ecosystem services provided by their natural capital, then upland management must maintain viable stocks of natural capital that allow ecosystem services to function viably. This suggests that to assess, for example, the sustainability of different potential adaptations to climate change, it is necessary to determine whether they threaten the long-term viability of ecosystem services, both singly and in conjunction with other potential adaptations. Indicators with thresholds have been developed for a range of ecosystem services to date and could be used to ensure that adaptation does not simply delay or create new problems. An example is nitrates in drinking water, which are strictly limited to concentrations below 11 mg l^{-1} N–NO_3. The concentration of nitrates in groundwater boreholes or surface water streams is therefore a good indicator with a clear threshold. For waters with levels above the limit, dilution with less nitrate-rich waters is the simplest solution. However, if all the gathering areas for water have high nitrate concentrations, then dilution cannot occur locally and the problem becomes very difficult to deal with. Solutions to the nitrate problem have therefore been to deal with the source of the problem on the landscape and limit its application as fertilizer or use other management techniques to limit leaching into water courses.

Such indicators can be used by stakeholders themselves (including regulatory bodies) or they can be used in conjunction with simulation models. To avoid creating new problems given the inherent interdependence of ecosystem services, the importance of identifying interactions and modelling trade-offs for different management scenarios cannot be emphasized enough. This is paramount for:

1 Avoiding irreversible losses of some services, which are not in use now, but which could be realized or re-established in the future. This is in line with precautionary principles (Ciracy-Wantrup, 1968) aimed at avoiding irreversible scenarios at any point. Avoiding irreversible actions is especially important if flexibility is to be maintained, given the ever-changing preferences and demands of society and the unknown needs of future generations.
2 Modelling trade-offs can indicate how the focus on one ecosystem service may affect other services. For example, it could show how managing the area for water purification could impair livestock production, increase flood control and influence carbon sequestration.
3 Modelling could also be used to forecast changes and to develop and assess future scenarios with regard to the availability and quality of ecosystem services. This could take into account land use and land management changes under, for example, conditions of changing societal demand based on changing preferences, market conditions and prices.

Thus, while implementing new management activities and learning from the results of these actions can sometimes take years to see the effects, formal computer models can be used to inform stakeholders about the implications of their actions in terms of their own economic situation and also environmental effects such as the impacts they are having on biodiversity, soil erosion, water quality and carbon fluxes (Hubacek and Reed, 2009). All environmental decisions produce a diversity of outcomes dependent on the conditions of the land, the current uses and opportunity costs and most importantly the perceptions of a wide range of stakeholders influenced by their own socio-cultural backgrounds.

Participatory adaptive management for dealing with trade-offs between ecosystem services

Payments for ecosystem services only provide partial solutions given the uncertainty of future ecosystem services. Participatory adaptive management is necessary to manage the trade-offs that will inevitably occur between ecosystem services for different users. Ecosystem services, especially regulating, cultural and supporting services can be conceptualized as public goods and are thus most often underprovided. This is based on the assumption that individuals acting in their own self-interest would under-provide services not in their immediate interest. Thus incentives need to be offered to reward property-rights owners for providing such services. Their public benefits are often not captured in market prices and thus, over the last few decades, a growing body of literature has emerged using sophisticated economic valuations to quantify costs and benefits for maintenance of usually single ecosystem services.

Markets for ecosystem services have been developed to enhance direct use and commercialize ecosystem services in the areas of water quality protection, biodiversity conservation, landscape protection/countryside management and carbon sequestration (Daily et al, 2000; Pagiola et al, 2002) and thus increase income opportunities especially in the developing world (MA, 2005). Consequently, incentive schemes for ecosystem services have been developed mainly focusing on single-value consideration of ecosystem services, single scale (temporal and spatial) of analysis, and pursuing efficiency and effectiveness objectives over legitimacy and equity objectives. However, there are limitations in applying a single (monetary) exchange value to ecosystem services for designing and implementing measures through markets (see, for example Corbera et al, 2007; Hubacek and Mauerhofer, 2008; O'Neill, 1996; Spash, 1994).

Firstly, scientists have only a limited understanding of the complex relationships between different ecological and biophysical processes and functions. The process of causation is complex and governed by the interaction of a series of variables that may affect one another. For example, the presence of peatlands in some upland areas is determined by the subtle interaction of key variables including: water, climate (temperature, rainfall), altitude, topography and the presence of specific types of vegetation. If the balance between peat, water or vegetation is disturbed, it can fundamentally change the nature of the peatland (Parish et al, 2007). In addition, there is not only considerable uncertainty that eludes the call of a firm scientific evidence base, but there is also a disparity between scientific understanding and public perceptions of systems interactions, making it even more difficult to unpack the complexity (Tognetti et al, 2004).

Secondly, markets are not very well suited to efficiently allocate resources governed by such complexity and uncertainties. The consequences of environmental degradation and benefits from environmental improvement are heterogeneous and therefore in principle incomparable. Neither the scope nor the tools of conventional economic analysis are adapted to these types of interdependencies and complex causal sequences, as: 'these interdependencies have nothing to do with market transactions or exchanges of any kind, nor are they the result of choices unless one is prepared to argue that they are caused by the deliberate action of private firms which in full knowledge of the consequences decide to shift part of their costs to third persons or to society' (Kapp, 1970, p839). This makes price setting for ecological services very hard to impossible. To assign prices based on an exact evaluation demands a precisely demarcated object, that is, one for which conceptual boundaries can be drawn and property rights attached. Yet, '[m]any environmental goods fail to conform to discrete units which can be broken into marginal changes for the purpose of economic valuation' (O'Neill and Spash, 2000, p527).

Another set of criticisms arises from the incommensurability of different qualities (i.e. they cannot rationally be compared) through their translation into monetary values. The aggregation of different types of values into one 'super-numeraire' hides rather than reveals underlying values and thus is an obstacle rather than a tool to support deliberative stakeholder processes (Hubacek and Mauerhofer, 2008; O'Neill, 1996). Relying on simple monetary evaluation is therefore inadequate to capture this complexity. Moreover, due to the public-good characteristics of ecosystem services, markets typically reward short-term values of natural resources to the detriment of long-term ecological health (Turner and Daily, 2008).

Thirdly, different stakeholders at different scales perceive benefits from the ecosystem services differently. Sometimes these perceptions can be complementary but frequently they might be in conflict (Turner and Daily, 2008). An example of this is the utilitarian value provided by the upland moorland system at the local level and national level to support the pastoral farming systems and national production, while at the same time there is an intrinsic value attached to the same resources and landscape by second-home owners or people seeking recreation or spiritual benefits. To pursue the utilitarian value (i.e. short-term productivity or profitability), natural capital can be substituted with other forms of capital, below the functional threshold of biodiversity or other ecological key functions of the system. Given the public-good characteristics of ecosystem services there is often little incentive for land managers to provide for unpaid ecosystem services and invest in them beyond their own perceived benefit. For example, grazing reduction in uplands might benefit national stakeholders valuing improvement in species richness or run-off attenuation, or potential increases in carbon sequestration benefitting global stakeholders, but this may be to the detriment of the land owner, who might incur higher costs or forego benefits through such measures.

While ecosystem services payments designed to compensate for such 'institutional and market failure' have frequently increased efficiency and effectiveness of environmental decisions (Chichilnisky and Heal, 1998), they are not necessarily able to address issues of equity and legitimacy – considerations that a number of authors have argued the need to explicitly incorporate. For example, Corbera et al (2007) argue that markets ignoring local contexts will reinforce existing power structures, inequities and vulnerabilities, and see this as a product of the nature of emerging markets. These markets are relatively new in comparison to established-good markets and lack a set of institutions that have evolved with them. These are usually promoted by national or international agencies committed to market-based conservation without proper acknowledgement of local socio-ecological contexts. Thus, markets for ecosystem services are seen as being limited in promoting more legitimate forms of decision-making and a more equitable distribution (Corbera et al, 2007). Similarly, Turner and Daily (2008) recognize that payments for ecosystem services need to incorporate local social, political, legal and cultural complexities into their design and practice: 'Economic incentives on their own are unlikely to transform local cultural, ethical and behavioral traits towards environmental stewardship and citizenship' (Turner and Daily, 2008, p29).

Finally, the anthropocentric nature of the ecosystem service concept means that services can only be defined in relation to the needs and priorities of those who benefit from the land. But who has the right and the power to decide what the land is used for? Despite the right for the public to be involved in environmental decision-making enshrined in EU law through the Aarhus Convention,[1] the public has little possibility for direct influence (Swales and Woods, 2008). But should we just accept that those who hold greatest power will shape future land use if this marginalizes groups of people who are significantly affected by the decisions taken but have little power to influence what happens (Swales and Woods, 2008), especially when exchanges happen under monetary market systems (Corbera et al, 2007)? Given the conflicting ways in which land is often used and valued by different stakeholders with no equivalent scales and recognizing an inequity problem, the management of ecosystem services must involve participation from the full range of stakeholders, working together to design and test policies and practices that minimize trade-offs, while creating and exploiting synergies between complementary ecosystem services where possible.

Conclusions

Ecosystem services may represent another shift in environmental policy paradigms. After a perceived failure (or sustained dislike) of the top-down command and control policies in the 1960s and 1970s, and various waves of subsidy schemes, the emergence of decentralized or 'light-touch' approaches have been observed. We are now starting to enter a phase of markets for ecosystem services where we aim to encourage owners of land or property rights to provide ecosystem services by paying them for it. We thus effectively replaced the earlier doctrine of the 'polluter pays' principle with the 'beneficiary pays' principle, in which the public pays for environmental benefits. In doing so, this redistributes property rights from the public to the land owner. In fact, the step was partly made already through granting subsidies. The shift from conventional production subsidies to Higher Level Stewardship (HLS) and other market incentives thus aims to replace the earlier, relatively unconditional support for provisioning services through payments for environmental benefits attained through promotion of regulatory, cultural and supporting services.

More recently the debate has widened to question the market philosophy of payments for ecosystems services based on notions of legitimacy, fairness and power. Thus it is important not only to look at how property-rights owners can be incentivized to provide protection for water quality or carbon stores, but also to recognize that in paying for such services the list of property rights has been extended from ownership of a parcel of land to providing vital public goods of importance for a global community.

In response to this, a process is required that attempts to integrate these concerns more fully, considering the social, economic and politico-cultural contexts of ecosystem services and the distributional effects payment for ecosystems services may have. Different stakeholders are likely to value ecosystem services differently and this may differ in different areas; thus there is a need to emphasize the importance of stakeholder perceptions, property rights and institutions in the management of ecosystem services. This in turn stresses the need for participatory approaches and greater decentralization of control over ecosystem service management.

Uplands around the world are facing significant socio-economic and environmental change and land managers and other decision-makers need to better understand the implications of their actions with regards to environmental systems if they are to adapt and maintain upland goods and services. The ecosystem service concept and associated tools such as mapping of ecosystem services, modelling of scenarios, and trade-offs and sustainability indicators are important components to providing the required information. While many upland regions have been coined severely disadvantaged areas with respect to provisioning ecosystem services, they should now be recognized for their significant potential to provide for regulating and cultural ecosystem services.

Notes

1 United Nations Economic Commission for Europe, Convention on Access to Information, Public Participation in Decision-Making and Access to Justice in Environmental Matters, http://www.unece.org/env/pp/) [29.07.2009]

References

Anderson, P. and Yalden, D.W. (1981) 'Increased sheep numbers and the loss of heather moorland in the Peak District, England', *Biological Conservation*, vol 20, pp195–213

Armstrong, A., Holden, J., Kay, P., Francis, B., Foulger, M., Gledhill S., McDonald A.T. and Walker, A. (in review) 'The impact of peatland drain-blocking on dissolved organic carbon loss and discolouration of water: results from a national survey', *Journal of Hydrology*

Bonn, A., Rebane, M. and Reid, C. (2008) 'Ecosystem services: a new rationale for conservation of upland environments', in Bonn, A., Allott, T., Hubacek, K. and Stewart, J. (eds) *Drivers of Environmental Change in Uplands*, Routledge, Abingdon, pp448–474

Brown, T.C., Bergstrom, J.C. and Loomis, J.B. (2007) 'Defining, valuing, and providing ecosystem goods and services', *Natural Resources Journal*, vol 47, pp329–376

Burt, T.P. and Haycock, N.E. (1992) 'Catchment planning and the nitrate issue: a UK perspective', *Progress in Physical Geography*, vol 16, pp379–404

Burt, T.P., Donohoe, M.A. and Vann, A.R. (1983) 'The effect of forestry drainage operations on upland sediment yields: The results of a storm-based study', *Earth Surface Processes and Landforms*, vol 8, pp339–346

Burton, R., Mansfield, L., Schwarz, G., Brown, K. and Convery, I. (2005) 'Social capital in hill farming. Report for the Upland Centre', International Centre for the Uplands, Penrith

Chambers, F.M., Mauquoy, D., Gent, A., Pearson, F., Daniell, J.R.G. and Jones, P.S. (2007) 'Palaeoecology of degraded blanket mire in South Wales: Data to inform conservation management', *Biological Conservation*, vol 137, pp197–209

Chan, K.M.A., Shaw, M.R., Cameron, D.R., Underwood, E.C. and Daily, G.C. (2006) 'Conservation planning for ecosystem services', *PLoS Biology*, vol 4, pp2138–2152

Chichilnisky, G. and Heal, G. (1998) 'Economic returns from the biosphere', *Nature*, vol 391, pp629–630

Ciracy-Wantrup, S.V. (1968) *Resource Conservation: Economics and Policies*, University of California Division of Agricultural Sciences, University of California Press, Berkeley

Condliffe, I. (2008) 'Policy change in the uplands', in Bonn, A., Allott, T., Hubacek, K. and Stewart, J. (eds) *Drivers of Environmental Change in Uplands*, Routledge, Abingdon, pp59–89

Corbera, E., Brown, K. and Adger, N.W. (2007) 'The equity and legitimacy of markets for ecosystem services', *Development and Change*, vol 38, pp587–613

Corbera, E., González Soberanis, C. and Brown, K. (2009) 'Institutional dimensions of payments for ecosystem services: An analysis of Mexico's carbon forestry programme', *Ecological Economics*, vol 68, no 3, pp743–761

Crowle, A. and McCormack, F. (2008) 'Condition of upland terrestrial habitats', in Bonn, A., Allott, T., Hubacek, K. and Stewart, J. (eds) *Drivers of Environmental Change in Uplands*, Routledge, Abingdon, pp156–170

Curry, N. (2008) 'Leisure in the landscape: rural incomes and public benefits', in Bonn, A., Allott, T., Hubacek, K. and Stewart, J. (eds) *Drivers of Environmental Change in Uplands*, Routledge, Abingdon, pp277–290

Daily, G.C. (1997) *Nature's Services: Societal Dependence on Natural Ecosystems*, Island Press, Washington, DC

Daily, G.C. (2000) 'Management objectives for the protection of ecosystem services', *Environmental Science & Policy*, vol 3, pp333–339

Daily, G.C., Soederqvist, T., Aniyar, S., Arrow, K., Dasgupta, P., Ehrlich, P.R., Folke, C., Jansson, A.M., Jansson, B.O., Kautsky, N., Levin, S., Lubchenco, J., Maeler, K.G. and Walker, B. (2000) 'The value of nature and the nature of value', *Science*, vol 289, pp395–396

Davies, S. (2006) 'Recreation and visitor attitudes in the Peak District moorlands', *Moors for the Future report*, no 12, Moors for the Future Partnership, Edale

Defra (2003) 'Review of the Hill Farm Allowance in England. Consultation Document', Department for the Environment, Food and Rural Affairs, London

Defra (2007a) *The Heather and Grass Burning Code*, Department for the Environment, Food and Rural Affairs, London http://www.defra.gov.uk/rural/pdfs/uplands/hg-burn2007.pdf

Defra (2007b) *Securing a Healthy Natural Environment: An Action Plan for Embedding an Ecosystems Approach*, Department for the Environment, Food and Rural Affairs, London, http://www.defra.gov.uk/wildlife-countryside/natres/pdf/eco_actionplan.pdf

Dodgshon, R.A. and Olsson, G.A. (2006) 'Heather moorland in the Scottish Highlands: The history of a cultural landscape, 1600–1880', *Journal of Historical Geography*, vol 32, pp21–37

English Nature (2001) *The Upland Management Handbook*, English Nature, Peterborough

Gardner, S.M., Waterhouse, T. and Critchley, C.N.R. (2008) 'Moorland management with livestock: The effect of policy change on upland grazing, vegetation and farm economics', in Bonn, A., Allott, T., Hubacek, K. and Stewart, J. (eds) *Drivers of Environmental Change in Uplands*, Routledge, Abingdon, pp186–208

Goldman, R.L., Thompson, B.H. and Daily, G.C. (2007) 'Institutional incentives for managing the landscape: Inducing cooperation for the production of ecosystem services', *Ecological Economics*, vol 64, pp333–343

Goldman, R.L., Tallis, H., Kareiva, P. and Daily, G.C. (2008) 'Field evidence that ecosystem service projects support biodiversity and diversify options', *Proceedings of the National Academy of Sciences*, vol 105, pp9445–9448

Gowdy, J.M. (1997) 'The value of biodiversity: Markets, society, and ecosystems', *Land Economics*, vol 73, pp25–41

Hein, L., Koppen, K. van, Groot, R.S. de and Ierland, E.C. van (2006) 'Spatial scales, stakeholders and the valuation of ecosystem services', *Ecological Economics*, vol 57, pp209–228

Holden, J. (2005) 'Peatland hydrology and carbon cycling: Why small-scale process matters', *Philosophical Transactions of the Royal Society A*, vol 363, pp2891–2913

Holden, J., Shotbolt, L., Bonn, A., Burt, T.P., Chapman, P.J., Dougill, A.J., Fraser, E.D.G., Hubacek, K., Irvine, B., Kirkby, M.J., Reed, M.S., Prell, C., Stagl, S., Stringer, L.C.,

Turner, A. and Worrall, F. (2007) 'Environmental change in moorland landscapes', *Earth-Science Reviews*, vol 82, pp75–100

Hubacek, K. and Mauerhofer, V. (2008) 'Future generations: Economic, legal and institutional aspects', *Futures*, vol 40, pp413–423

Hubacek, K. and Reed, M.S. (2009) 'Lessons learned from a computer-assisted participatory planning and management process in the Peak District National Park, England', in Allen, C. and Stankey, G. (eds) *Adapative Environmental Management: A practical guide*, Springer, Heidelberg

Hubacek, K., Dehnen-Schmutz, K., Qasim, M. and Termansen, M. (2008) 'Description of the upland economy: Areas of outstanding beauty and marginal economic performance', in Bonn, A., Allott, T., Hubacek, K. and Stewart, J. (eds) *Drivers of Environmental Change in Uplands*, Routledge, Abingdon, pp293–308

Jaarsveld, A.S. van, Biggs, R., Scholes, R.J., Bohensky, E., Reyers, B., Lynam, T., Musvoto, C. and Fabricius, C. (2005) 'Measuring conditions and trends in ecosystem services at multiple scales: The Southern African Millennium Ecosystem Assessment (SAfMA) experience', *Philosophical Transactions of the Royal Society B, Biological Sciences*, vol 360, pp425–441

Kapp, K.W. (1970) 'Environmental disruption and social costs: A challenge to economics', *Kyklos*, vol 23, pp843–847

Kremen, C., Daily, G.C., Klein, A.-M. and Scofield, D. (2008) 'Inadequate assessment of the ecosystem service rationale for conservation: Reply to Ghazoul', *Conservation Biology*, vol 22, pp795–798

Lane, S.N., Brookes, C.J., Kirkby, M.J. and Holden, J. (2004) 'A network-index-based version of TOPMODEL for use with high-resolution digital topographic data', *Hydrological Processes*, vol 18, pp191–201

Millennium Ecosystem Assessment (MA) (2005) *Millennium Ecosystem Assessment. Ecosystems and Human Well-Being: Synthesis*, MA and Island Press, Washington DC

Naidoo, R. and Ricketts, T.H. (2006) 'Mapping the economic costs and benefits of conservation', *PLoS Biology*, vol 4, pp2153–2164

Naidoo, R., Balmford, A., Costanza, R., Fisher, B., Green, R.E., Lehner, B., Malcolm, T.R. and Ricketts, T.H. (2008) 'Global mapping of ecosystem services and conservation priorities', *Proceedings of the National Academy of Sciences*, vol 105, pp9495–9500

Neff, J.C. and Hooper, D.U. (2002) 'Vegetation and climate controls on potential CO_2, DOC and DON production in northern latitude soils', *Global Change Biology*, vol 8, pp872–884

O'Neill, J. (1996) 'Cost-benefit analysis, rationality and the plurality of values', *The Ecologist*, vol 26, pp98–103

O'Neill, J. and Spash, C.L. (2000) 'Appendix: Policy research brief conceptions of value in environmental decision-making', *Environmental Values*, vol 9, pp521–536

Pagiola, S., Landel-Mills, N. and Bishop, J. (eds) (2002) *Selling Forest Environmental Services: Market-Based Mechanisms for Conservation and Development*, Earthscan, London

Parish, F., Sirin, A., Charman, D., Joosten, H., Minaeva, T. and Silvius, M. (eds) (2007) *Assessment on peatlands, biodiversity and climate change*, Global Environment Centre, Kuala Lumpur and Wetlands International, Wageningen

Pearce-Higgins, M.C., Grant, C.M., Beale, G., Buchanan, M. and Sim, I.M.W. (2008) 'International importance and drivers of change of upland bird populations', in Bonn,

A., Allott, T., Hubacek, K. and Stewart, J. (eds) *Drivers of Environmental Change in Uplands*, Routledge, Abingdon, pp209–227

Puhr, C.B., Donoghue, D.N.M., Stephen, A.B., Tervet, D.J. and Sinclair, C. (2000) 'Regional patterns of streamwater acidity and catchment afforestation in Galloway, SW Scotland', *Water, Air, and Soil Pollution*, vol 120, pp47–70

Reed, M.S., Arblaster, K., Bullock, C., Burton, R., Fraser, E.D.G., Hubacek, K., Mitchley, J., Morris, J., Potter, C., Quinn, C.H. and Swales, V. (2009) 'Using scenarios to explore UK upland futures', *Futures*, vol 41, pp619–630

Simmons, I.G. (1989) 'Prehistory and planning on the moorlands of England and Wales', *Landscape and Urban Planning*, vol 17, pp251–260

Sotherton, N., May, R., Ewald, J., Fletcher, K. and Newborn, D. (2008) 'Managing uplands for game and sporting interest: An industry perspective', in Bonn, A., Allott, T., Hubacek, K. and Stewart, J. (eds) *Drivers of Environmental Change in Uplands*, Routledge, Abingdon, pp241–260

Spash, C.L. (1994) 'Double CO_2 and beyond: Benefits, costs and compensation', *Ecological Economics*, vol 10, pp27–36

Swales, V. and Woods, A. (2008) 'Rising to the land use challenge: Key issues for policy-makers', Discussion Paper prepared for the Rural Economy and Land Use Programme, unpublished

Tharme, A.P., Green, R.E., Baines, D., Bainbridge, I.P. and O'Brien, M. (2001) 'The effect of management for red grouse shooting on the population density of breeding birds on heather-dominated moorland', *Journal of Applied Ecology*, vol 38, pp439–457

Tognetti, S., Mendoza, G., Ayward, B., Southgate, D. and Garcia, L. A. (2004) 'Knowledge and assessment guide to support the development of payments arrangements for watershed ecosystem services', Report prepared for the World Bank Environment Department, http://www.flowsonline.net/data/pes_assmt_guide_en.pdf accessed 29 July 2009

Troy, A. and Wilson, M.A. (2006) 'Mapping ecosystem services: Practical challenges and opportunities in linking GIS and value transfer', *Ecological Economics*, vol 60, pp435–449

Turner, R.K. and Daily, G.C. (2008) 'The ecosystem services framework and natural capital conservation', *Environmental Resource Economics*, vol 39, pp25–35

Wilby, R.L., Orr, H.G., Hedger, M., Forrow, D. and Blackmore, M. (2006) 'Risks posed by climate change to the delivery of Water Framework Directive objectives in the UK', *Environment International*, vol 32, pp1043–1055

Wünscher, T., Engel, S. and Wunder, S. (2008) 'Spatial targeting of payments for environmental services: A tool for boosting conservation benefits', *Ecological Economics*, vol 65, pp822–833

Part 2

Emerging Issues and New Perspectives

8

Adaptation of Biodiversity to Climate Change: An Ecological Perspective

John Hopkins

Introduction

Although the current episode of climate change is unique because it originates from human influence upon greenhouse gas levels in the atmosphere (IPCC, 2007a), climate change is not a new phenomenon. The past *c.*10,000 years of the Holocene, in which human activities, especially agriculture, have shaped the land have been a period of relative climatic stability (Burroughs, 2005). For much of the rest of the *approximate* two million years of the Quaternary Period the climate has been much less stable. This means that species in Britain today have developed traits which allowed them to respond to past change in climate, not least by dispersing to new areas with favourable climate conditions as change occurred. Those which did not have not survived (Tallis, 1991).

However, sole reliance upon this natural adaptive capacity is unlikely to result in a high level of late 21st-century biodiversity. In the past species responded to climate change in landscapes upon which human activity had little impact. Today the countryside has been transformed through simplification of its ecology and reduction in the extent and quality of many wildlife habitats. The way in which the land is managed in order to allow nature to respond will be a key determinant of future biological richness of Britain and is the subject of this chapter.

This task, however, must build upon a range of established conservation policy and practice. In the 20th century the conservation of biodiversity became an important policy consideration, reflected in a range of Government policy, UK and EU legislation and the setting up and funding of environmental agencies (Evans 1997; Sheail, 1998; Marren, 2002). This has resulted in a significant influence of conservation upon the countryside. At the time of writing approximately 10 per cent of the land surface of Great Britain is within an SSSI in order to protect its biological or geological interest (Joint Nature Conservation Committee, unpublished). In England €3.5 billion (approx. £2.9 billion) has been allocated between 2007 and 2013 to fund agri-environment schemes, very largely to conserve or enhance biodiversity (Defra, 2007).

To address the threats associated with climate change calls for *adaptation* of current conservation arrangements, that is, the adoption of policies and practices which minimize the adverse impacts of climate change (IPCC, 2007b). Thus far *mitigation*, that is, stabilizing or reducing the GHGs in the atmosphere, has received most attention. The need for adaptation cannot be avoided, because even if GHG emissions were to stop tomorrow we are already locked into climate change for decades to come due to lags in the climate system.

At the time of writing, a coherent body of policy and practice to address this challenge has yet to emerge and it is the principles that might underpin a new approach (Hopkins et al, 2007) which are discussed here in the context of underlying ecological processes.

We may also see large land and water use changes driven by climate, not least due to alteration of types of arable and pastoral farming (Mitchell et al, 2007). These could have a significant indirect impact upon biodiversity. However, to what degree such impacts have already occurred and how they might manifest themselves in the future is not well understood and they are not dealt with here, despite their undoubted importance.

Observed and projected climate change

Significant change to Britain's climate has already been observed, with temperature increases above the global average. In fact the pattern of climate change is especially well documented in Britain due to continuous recording since 1659 of the Central England Temperature (CET), the longest continual observation of daily temperature (Manley, 1974). Since 1950 the CET has risen by 1°C; 2006 was the warmest year in 365 years and 9 of the 15 warmest years have been since 1990 (Jenkins et al, 2007). The number of frost-free days and hot summer days has increased and the growing season has become longer (Hulme et al, 2002). The England and Wales Precipitation Record is of monthly mean precipitation since 1766. There is no evidence for significant change in annual rainfall in recent decades, although there is evidence of decreasing summer rainfall and increased winter rainfall, and rainfall is in heavier downpours (Hulme et al, 2002; Jenkins et al, 2007). Due to the combination of thermal expansion of the warming oceans and, to a lesser extent, melting of ice, sea level rose by one millimetre per year in the 20th century, although in the UK this is complicated by the fact that land is falling in the south-east and rising in the north-west so that rates above and below the global average value were encountered (Hulme et al, 2002).

Under the United Kingdom Climate Impact Programme (UKCIP) scenarios, which cover a range of future potential GHG emissions and so several climate projections, a rise in UK temperature of between 2°C and 3.5°C is projected before the end of the 21st century, with greater warming in the south-east than in the north-west. The last time these temperatures occurred was in the Ipswichian Interglacial *c.*120,000 years ago, when it was warm enough for hippopotamuses to roam widely in southern England.

By mid-century, current typical spring temperatures may occur between one and three weeks earlier and onset of winter could be delayed by a similar period, lengthening the growing season. Annual average precipitation may decrease slightly, winters become wetter and summers drier, particularly in the south-east, where summer precipitation may decrease by 50 per cent by the end of the century under the high scenario. Periods of heavy winter rainfall may become more frequent. By the end of the century sea levels in Scotland may be between 2 cm below and 58 cm above current level, with a possible rise of between 26 and 86 cm in south-east England (Hulme et al, 2002) and inevitable losses of intertidal and coastal land, especially on low-lying parts of the coast.

Response of species to climate change

Already a global 'fingerprint' of climate change response by species has been observed (Parmesan and Yohe, 2003; Root et al, 2003), which might be seen as the inherent adaptive capacity of wild nature. As with climate this biodiversity response is especially well documented in the UK (Root et al, 2003), due to the British genius for amateur natural history and biological recording.

Although it is important to note that many species do not follow the main trajectory of change, and a few show an opposite response, three general patterns of response have been shown by species:

1 *Phenological change*, that is, change to the timing of seasonal events. There has been a general trend of spring and summer events taking place earlier in the year (Menzel et al, 2006). These include earlier first-leafing dates of trees (Sparks et al, 1997), flight times of moths and butterflies (Woiwod, 1997; Roy and Sparks, 2000), egg-laying dates in birds (Crick et al, 1997; Crick and Sparks, 1999), first spawning of amphibians (Beebee, 1995), first appearance of hoverflies (Morris, 2000) and earlier summer fruiting (Menzel et al, 2006). Autumn events are more complex (Menzel et al, 2006), but delayed migration is reported for some bird species (Sparks, 1999). The conservation implications of such changes are starting to emerge. In addition to their implications for timing of management activities, discussed below, of greatest importance may be that interdependent species no longer have life cycles that are synchronized. There is, for example, good evidence that some Dutch populations of the pied flycatcher (*Ficedula hypoleuca*) are declining because birds are now breeding after the

time of peak caterpillar abundance, which has become earlier (Both et al, 2006). In other cases the pattern of change may be ecologically complex, for example great tits (*Parus major*) at Wytham Wood, in Central England, have changed their time of breeding to remain synchronized with food supply, but this response has not occurred in a population at a study site in the Netherlands (Lyon et al, 2008).

2 *Range change.* Many species are showing evidence of changes in their range. Wading birds and wildfowl such as knot (*Calidris canuta*) and sanderling (*C. alba*) which migrate to spend winter on the UK coast have shown a particularly rapid response. In recent years reduced numbers of waders have been found in the south and west of the UK, because warmer winters mean they are able to feed further north and east in the UK, nearer to their overseas breeding sites (Austin and Rehfish, 2005).

Non-migratory species which reach a northern limit of distribution in the UK are widely thought to be limited by climate, particularly temperature (Thomas 1993; Thomas et al, 1999). This assumption, which is not well tested by field experiment, is supported by the fact that many apparent thermophiles are expanding their range northwards with increasing temperatures. This general trend can be seen in birds (Thomas and Lennon, 1999), butterflies (Warren et al, 2001), dragonflies and damselflies (Hickling et al, 2005), while a study of 329 southern species (from 16 invertebrate and vertebrate animal groups) which reach their northern limit in Britain (Hickling et al, 2006) found average northwards expansion in 279 species, including species from all groups except amphibians and reptiles. Hickling et al (2006) have concluded that the average rate of northern expansion for the species they studied is in the range 12.5–19 km per decade, equivalent to a brisk walk every year. Not surprisingly this process has been associated with some continental species extending their range across the English Channel and establishing breeding populations in the UK, as with a bumble bee (*Bombus hypnorum*) and the small red-eyed damselfly (*Erythromma viridulum*).

In addition thermophilous species are expanding onto higher ground, including 227 of the 329 species studied by Hickling et al (2006) which increased their range uphill at between 4.7 and 10.7 m per decade, a phenomenon of greater adaptive significance than might at first appear, as discussed below.

A much less studied retreat to the north of cold-tolerant species that reach their southern limit in the UK is also occurring (Hickling et al, 2005; Franco et al, 2006).

3 *Changing habitat preference.* Many UK species, particularly plants and cold blooded animals, occupy a different and wider range of habitat conditions further south in Europe. In many cases this is likely to be due to warmer

climates (Perring, 1960; Thomas, 1993; Thomas et al, 1999). It can be expected that change to habitat occupation by some species will occur as climate changes. Such sensitivity to climate has already been demonstrated in southern England for the silver-spotted skipper butterfly (*Hesperia comma*). Previously this species mainly bred in short turf on south-facing slopes, where climate conditions are particularly warm. It now breeds in taller, cooler vegetation on slopes with a wider range of aspect and temperature (Davies et al, 2006). Climate warming may also account for the shift in larval food plant preference by the brown argus butterfly (*Aricia agestis*) by making food plants in colder habitats available (C.D. Thomas et al, 2001).

Community and habitat responses

Climate-induced changes at the level of community and habitat are less easily detected, as factors operating but not linked to climate may also drive ecological change. For example increased nitrogen deposition and decreased level of atmospheric sulphur deposition have occurred in recent decades at the same time as climate has warmed. These have been drivers of significant biodiversity change not clearly linked through causality to climate (NEGTAP 2001; Stevens et al, 2004, Hopkins and Kirby, 2007).

Long-term monitoring of butterflies (Roy et al, 2001; González-Megías et al, 2008) and moths (Conrad et al, 2004) has detected change in community composition correlated with climate. The analysis by González-Megías et al (2008) indicates a *c.*15 per cent turnover of butterfly species in 25 years, mainly due to the spread of southern and habitat-generalist butterflies. If typical of other insects this suggests major change to insect communities (and so the largest animal component of most ecosystems) has already occurred.

Much less is known about how climate change is impacting upon plant species and the results of vegetation monitoring have yielded inconsistent results, suggesting that there may be a great deal of context-specific change to vegetation. The studies of Dunnett et al (1998) and Kirby et al (2005) have detected changes in the relative proportions of plant species in road-side grasslands and woodlands correlated with climate change. However, 13 years of experimentation on long-established limestone grasslands in Derbyshire has shown a high stability of these grasslands in the face of simulated temperature and rainfall change, suggesting some habitat may be highly resistant to climate change (Grime et al, 2008). Despite this, similar treatment of grassland recently established on abandoned arable land showed a more rapid shift in species composition (Grime et al, 2000). The recently reported results of Countryside Survey 2007, a vegetation monitoring programme for the UK, similarly found no clear signal of climate impact upon vegetation, in part due to the difficulty of separating climate change effects from those caused by other factors (Carey et al, 2008).

It might be expected that vegetation impacts are particularly marked and rapid in communities which are dominated by one or a few plant species. Beech *(Fagus sylvatica)* frequently dominates the woodland in which it occurs and due to its shallow rooting is particularly susceptible to drought (Peterken and Mountford, 1996; Jump et al, 2006). It would appear plausible that in South-East England major change to woodlands will occur due to disappearance or reduction of beech (Harrison et al, 2001; Wesche et al, 2006). Similarly common heather (*Calluna vulgaris*) frequently dominates the vegetation in which it occurs, and warming experiments suggest under climate change a heather decline may occur due to increased insect herbivory (Peñuelas et al, 2004).

Because species respond individualistically to climate change, new types of vegetation may emerge unlike any seen today (Williams and Jackson 2007), although available evidence of a relatively slow rate of change suggests strikingly novel vegetation is unlikely to appear in the next few decades.

Adaptive management

Although for some climate trends there is a high level of confidence about the direction of change, large uncertainties surround the rate of change (IPCC, 2007a). Such uncertainty is compounded in attempts to model how biodiversity will change in response to climate by knowledge gaps about how climate and other factors impact upon species and ecosystems, and technical limitations of various modelling approaches (Botkin et al, 2007; Doak et al, 2008). Such uncertainty has profound implications for the approach to adaptation to be adopted.

A significant part of current conservation practice is founded upon a static paradigm, essentially the conservation of habitats and species where they occur today. This is perhaps clearest in the context of protected areas. In the UK the most important of these are SSSIs, Special Areas of Conservation (SACs), under the EU Habitats Directive, and Special Protection Areas (SPAs), under the EU Birds Directive. Despite the likelihood of unavoidable range change and even great uncertainty about how habitats will change, site selection for protected status is based upon species and habitats which occur at the time of designation (Nature Conservancy Council, 1989; Stroud et al, 2001; McLeod et al, 2005). It is likely that over time some of the features for which sites are selected will no longer be naturally sustainable. Other aspects of conservation planning, including monitoring (JNCC, 2004) in large part continue to be within the paradigm of a static view of nature. Potentially this could result in considerable wasted effort as increasingly expensive and unsuccessful interventions are called for to maintain the status quo. Although this problem, which might require legal changes to be introduced, is widely recognized, as yet no alternative approach to the designation process has been proposed.

Conversely, long-term planning based upon projections of how species may change their range and habitats change their ecological character, is unlikely to be efficient. Given the large uncertainties surrounding such projections, it is likely to result in either over-provision or under-provision of conservation effort.

A strategy would appear to be called for which adopts the basic tenets of 'adaptive management' (Holling, 1978) and from the outset explicitly recognizes both the inevitability of change and high levels of uncertainty. It should also incorporate learning from practice and research, so that plans and programmes can be modified in the light of new knowledge, through regular review.

Although the concept of adaptive management is not used explicitly in many conservation activities in the UK, in some areas of current practice there appear to be only limited barriers to its adoption. For example Habitat Action Plans and Species Action Plans, which have underpinned the UK Biodiversity Action Plan, comprise a series of time-limited targets amenable to periodic revision (UK BAP, 2006). At the time of writing, an ecosystem-focused approach to delivery of the UK BAP is being developed which is likely to retain the capacity for periodic review. Management plans for most nature reserves are again normally time limited and subject to update and review, allowing for an adaptive management approach (SNH, 2007; Natural England, in preparation).

Agri-environment schemes that reward farmers for adopting a set of environmentally beneficial practices are based upon time-limited agreements. There is potential for adjustment of the agreements over time to address emerging climate-driven change and incorporate new practices as the agreements become eligible for renewal. A major review of agri-environment and other land-based conservation schemes funded under the European Rural Development Programme (ERDP) will almost certainly occur at the end of the current ERDP in 2013, and we can anticipate that climate change adaptation will be a significant consideration at that time, if not before.

The role of protected areas and other wildlife habitat

Although designation as protected area will not prevent sites from being impacted by climate change, protected areas are likely to remain a cornerstone of conservation policy in both the short and long term. In the short term they will play a major role as 'arks' for biodiversity. They are chosen to encompass a wide range of biodiversity. This is especially true for species and habitats with a very restricted range or abundance, many of which are mainly or only found within protected areas. For example all viable breeding sites for species such as the swallow-tail (*Papilio machon*) and heath fritillary (*Melitaea athalia*) butterflies qualify for notification as SSSI (Nature Conservancy Council, 1998), and 68 per cent of all semi-natural grassland in England is within an SSSI (Natural England, 2008).

We can expect that there will be some European species which colonize Britain in response to climate change. For example 160 moth species which do not yet breed in Britain are regularly recorded (Sparks et al, 2005). However, it seems unlikely that such species will come to dominate British biodiversity, not least due to the isolating effects of our surrounding seas. It is a reasonable assumption that in the present century the biodiversity in the UK will be founded mainly upon the level of protection already afforded to species and habitats.

In many parts of lowland Britain (Fox, 1932) the protected site series accounts for a high proportion of all high-quality wildlife habitat, and its role in securing a future biodiversity will be particularly critical. However, by no means all high-quality habitat is included within protected areas. For example only 11 per cent of native broadleaved woodland and wood pasture is within a protected area in England (Natural England, 2008). Linear habitats such as road verges, railways and riverbanks, as well as small fragments of many other habitats, are mostly outside the SSSI series but are important species reservoirs in biologically impoverished landscapes (Smart et al, 2006). Action taken to protect and manage such areas through, for example, national policy frameworks (e.g. Forestry Commission England, 2005), Local Biodiversity Action Plans and agri-environment schemes is therefore essential. However, the collective effectiveness of the full suite of such measures outside protected areas is not well understood. Monitoring of a representative sample of 104 heathlands outside of protected areas found that none was in favourable condition, in contrast to 81 per cent in favourable condition within SSSIs (Hewins et al, 2007; Natural England, 2008). Monitoring grasslands outside of SSSIs revealed only about 20 per cent to be in favourable condition (Hewins et al, 2005; Natural England, 2008).

There are also ecological grounds for investing over the long term in protected areas and other high-quality wildlife habitats. Such areas have a set of characteristics such as low-fertility soils associated with high species richness (Grime, 1973), as well as varied hydrology, soils, geology and landform which result in high habitat diversity, the main determinant of species richness in many ecosystems (Rosenzweig, 1995; Whittaker and Fernández-Palacios, 2007). In other cases such areas have features such as high levels of heavy metals or shingle, which although not associated with high species richness are habitat conditions required by rare and local species narrowly adapted to such harsh environments (Rodwell, 2000; Shardlow, 2001).

It seems unlikely that the ecological conditions which confer biodiversity importance to most high-quality habitats can be easily or quickly recreated elsewhere. For example studies in southern England of comparable calcareous grasslands of different age indicate that the characteristic species composition of ancient grasslands takes many decades, possibly a century or more, to develop in sites which escaped intensive agriculture (Gibson and Brown, 1991). Today over quite large parts of the farmed landscape ploughing, drainage and nutrient enrichment are likely to have modified the soils and depleted residual biodiversity. In such cases restoration to conditions which support important biodiversity is no longer possible even after decades, a problem which may be compounded by non-native species invasion (Cramer et al, 2008).

Ecological range and variation

In order to ensure species survival in the face of climate change, Saxon (2003) has advocated that protected-area series and other conservation frameworks should encompass an element of redundancy, that is, include multiple representation of a species or habitat. This provides insurance in the face of uncertain climate-change impacts that the species or habitat will survive in at least some of its current localities. Added insurance value is provided by conserving the full geographical distribution and range of ecological situations in which given species or habitats occur (Hopkins et al, 2007). For example horseshoe vetch (*Hippocrepis comosa*) is a plant typical of lowland chalk and limestone grasslands in southern Britain. It has atypical high-altitude occurrences in Upper Teesdale, County Durham and parts of Cumbria (Elkington, 1978; Preston et al, 2002). It is plausible that over time these upland localities could become the main centre for the survival of this species, despite the fact that today it is more abundant in lowland southern England.

A significant level of adaptive insurance is likely to exist under current arrangements for protected areas. Guidelines for selection of SSSIs advocate the conservation of the full range of variation in most habitats, and for a range of rare species all occurrences judged to have viable breeding populations are recommended for inclusion. Further, with few exceptions the guidelines recommend that examples of important habitat types are selected within each 'area of search' where they occur (roughly equivalent to a small- or medium-sized county or major division of a large county or Scottish Region) (Nature Conservancy Council, 1989).

It is important not to overlook the fact that some once common species, such as the North American Passenger Pigeon (*Ectopistes migratorius*) have undergone extinction (Gaston and Fuller, 2008). To some degree the requirement to conserve common species is likely to be met by conserving a full range of semi-natural habitats within SSSIs. But there have been notable declines of species with highly anthropogenic associations, such as the house sparrow (*Passer domesticus*) (Hole et al, 2002), and such species may not be fully taken into account due the emphasis in protected area selection upon semi-natural habitats.

Conserved and enhanced landscape heterogeneity

Mathematical models have come to play a role in climate change policy unparalleled by any previous environmental issue. There has been much concern about the confidence which can be attached to the outputs from such models, reflected in the meticulous reporting of levels of confidence of model outputs and other scientific findings by the IPCC (2007a). The spatial scale of the models has received much less comment yet may result in considerable misunderstanding about the appropriate spatial scale for climate change adaptation.

Typically general circulation models of climate, and linked bio-climatic models which attempt to project species impacts of climate change are run and reported upon at large spatial scale (Burroughs 2001), for example as 50-km grid squares (2500 km^2) in the reports of Hulme et al (2002) and Walmsley et al (2007). This is not an especially meaningful scale of ecological analysis. Models use air temperatures not surface temperatures, and surface temperatures can vary over very short distances of metres and centimetres. It is this small-scale variation in microclimate which has an impact upon most organisms and most ecological processes, particularly given that the majority of multi-celled organisms are insects, too small to sense and respond to climate at large scales, and most plants and some animals are sessile. The British countryside should be seen as composed of a finely textured mosaic of microclimatic patches which controls the distribution and abundance of species and the ecological variability of habitats.

The ecological importance of such local variation to species' survival is common knowledge among successful gardeners, who will know within their own garden where there are warm locations for ripening summer fruits and frost-free patches for growing frost-tender species. In nature this mosaic means that many species can respond to climate change by small-scale movement and dispersal within the landscape, although of course those which already occupy the coldest patches, as on mountain tops, are still likely to be vulnerable to local extinction (Pauli et al, 2007).

Understanding how topography, hydrology, soils and vegetation influence local climate is therefore critical to climate change adaptation, particularly in the short to medium term. Armed with such knowledge there is also the possibility of managing the landscape to maintain and enhance desirable climatic features, an approach already being adopted in the urban context (Bonan 2008).

One of the most important correlates of local climate variation is elevation, more specifically a generally observed cooling of climate at higher altitude, although precipitation and the wind environment also vary with altitude and air movements can complicate this general pattern (Bonan 2008). An indication of the scale of the cooling effect on local climates can be gathered by consideration of the *adiabatic lapse rate*, that is the rate at which air cools as it rises. For dry air the *dry adiabatic lapse rate* is 9.8°C per 100 m; for air warmed by condensation the *wet adiabatic lapse rate* is approx. 0.6°C per 100 m.

Steeper altitudinal climate gradients occur in hyper-oceanic Northern and Western Isles of Scotland, with, on Orkney, agriculture on the coast and tundra vegetation at 420 m above sea level (Crawford, 2008).

This elevational effect means that species are more likely to successfully expand their range in step with global warming by moving uphill. Hopkins' Bioclimatic Law states that a 1°C temperature decrease equates to moving 100 km towards the pole, but only 130 m uphill (Hopkins, 1920; Kerr and Kharouba, 2007), that is, escape from the impacts of climate change is at a first approximation 1000 times easier by movement uphill than by changing latitude.

A number of studies now document uphill expansion of species into the mountains and other uplands of the world during the 20th and 21st centuries (Tryjanowski et al, 2005; Battisiti et al, 2006; Parolo and Rossi, 2007; Raxworthy et al, 2008). The study of Hickling et al (2006) have reported that 227 of 329 animal species showed an uphill expansion of range of between 2.8 and 10.1 m per decade in the late 20th century, an order of magnitude of change which may allow some of the species to track the approximately 1°C increase in temperatures reported in recent decades (Jenkins et al, 2007).

Recording of altitudinal change in plant species of the Alps by Grabherr et al (1994) provides more spatially precise results. To track climate change in the Alps, plants were expected to increase their altitudinal limit at a rate of 1 m $year^{-1}$. The observed rate of change was closer to 0.1 m $year^{-1}$, with the fastest species only expanding at a rate of 0.4 m $year^{-1}$. These results suggest range expansion is lagging behind the expansion of the favourable climate envelope and may illustrate a slower response to climate change by plants than that observed for animals. Despite the likely long-term vulnerability to extinction of cold-tolerant species found there, Britain's uplands are likely to increase in diversity and conservation importance in the 21st century.

A potential difficulty for species which move their distribution into the uplands is that, depending upon the convexity or concavity of the landform, with increased altitude the actual physical area available declines. The carrying capacity of summits will be lower and, for species that occupy large territories at low density, this may mean viable populations cannot be maintained. This same effect will be experienced as species move their range northwards due to the smaller surface area of the globe at high latitudes (Rosenzweig, 1995). Conservation programmes that expand the area of available land for conservation in the uplands may therefore be of critical importance. In addition the movement of species into the uplands may result in increased isolation of populations and decreased genetic exchange, which could also have deleterious effects and require conservation interventions to maintain genetic diversity.

Even in the absence of significant range of elevation, landscapes can be rich in local climatic variation due the effects of slope and aspect, which can result in climate variation within valleys and on hills equivalent to many degrees of latitude. Slope and aspect interact so that high levels of insolation and so warm climates occur on steep south-facing slopes and low levels of sunlight reach cold, steep north facing slopes, due to shading effects and low angles of sunlight incidence. Under a temperate climate in Germany, hill slopes of different aspect were found to vary by as much as 3.5°C (Bonan 2008).

The effects of such topographic variation on biodiversity are most clearly shown among Britain's insects, where a high proportion of rare and local species are warmth loving and at the northern edge of their range. They are mainly found on sites with southerly aspects and short vegetation maintained mainly by grazing and other human activities (Thomas 1993). Plant species similarly show patterns of distributions associated with specific ranges of slope and aspect in otherwise uniform chalk grassland vegetation due to local variation in climate (Perring, 1960). Small scale variation in slope and aspect associated with valley systems and hill slopes offers species opportunities for local adaptative movement as climate warms. This has been shown in the case of the silver-spotted skipper (*Hesperia comma*). As discussed above, in recent years this butterfly has spread in response to warming from south facing slopes to breed on a wide range of aspect in chalk grasslands of southern England (Davies et al, 2006).

It can be expected that in areas with a variety of slope and aspect, as for example in many valley systems, higher levels of current biodiversity will be maintained and their conservation will become of increased importance, although species already confined to cold habitat patches on north facing slopes may be extirpated. In the lowlands biodiversity loss may be greater on scarp slopes and other land forms with a uniform aspect and also in areas with flat topography, as in parts of East Anglia.

Although the topic has attracted much less attention among ecologists, soil can also have an important influence upon local climate. High soil moisture has an atmospheric cooling effect, and can lead to the creation of small scale climate mosaics (Bonan 2008). From a theoretical perspective at least it can be expected that variation in soil creates local climate patchiness of adaptive value.

An additional and more malleable factor is the influence of vegetation upon local climate. The shading effect of light captured by the canopy, coupled with transpiration of water from the leaves, have a combined local cooling influence during the day. This effect is commonly experienced by humans entering woodlands during hot weather, however, smaller scale variation is surprisingly large. For example summer measurement of ground surface temperatures under a canopy of the low growing horseshoe vetch (*Hippocrepis comosa*) with a range in height from one to ten centimetre, varied by 8°C (Thomas, 1990).

A range of possibilities exist to manage rural climates for biodiversity adaptation (Hopkins et al, 2007). This could be by:

- Designing engineered land forms such as quarries and roads to add or reinforce local topographic diversity.
- Removing or fragmenting blanket forests on hill slopes to allow valley species to spread gradually onto high ground through zones with intermediate climate properties.
- Restoring habitats in topographically complex system, such as valleys, to create habitat patches over a more complete microclimate range.
- Managing habitat to create a wider range of vegetation structure and so microclimate variation.

Dispersal and connectivity

Although a poleward expansion of range has been observed for many animal species which reach the northern edge of their range in Britain, not all southern species show this general trend. Out of 329 southern animal species which might be expected to expand their range polewards Hickling et al (2006) identified 52 which retreated southwards and 2 which remained stable. Out of 46 butterflies at their northern limit three-quarters failed to expand their range between 1970 and 1999, including 89 per cent of habitat specialist species (Warren et al, 2001). Given the majority of British butterflies are southern species it might be expected that overall the spread of species will result in increased diversity of butterfly communities, but only one-third of the expected diversification occurred between 1970 and 1982 (Menendez et al, 2006).

A range of factors may account for such responses. Some species may only survive in sub-optimal conditions and be unable to produce sufficient offspring for effective dispersal and colonization, the so-called 'mass effect' (Holyoak et al, 2005, pp 17–18). In other cases inbreeding, as reported for the natterjack toad *Bufo calamita* (Rowe et al, 1999), may have reduced genetic fitness of offspring and so dispersal and colonization.

Equally the widespread destruction of semi-natural habitats in the 20th century, including for example a 97 per cent loss of unimproved lowland grasslands between the 1930s and 1980s (Fuller, 1987), has resulted in a high level of fragmentation of habitat and for some species (perhaps many) decreased or even blocked dispersal due to the large distance between habitat patches and hostility of the countryside matrix between them. The clearest evidence of this is provided by several British butterflies, which have increased local abundance, but not expanded their range (Warren et al, 2001).

Increasing *connectivity*, that is, increasing the ease with which species are able to move within the landscape, has therefore been seen as an essential component of adaptation to climate change. Most often among conservation planners this appears to have been identified as a requirement to establish wildlife corridors (Saunders, 2007), that is physically continuous linear habitat linking existing habitat patches. There is, however, only weak evidence for the effectiveness of such habitat continuity. Among reasons for thinking this is not a sound approach for most species is the fact that corridors of, say, woodland habitat would preclude corridors for grassland and so the geometry of this approach would not work in landscapes with more than one fragmented habitat (Huntley, 2007). Further, in nature habitats are typically patchy and natural habitat corridors rare, so few species are adapted to disperse along linear routes but mainly 'jump disperse' between patches (Watkinson and Gill, 2002; Huntley 2007).

Although several books deal exclusively with this topic (Jongman and Pungetti 2004; Crooks and Sanjayan 2006; Hilty et al, 2006) only a limited consensus has emerged about what practical steps are required to restore connectivity (Lindenmayer et al, 2007). This is in large part due to the fact that the process of dispersal is complex and under control of factors other than landscape character. For example seed plants are variously dispersed passively by wind, water, animals and humans; vectors which impose quite different patterns of movement (Cousens et al, 2008). The active dispersal typical of animals imposes even greater complexity, including very commonly flight, which loosens the constraint of the landscape upon dispersal.

Life stage is also important in animals. For example, while mature dormice *Muscardinus avellanarius* stick tightly to hedgerows as they travel, and have been seen as a species which uses habitat corridors (Bright, 1998), immature dormice may range over hundreds of metres though open country (Büchner, 2008). Most insects, the largest animal group on land, are winged but small. However, they disperse at altitude in their millions, over tens and hundreds of kilometres, by a combination of passive and active flight (Chapman et al, 2003). Prevailing wind direction (Whittaker and Fernández-Palacios, 2007), storm frequency (Soons et al, 2004) and temperature (Sparks et al, 2005) are climatic factors which influence dispersal and may themselves change, further complicating the situation.

Connectivity is therefore not a single landscape property. A landscape which is freely permeable to one organism may be completely impermeable to another. Just as one would not expect to apply the same approach to management of sand dunes and upland bogs, restoring and managing connectivity at landscape and larger scales is unlikely to be amenable to a single formulaic solution.

Given the lack of off-the-shelf solutions, the large-scale nature of the land use change required, the implicit assumption that new habitat will need to be created and the long time-scales over which dispersal will need to be facilitated, Hopkins et al (2007) advocate a multi-disciplinary team is required for the design and implementation of schemes which address problems of connectivity. There would, however, appear to be a limited range of generic options to be considered, which will vary in their importance from place to place (Hopkins et al, 2007; Huntley, 2007) but include:

- Conservation of core areas of existing habitat, particularly protected areas, as these will have healthy breeding populations and so fuel the dispersal process.
- The establishment of buffer areas around existing habitat to reduce the impact of adverse surrounding land use.
- Restoration of degraded habitat and creation of new habitat to re-enforce the patchwork of habitats as 'stepping stones' for dispersal.
- Reducing the intensity of land use in the matrix between habitat patches to facilitate dispersal.
- Building functional corridors such as under- and overpasses if the dispersal of target species such as flightless mammals is impeded by barriers such as major road networks.

One should also not forget that it is not just dispersal but also establishment and breeding at new sites which is critical to restoring connectivity. The ecological quality of habitat patches, not just their spatial arrangement is important (J.A. Thomas et al, 2001).

Taking measures which improve connectivity may not, however, be adequate for some less mobile species, or where favourable future habitat is separated by long distances. For example calcareous soils are rare over a large part of Central England and this could prevent some species found on chalk and limestone in the south from spreading to the limestones of the Peak District and further north. In such cases transplantation may be the only way of conserving some species (Hoegh-Guldberg et al, 2008).

Land management

The way in which land and water are managed for biodiversity may also need to change. An earlier growing season, earlier flowering of plants and breeding of animals might mean that it is already appropriate to bring forward cutting dates in traditional hay meadows. More frequent summer drought may require a planned removal of livestock in summer from some areas and additional water supplies may be needed where ponds and streams are relied upon for watering of stock. Change of livestock breeds to ones better adapted to new climates may also become appropriate. In low lying areas both inland and on the coast planning to accommodate more frequent flooding may be needed.

One consequence of a projected increase in drought frequency would be a greater risk of summer wildfire. In addition to the requirement for additional infrastructure to fight fire, strategic land management options include grazing to prevent litter build-up, firebreak construction and the planting of less fire prone species.

The timing of a range of farming operations across Europe has already changed in response to climate (Menzel et al, 2006) and to accommodate earlier breeding of ground nesting birds the Forestry Commission has modified the calendar for heathland management on the New Forest (David Morris, personal communication). Similar ad hoc decisions to accommodate changing species phenology may have been taken by other conservation managers and would be helped by further research on this topic.

Conclusion

The last 10,000 years in which Britain's countryside developed has been a period of relative climate stability. For much of the rest of the approximate two million years of the Quaternary Period the climate has been much less stable. Our flora and fauna evolved to respond to repeated, sometimes rapid, climate oscillations (Overpeck et al, 2005). Through plasticity of traits such as phenology, dispersal and habitat choice, species evolved to cope with climate change, or became extinct. However, humankind has now modified the countryside through simplifying its ecology and fragmenting the semi-natural habitats on which many species depend. In the current climate change episode simply conserving existing biodiversity so that it can respond naturally, although essential, is unlikely to be enough.

An unanswered question remains how much habitat is needed to allow for adaptation of all species. As species change their range they are likely to need to occupy a larger area in the period of transition, including an established source area and new recipient areas. The process of change may mean that in parts of the extended range the species is at low densities, as new populations establish, so that a larger area is needed to hold viable populations. Recent research in Western Europe (Hannah et al, 2007) concludes that to most efficiently capture both present and potential future distribution of plant species would require a 35 per cent area increase of the protected site network. The degree to which this figure is an under or over estimate in the UK is unclear, as is the degree to which such habitat lies outside of protected areas, or might be accommodated by the habitat expansion proposed under the UK Biodiversity Action Plan (UK BAP, 2006).

References

Austin, G.E. and Rehfish, M.M. (2005) 'Shifting distributions of migratory fauna in relation to climatic change', *Global Change Biology*, vol 11, pp31–38

Battisiti, A., Stastny, M., Buffo, E. and Larsson, S. (2006) 'A rapid altitudinal range expansion in the pine processionary moth produced by the 2003 climatic anomaly', *Global Change Biology*, vol 12, pp622–671

Beebe, T.J.C. (1995) 'Amphibian breeding and climate', *Nature*, vol 374, pp219–220

Bonan, G.B. (2008) *Ecological Climatology: Concepts and Applications*, Cambridge University Press, Cambridge

Both, C., Bouwhuis, S., Lessells, C.M. and Visser, M.E. (2006) 'Climate change and migratory population declines in a long-distance migratory bird', *Nature*, vol 441, pp81–83

Botkin, D.B., Saxe, H., Araújo, M.B., Betts, R., Bradshaw, R.H.W., Cedhagen, T., Chesson, P., Dawson, P., Etterson, J.R., Faith, D.P., Ferrier, S., Guisan, A., Skjoldborg-Hansen, A., Hilbert, D.W., Loehle, C., Marguiles, C., New, M., Sobel, M.J. and Stockwell, D.R.B. (2007) 'Forecasting the effects of climate change on biodiversity', *BioScience*, vol 57, pp227–236

Bright, P.W. (1998) 'Behaviour of specialist species in habitat corridors: Arboreal dormice avoid corridor gaps', *Animal Behaviour*, vol 56, pp1485–1490

Büchner, S. (2008) 'Dispersal of common dormice *Muscardinus avellarius* in a habitat mosaic', *Acta Theriologica*, vol 53, pp259–262

Burroughs, W.C. (2001) *Climate Change: A Multidisciplinary Approach*, Cambridge University Press, Cambridge

Burroughs, W.C. (2005) *Climate Change in Prehistory: The End of the Reign of Chaos*, Cambridge University Press, Cambridge

Carey, P.D., Wallis, S.M., Maskell, L.C., Murphy, J., Norton, L.R., Simpson, I.C. and Smart, S.S. (2008) *Countryside Survey: UK Headline Messages from 2007*, Centre for Ecology and Hydrology, Wallingford

Chapman, J.W., Reynolds, D.R. and Smith, A.D. (2003) 'Vertical-looking radar: a new tool for monitoring high-altitude insect migration', *BioScience*, vol 53, pp 503–511

Conrad, K.F., Woiwod, I.P., Parsons, M., Fox, R. and Warren, M.S. (2004) 'Long-term population trends in widespread British moths', *Journal of Insect Conservation*, vol 8, pp119–136

Cousens, R., Dytham, C. and Law, R. (2008) *Dispersal in Plants: A Population Perspective*, Oxford University Press, Oxford

Cramer, V.A., Hobbs, R.J. and Standish, R.J. (2008) 'What's new about old fields? Land abandonment and ecosystem assembly', *Trends in Ecology and Evolution*, vol 23, pp104–112

Crawford, R.M.M. (2008) *Plants at the Margin: Ecological Limits and Climate Change*, Oxford University Press, Oxford

Crick, H.Q.P. and Sparks, T.H. (1999) 'Climate change related to egg laying trends', *Nature*, vol 399, pp423–424

Crick, H.Q.P., Dudley, C., Glue, D.E. and Thomson, D.L. (1997) 'UK birds are laying eggs earlier', *Nature*, vol 388, p526

Crooks, K.R. and Sanjayan, M. (eds) (2006) *Connectivity Conservation*, Cambridge University Press, Cambridge

Davies, Z.G., Wilson, R.J., Coles, S. and Thomas, C.D. (2006) 'Changing habitat associations of a thermally constrained species, the silver spotted skipper butterfly, in response to climate warming', *Journal of Animal Ecology*, vol 75, pp247–256

Defra (2007) *The Rural Development Programme for England 2007–2013*, Defra, London http://www.defra.gov.uk/rural/rdpe/progdoc.htm, accessed 12 December 2008

Doak, D.F., Estes, J.A., Halpern, B.S., Jacob, U., Lindberg, D.R., Lovvorn, J., Monson, D.H., Tinker, T.M., Williams, T.M., Wootton, J.T., Carroll, I., Emmerson, M., Micheli, F. and Novak, M. (2008) 'Understanding and predicting ecological dynamics: are surprises inevitable?', *Ecology*, 89, pp952–961

Dunnet, N.P., Willis, A.J., Hunt, R. and Grime J.P. (1998) 'A 38 year study of relations between weather and vegetation dynamics in road verges near Bibury, Gloucestershire', *Journal of Ecology*, vol 86, pp610–623

Evans, D. (1997). *A History of Nature Conservation in Britain*, 2nd edn, Routledge, London & New York

Elkington, T.T. (1978) 'Phytogeography, variation and evolution', in Clapham, A.R. (ed) *Upper Teesdale, the area and its natural history*, Collins, London

Forestry Commission England (2005) *Keepers of time: a statement of policy for England's ancient and native woodlands – Action Plan 2005–2007*, Forestry Commission England, Cambridge

Fox, C.F. (1932) *The Personality of Britain*, 2nd edn, National Museum of Wales, Cardiff

Franco, A.M.A., Hill, J.K., Kitchke, C., Collingham, Y.C., Roy, D.B., Fox, R., Huntley, B. and Thomas, C.D. (2006) 'Impacts of climate warming and habitat loss on extinction of species: Low-latitude range boundaries', *Global Change Biology*, vol 12, pp1545–1553

Fuller, R.M. (1987) 'The changing extent and conservation interest of lowland grasslands in England and Wales: a review of grassland surveys 1930–84', *Biological Conservation*, vol 40, pp281–300

Gaston, K.J. and Fuller, R.A (2008) 'Commonness, population depletion and conservation biology', *Trends in Ecology and Evolution*, vol 23, pp14–19

Gibson, C.W. and Brown V.K. (1991) 'The nature and rate of development of calcareous grassland in Southern Britain,' *Biological Conservation*, vol 58, pp297–316

González-Megías, A., Menédez, R., Roy, D., Brereton, T. and Thomas, C.D. (2008) 'Changes in the composition of British butterfly assemblages over two decades', *Global Change Biology*, vol 14, pp1464–1474

Grabherr, G., Gottfried, M. and Pauli, H. (1994) 'Climate effects on mountain plants', *Nature*, vol 369, p448

Grime, J.P. (1973) 'Competitive exclusion in herbaceous vegetation', *Nature*, 242, pp242–347

Grime, J.P., Brown, V.K., Thompson, K., Masters, G.J., Hillier, S.H., Clarke, I.P., Askew, A.P., Corker, D. and Kielty, J.P. (2000) 'The response of two contrasted grasslands to simulated climate change', *Science*, vol 289, pp762–765

Grime, J.P., Fridley, J.D., Askew, A.P., Thompson, K., Hodgson, J.G. and Bennett, C.R. (2008) 'Long-term resistance to simulated climate change in an infertile grassland', *Proceedings of the National Academy of Science*, vol 105, pp10028–10032

Hannah, L., Midgley, G., Andelman, S., Araújo, M., Hughes, G., Martinez-Meyer, E., Pearson, R. and Williams, P. (2007) 'Protected area needs in a changing climate', *Frontiers in Ecology and the Environment*, vol 5, pp131–138

Harrison, P.A., Berry, P.M. and Dawson, T.P. (eds) (2001) *Modelling Natural Resource Responses to Climate Change (MONARCH)*, UKCIP Technical Report, Oxford

Hewins, E., Pinches, C., Arnold, J., Lush, M., Robertson, H. and Escott, S. (2005) 'The condition of lowland BAP priority grasslands: Results from a sample of non-statutory stands in England', English Nature Research Report No. 636, English Nature, Peterborough

Hewins, E., Toogood, T., Alonso, I., Glaves, D.J., Cooke, A. and Alexander, R. (2007) 'The condition of lowland heathland: results form a sample survey of non-SSSI stands in England', Natural England Research Report No. 2, Natural England, Sheffield

Hickling, R., Roy, D.B., Hill, J.K. and Thomas C.D. (2005) 'A northwards shift of range margins in British Odonata', *Global Change Biology*, vol 11, pp520–526

Hickling, R., Roy, D.B., Hill, J.K., Fox, R. and Thomas, C.D. (2006) 'The distributions of a wide range of taxonomic groups are expanding northwards', *Global Change Biology*, vol 12, pp450–455

Hilty, J.A., Lidicker, W.Z. and Merenlender, A.M. (2006) *Corridor Ecology: The Science and Practice of Linking Landscapes for Biodiversity Conservation*, Island Press, Washington DC

Hoegh-Guldberg, O., Hughes, L., McIntyre, S., Lindenmayer, D.B., Parmesan, C., Possingham, H.P. and Thomas, C.D. (2008) 'Assisted colonization and rapid climate change', *Science*, vol 321, pp345–346

Hole, D.G., Whittingham, M.J., Bradbury, R.B., Anderson, G.Q., Lee, P.L., Wilson, J.D. and Krebs J.R. (2002), 'Widespread local house-sparrow extinctions', *Nature*, 418, pp931–932

Holling, C.S. (ed.) (1978) *Adaptive Environmental Assessment and Management*, John Wiley & Sons, Chichester

Holyoak, M., Leibold, M.A., Mouquet, N.M., Holt, R.D. and Hoopes, M.E. (2005) 'A framework for large-scale community ecology', in Holyoak, M., Leibold, M.A., Mouquet, N.M. and Holt, R.D. (eds) *Metacommunities: Spatial Dynamics and Ecological Characteristics*, Chicago University Press, Chicago

Hopkins, A.D. (1920) 'The bioclimatic law', *Monthly Weather Review*, vol 48, p355

Hopkins, J.J. and Kirby, K.J. (2007) 'Ecological change in British broadleaved woodland since 1947', *Ibis*, vol 149, suppl 2, pp29–40

Hopkins, J.J., Allison, H.M., Walmsley, C.A., Gaywood, M. and Thurgate, G. (2007) *Conserving Biodiversity in A Changing Climate: Guidance on Building Capacity to Adapt*, Defra, London

Hulme, M., Jenkins, G.J., Turnpenny, J.R., Mitchell, T.D., Jones, R.G., Lowe, J., Murphy, J.M., Hassell, D., Boorman, P., McDonald, R. and Hill, S. (2002) *Climate Change Scenarios for the United Kingdom: The UKCIP02 Scientific report*, Tyndall Centre for Climate Change Research, School of Environmental Sciences, University of East Anglia, Norwich

Huntley, B. (2007) 'Climatic change and the conservation of European biodiversity: Towards the development of adaptation strategies', Council of Europe, Strasbourg, http://www.coe.int/t/dg4/cultureheritage/conventions/bern/T-PVS/sc27_inf03_en.pdf, accessed 20 December 2008

IPCC (2007a) 'Summary for policymakers', in Solomon, S., Qin, D., Manning, M., Chen, Z., Marquis, M., Avery, K.B., Tignor, M. and Miller, H.L. (eds) *Climate Change 2007: The Physical Science Basis*, Contribution of Working Group I to the Fourth Assessment Report of the Intergovernmental Panel on Climate Change, Cambridge University Press, Cambridge

IPCC (2007b) 'Summary for policymakers', in Parry, M.L., Canziani, O.F., Palutikof, J.P., Linden P.J. van der and Hanson, C.E. (eds) *Climate Change 2007: Impacts, Adaptation and Vulnerability*, Contribution of Working Group II to the Fourth Assessment Report of the Intergovernmental Panel on Climate Change, Cambridge University Press, Cambridge

Jenkins, G., Perry, M. and Prior, J. (2007) *The Climate of the United Kingdom and Recent Trends*, Hadley Centre, Met Office, Exeter

Joint Nature Conservation Committee (2004) 'Common standards monitoring: Introduction to the guidance manual', JNCC, Peterborough. http://www.jncc.gov.uk/pdf/CSM_introduction.pdf, accessed 15 December 2008

Jongman, R. and Pungetti, G. (eds) (2004) *Ecological Networks and Greenways: Concept Design and Implementation*, Cambridge University Press, Cambridge

Jump, A.S., Hunt, J.M. and Peñuelas, J.P. (2006) 'Rapid climate change related growth decline at southern range edge of *Fagus sylvatica*', *Global Change Biology*, vol 12, pp2163–2174

Kerr, J.T. and Kharouba, H.M. (2007) 'Climate change and conservation biology', in May, R.M. and McLean, A.R. (eds) *Theoretical Ecology: Principles and Applications*, Oxford University Press, Oxford

Kirby, K.J., Smart, S.M., Black, H.I.J., Bunce, R.G.H., Corney, P.M. and Smithers R.J. (2005) 'Long term ecological change in British woodland (1971–2001)', English Nature Research Report No. 653, English Nature, Peterborough

Lindenmayer, D., Hobbs, R.J., Montague-Drake, R., Alexandra, J., Bennett, A., Burgman, M., Cale, P., Calhoun, A., Cramer, V., Cullen, P., Driscoll, D., Fahrig, L., Fischer, J., Franklin, J., Haila, Y., Hunter, M., Gibbons, P., Lake, S., Luck, G., MacGregor, C., McIntyre, S., MacNally, R., Manning, A., Miller, J., Mooney, H., Noss, R., Possingham, H., Saunders, D., Schmiegelow, F., Scott, M., Simberloff, D., Sisk, T., Tabor, G., Walker, B., Wiens, J., Woinarski, J. and Zavaleta, E. (2007) 'A checklist for ecological management of landscapes for conservation', *Ecology Letters*, vol 11, pp78–91

Lyon, B.E., Chaine, A.S. and Winkler, D.W. (2008) 'A matter of timing', *Science*, vol 321, pp1051–1052

Manley, G. (1974) 'Central England temperatures: Monthly means 1659 to 1973', *Quarterly Journal of the Royal Meteorological Society*, vol 100, pp389–405

Marren, P. (2002) *Nature Conservation*, HarperCollins, London

McLeod, C.R., Yeo, M., Brown, A.E., Hopkins, J.J. and Way, S.F. (2005) *The Habitats Directive: Selection of Special Areas of Conservation in the UK*, 2nd Edn, Joint Nature Conservation Committee, Peterborough

Menéndez, R., González-Megias, A., Hill, J.K., Braschler, B., Willis, S.G., Collingham, Y., Fox, R., Roy, D.B. and Thomas, C.D. (2006) 'Species richness changes lag behind climate change', *Proceedings of the Royal Society Series B*, vol 273, pp1465–1470

Menzel, A., Sparks, T.H., Estella, N., Koch, E., Aasa, A., Ahas, R., Alm-Kubler, K., Bissolli, P., Braslavská, O., Briede, A., Chmielewski, F.M., Crepinsek, Z., Curnel, Y., Dahl, Å., Defila, C., Donnelly, A., Filella, Y., Jatczak, K., Måge, F., Mestre, A., Nordli, Ø., Peñuelas, J., Pirinen, P., Remišova, V., Scheifinger, H., Striz, M., Susnik, A., Van Vliet, A.J.H., Wielgolaski, F-E., Zach, S. and Zust, A. (2006) 'European phenological response to climate change matches the warming pattern', *Global Change Biology*, vol 12, pp1969–1976

Mitchell, R.J., Morecroft, M.D., Acreman, M., Crick, H.Q.P., Frost, M., Harley, M., Maclean, I.M.D., Mountford, O., Piper, J., Pontier, H., Rehfisch, M.M., Ross, L.C., Smithers, R.J., Stott, A., Parr, T., Walmsley, C., Watt, A.D., Watts, O. and Wilson, E. (2007) 'England Biodiversity Strategy – Towards adaptation for climate change', Final Report to Defra contract CR0327. Defra, London

Morris, R.K.A. (2000) 'Shifts in the phenology of hoverflies in Surrey: Do these reflect the effects of global warming?', *Dipterists Digest*, vol 7, pp103–108

Natural England (2008) *State of the Natural Environment 2008*, Natural England, Sheffield

Natural England (in preparation) *NNR Management Plans: A Guide*, Natural England, Sheffield

National Expert Group on Transboundary Air Pollution (2001) *Transboundary Air Pollution Acidification, Eutrophication and Ground Level Ozone in the UK*, Centre for Ecology and Hydrology, Edinburgh

Nature Conservancy Council (1989) *Guidelines for the Selection of Biological SSSIs*, Nature Conservancy Council, Peterborough

Overepeck, J., Cole, J. and Bartlein, P. (2005) 'A "paleoperspective" on climate variability and change', in Lovejoy, T.E. and Hannah, L. (eds) *Climate Change and Biodiversity*, Yale University Press, New Haven and London

Parmesan, C. and Yohe, G. (2003) 'A globally coherent fingerprint of climate change impacts across natural systems', *Nature*, vol 421, pp37–42

Parolo, G. and Rossi, G. (2008) 'Upward migration of vascular plants following a climate warming trend in the Alps', *Basic and Applied Ecology*, vol 9, pp100–107

Pauli, H., Gottfried, M., Reiter, K., Klettner, C. and Grabherr. G. (2007) 'Signals of range expansions and contractions of vascular plants in the high Alps: Observations (1994–2004) at the GLORIA master site, Schrankogel, Tyrol, Austria', *Global Change Biology*, vol 13, pp147–156

Peñuelas, J., Gordon, C., Llorens, L., Nielsen, T., Tietema, A., Beier, C., Bruna, P., Emmett, B., Estiarte, M. and Gorissen, A. (2004) 'Nonintrusive field experiments show different plant responses to warming and drought among sites, seasons and species in a north–south European gradient', *Ecosystems*, vol 7, pp598–612

Perring, F. (1960) 'Climatic gradients of chalk grassland', *Journal of Ecology*, vol 48, pp415–422

Peterken, G.F. and Mountford, E.P. (1996) 'Effect of drought on beech in Lady Park Wood, an unmanaged mixed deciduous woodland', *Forestry*, vol 69, pp117–128

Preston, C.D., Pearman, D.A. and Dines T.D. (2002) *New Atlas of the British and Irish Flora*, Oxford University Press, Oxford

Raxworthy, C.J., Pearson, R.G., Rabibison, N., Rakotondrazafy, A.M., Ramanamanjato, J-B., Raselimanana, A.P., Wu, S., Nussbaum, R.A. and Stone, D.A. (2008) 'Extinction vulnerability of tropical montane endemism from warming and upslope displacement: a preliminary appraisal for the highest massif in Madagascar', *Global Change Biology*, vol 14, pp1703–1720

Rodwell, J.S. (2000) *British Plant Communities Volume 5: Maritime communities and vegetation of open habitats*, Cambridge University Press, Cambridge

Root, T.L., Price, J.T., Hall, K.R., Rosenzweig, C. and Pounds, J.A. (2003) 'Fingerprints of global warming on wild animals and plants', *Nature*, vol 421, pp57–60

Rosenzweig, M.L. (1995) *Species Diversity in Space and Time*, Cambridge University Press, Cambridge

Rowe, G., Beebee, T.J.C. and Burke, T. (1999) 'Microsatellite heterozigosity, fitness and demography in natterjack toads *Bufo calamita*', *Animal Conservation*, vol 2, pp85–92

Roy, D.B. and Sparks, T.H. (2000) 'Phenology of British butterflies and climate change', *Global Change Biology*, vol 6, pp407–416

Roy, D.B., Rothery, P., Moss, D., Pollard, E. and Thomas, J.A. (2001) 'Butterfly numbers and weather: Projecting historical trends in abundance and their future effects of climatic change', *Journal of Animal Ecology*, vol 70, pp201–217

Saunders, D.A. (2007) 'Connectivity, corridors and stepping stones' in Lindenmayer, D.B. and Hobbs, R.J. (eds) *Managing and Designing Landscapes for Conservation: From Perspectives to Principles*, Blackwell, Oxford

Saxon, E.C. (2003) 'Adapting ecoregional pans to anticipate the impacts of climate change', in Groves, C.R. (ed) *Drafting a Conservation Blueprint: A Practitioners Guide to Planning for Biodiversity*, Island Press, Washington DC

Scottish Natural Heritage (2007) 'Local nature reserve management planning guidance: The process and the plan', SNH, Clydebank, http://www.snh.org.uk/pdfs/lnr/ManPlanGuidFeb07.pdf, accessed 12 December 2008

Shardlow, E.A. (2001) 'A review of the conservation importance of shingle habitat for invertebrates in the United Kingdom (UK)' in Packham, J.R., Randall, R.E., Barnes R.S.K. and Neal, A. (eds) *Ecology & Geomorphology of Coastal Shingle*, Westbury Academic and Scientific Publishing, Otley, UK

Sheail, J. (1998) *Nature Conservation in Great Britain – The Formative Years*, The Stationery Office, London

Smart, S.M., Marrs, R.H., Le Duc, M.G., Thompson, K., Bunce, R.G.H., Firbank, L.G. and Rossall, M.J. (2006) 'Spatial relationships between intensive land cover and residual plant species diversity in temperate farmed landscapes', *Journal of Applied Ecology*, vol 43, pp1128–1137

Soons, M., Heil, G.W., Nathan, R. and Katul, G.C. (2004) 'Determinants of long-distance dispersal by wind in grasslands', *Ecology*, vol 85, pp3056–3068

Sparks, T.H. (1999) 'Phenology and the changing pattern of bird migration in Britain', *International Journal of Biometeorology*, vol 42, pp134–138

Sparks, T.H. and Yates, T.J. (1997) 'The effect of spring temperature on the appearance dates of British butterflies 1883–1993', *Ecography*, vol 20, pp368–374

Sparks, T.H., Carey, P.D. and Combes, J. (1997) 'First leafing dates of trees in Surrey between 1947 and 1996', *The London Naturalist*, vol 76, pp15–20

Sparks, T.H., Roy, D.B. and Dennis, R.L.H. (2005) 'The influence of temperature on migration of Lepidoptera into Britain', *Global Change Biology*, vol 11, pp507–514

Stevens, C.J., Dise, N.B., Mountford, J.O. and Gowing, D.J. (2004) 'Impact of nitrogen deposition on the species richness of grasslands', *Science*, vol 303, pp1876–1879

Stroud, D.A., Chambers, D., Cook, S., Buxton, N., Fraser, B., Clement, P., Lewis, P., McLean, I., Baker, H. and Whitehead, S. (2001) *The UK SPA Network: Its Scope and Content, Volume 1: Rationale for the Selection of Sites*, JNCC, Peterborough

Tallis, J.H. (1991) *Plant Community History*, Chapman and Hall, London

Thomas, C.D. and Lennon, J.J. (1999) 'Birds extend their ranges northwards', *Nature*, vol 399, p213

Thomas, C.D., Bodsworth, E.J., Wilson, R.J., Simmons, A.D., Davies, Z.G., Musche, M. and Conradt, L. (2001) 'Ecological and evolutionary processes at expanding range margins', *Nature*, vol 411, pp 577–581

Thomas, J.A. (1990) 'The conservation of Adonis blue and Lulworth skipper butterflies – Two sides of the same coin', in Hillier, S.H., Walton, D.W.H. and Wells, D.A. (eds) *Calcareous Grasslands Ecology and Management*, Bluntisham Books, Huntingdon

Thomas, J.A. (1993) 'Holocene climate change and warm man-made refugia may explain why a sixth of British butterflies inhabit unnatural early-successional habitats', *Ecography*, vol 16, pp278–284

Thomas, J.A., Rose, R.J., Clarke, R.T., Thomas, C.D. and Webb, N.R. (1999) 'Intraspecific variation in habitat availability among ectothermic animals near their climatic limits and their centres of range', *Functional Ecology*, vol 12, suppl 1, pp55–64

Thomas, J.A., Bourn, N.A.D., Clarke, R.T., Stewart, K.E., Simcox, D.J., Pearman, G.S., Curtis, R. and Goodger, B. (2001) 'The quality and isolation of habitat patches both determine where butterflies persist in fragmented landscapes', *Proceedings of the Royal Society of London Series B*, vol 268, pp1791–1796

Tryjanowski, P., Sparks, T.H. and Profus, P. (2005) 'Uphill shifts in the distribution of the white stork *Ciconia ciconia* in southern Poland: The importance of nest quality'. *Diversity and Distributions*, vol 11, pp219–224

UK Biodiversity Action Plan (2006) 'Revised BAP Targets 2006', http://www.ukbap.org.uk/BAPGroupPage.aspx?id=98, accessed 22 December 2008

Walmsley, C.A., Smithers, R.J., Berry, P.M., Harley, M., Stevenson, M.J. and Catchpole, R. (2007) *MONARCH Modelling Natural Resource Change: a Synthesis for Biodiversity Conservation*. UK Climate Impacts Programme, Oxford

Warren, M.S., Hill, J.K., Thomas, J.A., Asher, J., Fox, R., Huntley, B., Roy, D.B., Telfer, M.G., Jeffcoate, S., Harding, P., Jeffcoate, G., Willis, S.G., Greatorex-Davies, J.N., Moss, D. and Thomas, C.D. (2001) 'Rapid response of British butterflies to opposing forces of climate and habitat change', *Nature*, vol 414, pp65–68

Watkinson, A.R. and Gill, J.A. (2002) 'Climate change and dispersal', in Bullock, J.M., Kenward, R.E. and Hails, R.S. (eds), *Dispersal Ecology*, Blackwell, Oxford

Wesche, S., Kirby, K.J. and Ghazoul, J. (2006) 'Plant assemblages in British beech woodlands within and beyond native range: Implications of future climate change for their conservation', *Forest Ecology and Management*, vol 236, pp385–392

Whittaker, R.J. and Fernández-Palacios, J.M. (2007) *Island Biogeography: Ecology, Evolution and Conservation*, 2nd edn, Oxford University Press, Oxford

Williams, J.W. and Jackson, S.T. (2007) 'Novel climates, non-analog communities and ecological surprises', *Frontiers in Ecology and the Environment*, vol 5, pp475–482

Woiwod, I.P. (1997) 'Detecting the effects of climate change on Lepidoptera', *Journal of Insect Conservation*, vol 1, pp149–158

9

Public Engagement in New Productivism

Neil Ravenscroft and Becky Taylor

Introduction

In this chapter we argue that the relationship between the public and the land has been changing, in a paradigmatic sense, since the 1970s. Until then, the relationship was an essentially consumptive one, informed by a rights agenda that invoked the 18th and 19th century enclosures as evidence of landowners assuming powers that were not theirs to assume (Shoard, 1987; Harrison, 1991; Ravenscroft, 1995, 1998). Following 50 years of agitating for greater rights of recreational access to the countryside, many walkers and their representative organizations believed that the time had come for this to change (Ramblers' Association, 1993). The Countryside and Rights of Way Act 2000 confirmed this belief, with many feeling that the balance had been redressed and that the public had now regained their rights to roam at will over open country (Parker and Ravenscroft, 2001). Yet even in the mid-1990s there was no evidence to suggest that the acquisition of these rights would lead to new consumption (House of Commons Environment Committee, 1995), while by 2005 it had become patently clear that the consumption of day visits to the countryside was declining steadily (Natural England, 2006).

Rather, what we have witnessed in the last two decades is a shift away from consumptive leisure, of the type typified by access on foot to private land, towards more productive forms of engagement with the land and rural environment. By 'productive' we mean two related concepts: on the one hand, a return to a more 'traditional' form of agriculture based on producing food (recognizing here that there are, equally, new forms of productivism relating to non-food crops); and, on the other, a deeper and more sustained relationship between people and land, in which people increasingly produce their own leisure (perhaps through specialist interests, or conservation work or gardening) as part of a lifestyle shift towards more ethical and environmentally friendly practices. This latter construct is akin to Stebbins' (1992, 1997) construct of 'serious leisure', in which people pursue non-work interests – and take on non-work identities – in ways more conventionally associated with work. These 'serious leisure' practices include recycling, sourcing local and organic foods and reducing 'unnecessary' leisure travel.

For some people, this commitment to produce new lifestyles is enacted largely through conventional liberal market regimes (albeit shifting from, say, the supermarket to the farmers' market – what Hegarty (2007), terms 'green-shifting') and is, therefore, intimately tied to established social structures such as class (see, for example, Ilbery and Maye, 2006; London Food Link, 2007; Sustain, 2008). An example of this commitment is identity formation through attachment to specific landscapes and ways of living (see Marsden, et al, 2003; Carnegie Commission for Rural Community Development, 2007). This has led to significant numbers of people relocating to rural areas (see Halfacree, 1995) and taking up lifestyle and consumption practices which they see as more sustainable (noting here Born and Purcell's construct of the 'local trap' in which, without evidence, people believe that locally sourced food is more ethically sound and environmentally sustainable than supermarket alternatives; Born and Purcell, 2006, p195).

For increasing numbers of others, however, the shift is more fundamental, with a commitment that extends beyond the market to encompass new approaches to lifestyle based on a 'back-to-the-land' creative (re)connection with the land and environment in a positive, productive way (Halfacree, 2001). The ultimate form of engagement in this reconnection with the land is, we argue, through farming and food production (Halfacree (2001), found examples of people relocating to rural areas to run smallholdings). While few people have the skills and will to actively take up farming (or even smallholding), increasing numbers of people are getting involved in farming through the membership of co-operatives and various forms of community-supported agriculture (CSA). For example, over 8000 people contributed financially to a trust established to save Fordhall Farm in the English Midlands (see Hollins and Hollins, 2007); over 400 people are shareholders in Tablehurst and Plaw Hatch Community Farms in Sussex (Countryside Agency, 2005; Soil Association, 2005a); and, in the USA and the UK, there are many CSAs in which individual people make annual financial pledges to support small, usually organic farms (McFadden, 2003a, 2003b; Soil Association, 2005a).

While there are various theories about why this has occurred, including concerns about health (Large, 2004), food security and climate change (McFadden, 2003b; Bjune and Torjusen, 2005; Parker, 2005), a wish to support local communities (Fieldhouse, 1996; Lea et al, 2006), a commitment to buy local food (O'Hara and Stagal, 2001; Lang, 2005) and 'green care' motives (Hegarty, 2007), we seek to argue that the underlying impulse has been primarily about establishing new forms of citizenship (see Parker, 2002), sometimes referred to as 'food citizenship' (Stevenson, 1998; Parker, 2005). For us, the significance of the trend is less its connection to food per se and more its role in expressing what Rojek (2001) has termed a new 'life politics', in which people seek 'civil labour' as a primary means of expressing their identity. In contrasting civil labour with Marx's 'necessary labour', Rojek (2001) argues that, faced with the breakdown of many core social institutions such as work, religion and class, increasing numbers of people are seeking new avenues to assert their identity, by using their 'leisure time' to undertake non-essential forms of work – such as joining and supporting a community farm initiative.

In this chapter we seek to set out the context for this changing relationship between the public and the land, and to describe and analyse some of the key forms of farming relationship that are in the vanguard of this movement. We will conclude the chapter by considering the potential implications of this shift from consumptive to productive leisure, particularly in terms of how we might reconstruct the relationship between the public and the land.

New citizenships in the land

In 2005, following an intensive media appeal, over 8000 people from 24 countries donated money (mostly in £50 units of not-for-profit shares) to a Community Land Trust to buy Fordhall Farm, a small organic farm in the English Midlands (Hollins and Hollins, 2007) faced with conversion to housing. They now collectively own the farm and let it to the tenants who would have been displaced by the sale. There was no profit motive, with most people buying the shares because they were inspired by the appeal and the thought of saving the farm from the developers, and by contributing to the continuation of small-scale organic farming. In his analysis of the buyers' motives, Hegarty (2007) has identified what he terms a 'green-care' movement, with people interested in sustainable farming and willing to pay to see it maintained. At Fordhall, this is largely about community land ownership (see Stroud Common Wealth, 2007). Elsewhere – Forest Row in Sussex, for example – this same impulse has led to 400 people putting up the share capital for a community farming company that owns two farm businesses (see Soil Association, 2005b), thus becoming what Cone and Kakaliouras (1995) have termed 'non-active' farmers (in the argot of the Forest Row community, the shareholders are referred to as 'farm partners'). Similarly, in the USA, at community-supported Caretaker Farm, the community of consumers are called 'members' rather than customers (Smith, 1997).

What these and countless other examples illustrate is an increasing wish on the part of many people to establish new relationships with farming and the land. Where once they were happy to exercise public rights to walk on the land – and sometimes to buy from farm shops and farmers markets – they are now seeking a deeper connection that is indicative (and supportive) of new life styles and life politics (Rojek, 2001). In his work, Parker (2002, 2005) suggests that this amounts to a new construct of citizenship, in which contemporary forms of consumer citizenship are taken to a new level of sustained commitment and involvement – what he and Stevenson (1998) have termed 'food citizenship'. For Stevenson (1998, p201) this is about the ways in which people participate consciously within (a localized) food system, such that they provide a 'human infrastructure for negotiating alternative agrifood systems'. Stevenson (1998) suggests that food citizens develop a suite of competences in order to participate in the new order:

- *Analytical competencies:* making connections and evaluating contradictions in the new food chains
- *Relational competencies:* focusing on new forms of organizational relationships between actors in the food chain
- *Ethical competencies:* valuing of non-market goods and the linkages between ethics and emotions
- *Aesthetic and spiritual competencies:* connecting agriculture and food with beauty and with what he calls sacramental living.

See also Pretty (2007), who adds *physical competency* as an attribute enjoyed by some new 'food citizens'.

It is this new form of citizenship, with its new competences and institutional formations, that is at the core of the shifting relationship between the public and the land. For Princen et al (2002), Parker (2005) and Winch (2005), this is essentially about shortening the food chain (*combating distance*), so that citizens gain better knowledge of both the production processes and the actual producer of the food. Indeed, in some community-supported agricultural systems, the citizens become the producers, which, when combined with highly localized farming systems such as biodynamic agriculture (Van En, 1988; McFadden, 2003a, 2003b; Vogt, 2007), can make redundant the very notion of the food chain. Thus, 'combating distance' is essentially an amalgam of geographic and cultural motivations, in recognizing that distance separates rights from responsibilities (Princen et al, 2002). And, of course, such rights and responsibilities flow in both directions: the right of the citizen to eat wholesome food that has been grown responsibly by the farmer; and the responsibility of the citizen to uphold the farmer's right to a fair price for the food. This is captured in Lockridge's (2005) epithet that the impulse is about 'putting the farmer's face on the food'. For Hudson, this approach represents

> A whole new food system, one that uses dollars but is not ruled by them... Here a new economics is being practiced, economics, as if ... people mattered. As if the land mattered. As if food were more than a commodity' (Hudson 2005, p12).

Thus, as Hudson (2005) argues, this shift in emphasis is less about forming new or alternative markets and more about decommodifying food. As such, it is less a case of green-shifting, as depicted by Hegarty (2007) and Ilbery and Maye (2006), and much more an example of Rojek's (2001) new life politics. This combination of localization and citizenship has been termed '*civic agriculture*' (DeLind, 2002; and Lyson and Guptill, 2004), in which the underlying philosophy is the (re)integration of people and land (what DeLind, 2002, has termed 'inhabitation' of the land), such that new forms of (food) democracy (Hassanein, 2003) emerge in ways that allow (food) citizens to determine agri-food policies at the local and even regional level. This is very much what the Policy Commission on the Future of Food and Farming (2002) had in mind when it called for food producers to 'reconnect' with consumers and markets (Sumberg, 2008).

As Parker (2005) has identified, this is a process that is occurring simultaneously in some of the most advanced international economies, including Japan and the USA, although often more in terms of shortening rather than removing the food chain. Thus, for example, although the Teikei system of co-operation in Japan has been widely claimed as the forerunner of the CSA movement (see McFadden, 2003a), Parker (2005) argues that it is better understood as a co-operative attempt to develop an alternative food network in which consumers have more control over the sourcing (and thus the methods of production and distribution) of their food. Many box and franchise schemes fulfill a similar function in other parts of the world. What they do not do, however, is offer a paradigm shift away from conventional food chains (and orthodox market mechanisms). Nor do they necessarily centre on shortening the cultural distance between producers and consumers.

And these factors – a paradigm shift in thinking about food chains allied to new cultural relationships between farmers and their local communities – are at the core of our arguments about the new relationships being forged between the public and the land. As Trauger Groh (Groh and McFadden, 1990), one of the founders of the CSA movement in the USA, has observed, the basis of the new relationship between the public and the land is based on care: the community supporting farming and the farms supporting the community (this is very much the sum of Stevenson's competences; Stevenson, 1998). A similar approach has been proposed by Braastad and Bjornsen (2006, p2), when they refer to 'the utilization of agricultural farms ... as a base for promoting mental and physical health, as well as quality of life, for a variety of client groups'.

What lies at the heart of this relationship is the generation of a new way of living *with* (as opposed to *from*) the land, in which people – the community – take explicit steps to become more responsible for the food that they eat and the care that is shown to the land and those who cultivate it. 'Local' is one element of this relationship, in reducing the spatial and cultural distance between the production and consumption of food. Sustainability of production is another. But these are not enough on their own; indeed, in possibly underpinning the local trap (Born and Purcell, 2006) they individually may do little more than shift the conventional market place ever so slightly away from supermarket shelves and into farmers' markets and farm shops. Rather, the glue that binds localization and sustainability into a new relationship between the public and the land is community action ('civil labour'; Rojek, 2001): the paradigm shift from consumer to (quasi) producer through which groups of people commit to sharing the risk and responsibility for producing local food from local land for consumption by local people. This is the basis of what McFadden (2003b) has referred to as 'regenerative agriculture', a form of enterprise that seamlessly incorporates a community of people engaged in civil labour to produce and consume the food (and land, landscape and amenity) that they collectively decide to grow.

Common forms of regenerative agricultural enterprise

As we have suggested, there are many different forms that regenerative agriculture can take (see Soil Association, 2005b). The most generically common of these are described and set out in Table 9.1 below. They comprise: community 'share' farming; community-owned social enterprise; and land trusts and community land trusts.

Community 'share' farming

The conventional understanding of share farming is that it involves the landowner and farmer 'sharing' the risks and rewards of farming, by each contributing capital (fixed and working) and each taking a share of the production (see Lastarria-Cornhiel and Melmed-Sanjak, 1999; Ravenscroft, 1999; Winter and Butler, 2008). This is reinterpreted in the case of many community farms, where the risk and rewards are shared by a community of subscribers. Classically, as set out by Van En (1988) and Groh (1990), this involves a group of people (the community) committing to pay the annual working capital for a farm and receiving all the produce from the farm, which they share out according to whatever they have agreed (usually needs-related). The farmer is a member of the community for this purpose, usually working with a core community group to plan the farm and its budget (effectively active and non-active farmers working together). As in conventional share farming, there may be provision for the farmer to sell excess produce, as a 'return' to the community. The payments by the community members may be in the form of 'pledges' based on their individual abilities to pay (with the total pledged matching the budget), or

Table 9.1 *A matrix of regenerative agricultural forms*

Enterprise type	Core business and objectives	Type of community involvement	Key stakeholders	Examples
Community 'share' farming (could be a partnership or other unincorporated association)	Small scale; community provides the finance and the farmer provides the skill	A defined group of people provide up-front working capital and take a share of the farm produce	A 'closed' community group (i.e. a fixed membership) and a farm team	Indian Line Farm; Temple Wilton Farm; Caretaker Farm (all eastern USA); Stroud Community Farm, Gloucestershire
Community-owned social enterprise (could take the form of a community interest company or other incorporated or unincorporated form)	Large scale with a formal legal structure (often an Industrial & Provident Society (IPS) with shares owned by many individuals)	People buy shares in the farming business (thus providing working capital), but do not receive dividends or other benefits. The produce is usually sold commercially to any customers	The farm team is central, usually reporting to a committee of shareholders representing the IPS	Tablehurst Farm & Plaw Hatch Farm, East Sussex, UK
Community Land Trusts (CLTs)	Charitable trusts established through multiple donations from individual people, with the purpose of owning and letting land to farming tenants	Primarily through financial donations to the charitable trust. In this case the 'community' could be widespread and not necessarily farm customers	Donors; trustees and farming tenants	Fordhall Farm, Shropshire, UK; Stroud Common Wealth, Gloucestershire, UK
Land trusts	Similar objectives to CLTs, but normally funded through a single major bequest	Depends upon the Trust, but usually limited	Trustees	See the Land Trusts Association (www.landtrusts.org)

there may be fixed shares (although sometimes there is a differential between 'full' and 'reduced' shares). As the Soil Association (2005b) sets out, there are a number of legal forms that such share farming arrangements can take, all of which are likely to be relatively informal and unincorporated (that is, outside company law).

As Groh (1990) has argued, this form of community-farm arrangement is centrally about mutual support: the 'community' supporting the farm, the farm supporting the community and, of course, the community supporting itself by matching individual annual financial contributions and produce shares to people's ability to pay and need for food. Thus, as long as the community supplies the annual working capital, it is of no consequence to the farm whether this is provided through equal or differential pledges (and whether the community members take equal or differential shares of the produce). The corollary to this is that such communities tend to be small and closed to outsiders, so that a level of trust can be built up about matching financial contributions to the needs of the farm and the individuals' relative abilities to pay, and so that there is also trust in the distribution of the produce. For example, Stroud Community Agriculture extends to 50 acres and involves a community of 180 people, while Goddard Farm CSA involves one farmer and a box scheme for 30 members (see Soil Association, 2005b).

Community-owned social enterprise

A number of legal vehicles exist to facilitate individual people making financial commitments to social enterprises. Typically this involves establishing a charity or an Industrial and Provident Society (IPS) through which people can make donations or buy non-tradable shares, with the proceeds used to finance the enterprise. It is equally possible to use an incorporated company structure or a community-interest company approach (see Soil Association, 2005b). Tablehurst and Plaw Hatch Community Farms are a typical example of the use of an IPS structure: shares are available at £100 each, for which the shareholders become members of the IPS, with voting rights for officers and a say in the overall strategic vision for the farms. The businesses themselves are run by boards of directors in much the same way as any social enterprise, with the farm work being undertaken by teams of farmers and workers. The produce from the farms is sold through farm shops, with the businesses aiming to create small surpluses for reinvestment in tenant capital. Generally, neither the farms nor the IPS own land or buildings, instead leasing them from a variety of landowners.

Community, in this sense, is about long-term investment in a social enterprise that benefits anyone who wishes to buy and consume the farm produce. The initial capital outlay is unlikely to be repaid, and the shareholders do not get any individual benefits from their investment. However, this is the extent of their exposure to risk. They do not put up annual working capital and they are not bound to take a share of the produce. Instead they can buy what they wish, when they wish, regardless of the farm's budgetary requirements. The farms are, effectively, commercial not-for-profit businesses that have to balance their social objectives with a financial imperative to earn enough to survive and reinvest. In a climate where few farms are able to achieve financial security, this is a hard task, and not one that is necessarily shared with the community. Rather, these are farms driven by the philosophy of their management teams and directors (underpinned by community support) to produce food in certain ways and to certain standards and, crucially, to make it available to everyone, not just the members of the co-operative or CSA.

Land trusts and community land trusts

In common with social enterprise schemes, land trusts are legal vehicles that allow individuals to leave their capital in trust for the public good; the only difference being that land trusts purchase and hold land, rather than providing working capital. It is also usually the case that a land trust is established by an individual, partly at least to ensure the preservation, use and development of their property (typically a landed estate). Where such a trust is established, the settlor relinquishes all rights over the land. These rights are given to the trustees of the charity who are required to manage the trust in accordance with the terms of the trust deed, which must be for specified charitable purposes.

Community Farm Land Trusts, in contrast, are established with the primary aim of supporting community-based sustainable agriculture and horticulture. They are generally controlled by, and run for the benefit of, local communities, through those people providing both the capital for the trust and some of the trustees to run the trust. The trusts themselves seek to let their property to community-supported farms, with the lettings often containing clauses about the types of farming and farm enterprise that are allowed. According to Large and Pilley (2007), a Community Farm Land Trust

- provides a mechanism for the democratic ownership of farm land and related assets by the community
- ensures permanently affordable access to farms for farmers
- retains farmland for farming, horticulture and related enterprise
- allows community access, and a range of benefits.

While it is clear that all these claims can be realized through community land trusts, it is important to understand that owning and letting land is far removed from the farming practices that take place on the land. In such cases 'community' has a meaning some way removed from the share-farming principle and is much more concerned with long-term political goals with respect to land. This is, however, not to deny the possibility of 'doing community' through community land trust ownership, as is the case with the vibrant community of investors who bought Fordhall Farm.

New institutional formations

There are many versions of 'community' farming and many ways in which individuals can 'do community', all of which are based on the three cornerstones of

- *New forms of co-operation*, with a network of human relations replacing the old hierarchical order of employers and employees.
- *New forms of economy*, founded on association, not competition, with a concomitant shift in focus from profit to the needs of the land and of the people involved in the enterprise. There is also a shift in addressing risk, from pledging material security to commercial banks, to distributing financial responsibility to the community.
- *New forms of property ownership*, in which land is held in common to ensure long-term financial and tenurial security for the farm.

While not all of these cornerstones may necessarily be found in all community farming ventures, they reflect the ideal type for long-term, stable community involvement in farming, one that is based on new forms of people–environment engagement characterized by the reduction or removal of hierarchy. What is apparent here is that very different forms of institutional relationship are at play with respect to community-supported farming. Central to the share arrangement is the dynamic of a group of individuals pledging financial contributions on the basis that they should give as much as they can afford, not as little as they can get away with. Equally important is the separation of the financial pledge from the receipt of the produce, such that the person who pledges the least could conceivably have the greatest need for the produce. While less dramatic than the share dynamic, social enterprise farming still involves suspending conventional economic and social logic, by members of the public contributing financially for no pecuniary gain in a situation where 'free-riders' (those who have not contributed financially) can buy the same produce at the same prices as those who have contributed. Arguably, the community land trust model makes this separation even more stark, in that the financial contributors do not, through their investment, have any direct or deliberative relationship with the farmers who rent the land.

Yet these relationships are increasingly common, with 'trust' and 'communality' replacing the competitive and contractual nature of the market. And they demand new institutional relationships that allow citizenship to flourish rather than be crowded out (Ostrom, 2000). These are not the neat, orderly, hierarchical systems that underpin conventional liberal approaches to enterprise. Rather, they are flexible and polycentric, developed around the idea that all people can play an effective role in the governance of the institution. For Parker (2005), problematizing institutional design in this way highlights how different relational formats can assist or constrain the ways in which people participate in the food chain; at their best, these new forms blur the traditional binary divide between production and consumption, replacing it with a mutual production project. The following extract from a memoir written by Woody Wodraska, a one-time member of the Temple-Wilton Community Farm in the USA (where Trauger Groh and his colleagues farmed), illustrates this proposition

> The farmers passed out copies of the proposed budget and discussed it briefly. I remember that it was in the neighborhood of $85,000 that year. There was general consensus that it was a reasonable budget. Trauger gave a high-minded little talk reminding people of the significance of the occasion – that they were engaging in a very unusual activity here, the application of principles of brotherhood in the sphere of economics, a radical departure from the ideology of the Wal-Mart store – cheap goods at the lowest possible price … It was a thrill to be part of this, to put the 'I' in Idealism and vote with our pocketbooks for food of the highest quality, grown by farmers who were first of all stewards of their land. Everyone, I think, was stoked, but most had been through the process before; this was my first time. For me, to speak aloud before this circle of strangers to tell what I could afford, to undergo such an exposure of private matters … this was exhilarating, humiliating and mighty disconcerting, all at the same time (Wodraska, 2008).

This example of rejecting the liberal market mechanism is an example of Ostrom's (2000) ideas; not only is it about new producer–consumer relations, but it is also about rejecting market orthodoxy – and in this way it goes beyond the type of relationship imagined by Ostrom (2000) and Parker (2005). In this example the conventional competitive market has gone, to be replaced by a form of co-operation built on mutuality and trust, as a further excerpt illustrates

> It was clear that the average pledge was going to have to be a little more than $1000 and it was understood that some could afford to offer more to cover the shortfall that was generated by some who could only give less, like me. I was prepared to pledge $750 for the year's share, all we

> could afford. Most of the others in the room had kids in Waldorf school too and were sorely burdened by the tuition as well as the high cost of living in the region, but many of them were employed by high tech firms in the western Boston suburbs and making three times what I did in my social service work, or they were professionals of one kind or another. Then there were the farmers, total masters of their craft, who were taking only a pittance for themselves … And so the protocol went: Each of us around the room, one by one, spoke out our pledge, without too much editorializing or justifying, to keep the process moving along. As each firm pledge figure escaped a shareholder's lips, a dozen poised fingers homed in on calculator keys, and a running total was kept. First round, $29,000 short; second round, $12,000 short; third round, DONE DEAL! We applauded lightly, many grinning faces and nodding heads. It was late. We congratulated ourselves and headed home (Wodraska, 2008).

While it is clear that there remains a cultural and economic distinction between the farmers and the community members (elsewhere in his book, Wodraska refers to the Temple-Wilton farmers as 'heroes'), there is no doubting the common cause in which they are engaged. This is no market, and the relationship between the community members is one of trust: they put forward what they feel that they can afford and no-one questions whether or not this is reasonable. Rather, the gesture is one of co-operation

> To run a CSA successfully, farmers must produce adequate, nutritious and attractive food. That's a baseline. But they and the people around them also have to know how to engage one another creatively and to weave themselves together into a modern community. Co-operation has been a key for those CSAs that have hung together and matured over a number of years. (McFadden, 2003b).

While the drama of the pledge process may be absent from other institutional forms of community-supported agriculture, the presentation of budgets and work plans remains central, underlining the proximity of the producer–consumer relationship. In these contexts the examination of the budget may be slightly less personal (to the extent that the community members do not have to articulate a pledge, nor necessarily buy the produce), but it may be more political or philosophical. Rather than 'just' taking on the responsibility for producing the farm outputs (as under the share situation), the farmers in the social enterprise situation may also be taking responsibility for delivering a particular version of farming, and predominantly at their – or at least the farming company's – risk.

And it is the risk/reward balance that is the key to the governance of these new institutional forms. The intention of those subscribing to these new institutions is to become active and share risk (essentially understood in financial terms) according to the means of each individual (whether through buying more shares or pledging large sums annually). Equally, to share in the proceeds of the farming endeavours, either through receiving a direct share of the output or by sharing knowledge, underpins a fundamentally new relationship between themselves and the land upon which the food is grown. And this remains the case where the community action is expressed through land ownership.

Between the acceptance of risk and the receipt of reward comes governance: 'doing community support'. Under the share institution, governance is performed through dialogue between farmers and members, and formally through the presentation and acceptance of the budget, with the farmers as the de facto leaders of the project. Under the social enterprise, governance is performed in legal and personal ways, primarily through the election of shareholder committees and, at the farm level, the appointment of directors. While the farmers may be ex officio members of the committees and may hold directorships of the farming companies, neither need be the case. Thus, while hierarchy may be evident, it remains a flat structure in which the enterprise is carried out in common, between the farmer producers and the 'non-active' producers who have put up the capital to fund the project. Although more at a distance, the same impulse and relational elements remain central in the operation of community land trusts, particularly in developing a vision for the land and working with their chosen tenants to achieve this vision.

Discussion

In its classic 'share' form, community-supported agriculture can be a genuinely joint venture in which the public can engage with farms and farmers to an extent well beyond the co-operatives and box schemes so often held up as archetypes of localism and community. The community share approach to farming reduces the distance between producer and consumer to virtually nil: without the community (or at least their money) there would be no production; and without a good farmer there could be little production to consume. This relationship, so well described by community farm pioneer Robyn Van En (1988) in her book about establishing Indian Line Community Farm in the USA, illustrates how the shift from a consumerist to a productionist focus can meet what Pretty (2007) has described as a need that people have for natural and social connections with one another and with the land and environment. In this relationship, food is more than a commodity; it is the common cause between those involved in the enterprise; it has a value, but this is related to the needs of the land, not the discipline of the market.

It is this that gives expression to the combination of competencies identified by Stevenson (1998) and Pretty (2007): people evaluate conventional food chains and find them lacking, largely for ethical reasons; they recognize the need for new institutional forms to reduce (or remove) distance; and they find that, through community-supported agriculture, they can connect with farming and the land, sometimes even in a physical way. These are the qualities that people associate with community-supported farming: striking a new relationship with the land that provides access to a different form of citizenship that is expressed through active participation in social enterprise.

Yet, while such an active affirmation of community may contribute to decommodifying food, the producer–consumer bifurcation remains, even if it has narrowed (especially at Indian Line Farm and some other CSAs, where members of the community help with harvests and at other busy times). The economics may be associative, to the extent that the community seeks to match a price, not undermine it, but there is no escaping the reality that 'active' community is largely practised through financial transactions. As Sam Smith (1997), another CSA pioneer (at Caretaker Farm in the USA) has observed, class remains an issue, even if a community farm is located in an area where there are different class fractions. Rather than representing an opportunity for people from different classes and social contexts to do community in ways not normally open to them, Smith (1997) suggests that CSAs are generally closed and socially narrow communities of people who share a lot more than merely living in close proximity to one another. Indeed, as the earlier quotes from Woody Wodraska suggest, one common factor that draws like-minded people together is the presence of a Waldorf school, where the teachings of Rudolf Steiner permeate both the classrooms and the biodynamic foundations of the local community farms.

While the term 'community' might denote openness, therefore, Smith (1997) argues that community farms tend to function in isolation, being small enough not to compete with each other, nor to have a major impact on the conventional local food chain. This is, of course, less the case with social enterprise, where not belonging to the farm community does not deny access to the food. Yet it remains the case that the shift from a consumptive to a productive relationship with land and farming remains largely the preserve of those with sufficient cultural and economic capital to 'afford' this form of civil labour.

There is an interesting juxtaposition here with the identity of the farmer and the nature of the farming itself. One of the central features of the community farm relationship is the disposition of the farmer: not only must the farmer be technically able, but they must also be able to manage people (and those people's expectations) – whether through the annual pledging process, or when those same shareholders want to undertake voluntary work on the farm or when they seek to question the farming methods (see Van En, 1988; Groh, 1990). Drawing up an acceptable budget in such circumstances requires an acute sense of balance: between keeping the cost at a minimum and seeking to invest and make changes to the farm. Circumstantially, it appears that cost minimization is paramount, with the farmers taking small wages (thus subsidizing the community) and eschewing opportunities to invest for future growth (potentially undermining the economic institution of determining the farm finances according to the needs of the land, not the business). These farms are also the least likely to move away from growing the main staples.

For Smith (1997), Van En (1988) and others, this again highlights a fundamental 'truth' about community-supported farming: that while there may be a rhetoric of putting the farmer's face on the food, it has to be the 'right' face – certainly in cultural terms. Doing community thus involves a much more complex set of relationships than simply shifting from walking on land to actively caring for it. And at the core of these relationships is 'cultural fit': that the values underlying the farming processes align with those of the individual; that active involvement makes a connection with the ways in which people want to see themselves – and be seen by others. Having an articulate farmer able to represent the farm and community in a favourable light comes high on the list (see Hegarty, 2007, with respect to the Hollins family at Fordhall Farm); being organic (at least) and promoting high standards of animal welfare is equally important; and steering clear of controversial issues, whether related to GM or to biofuels and other non-food products, is a given.

Conclusions

That there is a new social relationship emerging between people and the land seems clear. Evidence indicates that the supply of consumptive opportunities involving land has largely been met. Rather, increasing numbers of people are seeking new ways to use their leisure to express broader life choices, particularly with respect to their relationship with land and food. For many, expressing their preferences by buying local food – through box schemes, visits to farm shops or farmers' markets – satisfies this aspect of their lives. Their food choices are made more explicitly and ethically than might be the case with a trip to the supermarket, but their purchase decisions remain within a conventional economic and social paradigm. For others, this is not enough. In responding (explicitly or not) to the local trap of believing that local food is necessarily better than that which is transported in, these people seek to break the distance between food producer and consumer to such an extent that they take on (partially at least) a producer identity, which renders the food chain irrelevant.

The apotheosis of this intervention is CSA, in which new forms of land holding, co-operation and economy combine to offer an approach to food production that is far removed from conventional producer–consumer relationships, even where these are conducted locally. The cornerstone of CSA relationships is the replacement of social hierarchies with much flatter structures characterized by dense networks of human relations. Through these networks, individual people can work co-operatively to generate sustainable agricultural systems in which the core economic question is no longer about profit, but about the needs of the land and the people involved in the farming enterprise. As the excerpts from Woody Wodraska's book illustrate, these new relationships are, at once, complex, exposing and supportive; they offer new ways of working to achieve common goals (Rojek's life politics) but demand a level of individual authenticity that is quite at odds with most conventional social and economic relations.

There is no doubt that these relationships work, in certain social contexts, and that, as a result, individual human actions catalyze significant new and different farming enterprises and people–environment relationships. And it is these new relationships that form the core of our argument about the changing ways in which the public engage with land. In a conventional – global – environment in which there is increasing distance between food producers and consumers, active (community) engagement in local food offers many opportunities. Most basically it offers the ultimate food and bio-security: those involved know all there is to know about the food and its provenance and – environmental factors allowing – they know that it will be available to them regardless of events in the wider world. Beyond this, lending their support to community farming initiatives gives people a new connection with the land: they recognize that there is less need to consume fine or vulnerable landscapes because they begin to understand the unique productive capacity of the land in their locality – that, in Lefebrve's (1971) terms, the interaction of people and land makes for strong place identity.

And ultimately it is place identity that marks the new territorial grounds upon which the relationship between the public and the land rests. Fordhall Farm became 'real' for many people who had never heard of it or visited it, because they could sense the drama of a long-term farming family losing their home and land and, in the process, local people losing access to local food. 'Local' thus meant different things to different people, with each forming an attachment to the land – Fordhall Farm – that transcended any need to consume it (in the sense of walking on it for recreation). Rather, consumption was shifted from access to land per se, to ownership of the land in common with others to support the continuation of the land as a production surface.

It is this transition that underlines our argument. The public – or, at least, class-defined fractions of the public – recognize that it is possible to have increasingly complex and multi-layered relationships with the land: to be a landowner in common with others, but not to profit in a conventional sense; to be a farmer, albeit a 'non-active' one; to be simultaneously conservative, environmentally conscious and 'public-spirited'; and to be at once a producer and a consumer – of food, lifestyle and community. Thus it is, we argue, that the newly emergent food citizenship that leads people to invest time, money and emotion into farming has come to challenge the primacy of the consumer citizenship that formerly undermined the production focus of many of the small farms that are now in the vanguard of the community-land movement. And a core part of this food citizenship is, we argue, the attachment to place that follows from identity with that place: land ceases to be merely land, instead becoming Fordhall Farm, or Tablehurst Farm, or Temple-Wilton Farm, or any other of the thousands of community farms that have been established.

References

Bjune, M. and Torjusen, H. (2005) *Community Supported Agriculture (CSA) in Norway – A Context for Shared Responsibility*, Paper presented at the Second CCN (Consumer Citizenship Network) International Conference 'Taking responsibility', 26–27 May 2005, University of Economics, Bratislava, Slovakia. Available at www.andelslandbruk.no/Default.asp?WCI=file&WCE=463 accessed 30 July 2009

Born, B. and Purcell, M. (2006) 'Avoiding the local trap: Scale and food systems in planning research', *Journal of Planning Education and Research*, vol 26, pp195–207

Braastad, B. and Bjornsen, B. (2006) Proposal: COST Action 866: *Green Care in Agriculture*, Technical Annex EU Framework Programme, European Science Foundation, Aas, Norway: Norwegian University of Life Sciences.

Carnegie Commission for Rural Community Development (2007) *A Charter for Rural Communities*, Carnegie UK Trust, Dunfermline

Cone, C.A. and Kakaliouras, A. (1995) 'Community supported agriculture: Building moral community or an alternative consumer choice', *Culture and Agriculture*, vol 51/52, pp28–31

Countryside Agency (2005) *Capturing value for rural communities: Community land trusts and sustainable rural communities*, Publication CA209, Countryside Agency, Cheltenham

DeLind, L. (2002) 'Place, work and civic agriculture: Common fields for cultivation', *Agriculture and Human Values*, vol 19, pp217–224

Fieldhouse, P. (1996) 'Community shared agriculture', *Agriculture and Human Values*, vol 13, no 3, pp43–47

Groh, T. and McFadden, S.S.H. (1990) *Farms of tomorrow: Communities supporting farms, farms supporting communities*, Steiner Books, London

Halfacree, K. (1995) 'Talking about rurality: Social representations of the rural as expressed by residents of six English parishes', *Journal of Rural Studies*, vol 11, no 1, pp1–20

Halfacree, K. (2001) 'Going 'back-to-the-land' again: Extending the scope of counterurbanisation', *Espace, Populations, Sociétés*, vol 1, no 2, pp161–170

Harrison, C, (1991) *Countryside Recreation in a Changing Society*, TMS Partnership, London

Hassanein, N. (2003) 'Practicing food democracy: A pragmatic politics of transformation', *Journal of Rural Studies*, vol 19, no 1, pp77–86

Hegarty, J.R. (2007) 'How to be a care-farmer for €73: A shareholder survey of the Fordhall Community Land Initiative', Paper presented at the COST Conference '*Green care: Health effects, economics and policies*', a part of COST Action 866 – Green Care in Agriculture, hosted by Austrian Horticultural Society, Vienna, Austria, 20–22 June 2007, Keele: School of Psychology, Keele University

Hollins, B. and Hollins, C. (2007) *The Fight for Fordhall Farm*, Hodder and Stoughton, London

House of Commons Environment Committee (1995) *The environmental impact of leisure activities*. HC 246-1of Session 1994-1995. London: House of Commons.

Hudson, A. (2005) 'CSAs cropping up around the globe', *Critique: A Worldwide Journal of Politics*, Spring 2005

Ilbery, B. and Maye, D. (2006) 'Regional economies of local food production: Tracing food chain links between "specialist" producers and intermediaries in the Scottish–English borders', *European Urban and Regional Studies*, vol 13, pp337–354

Lang, B.K. (2005) 'Expanding our understanding of community supported agriculture (CSA): An examination of member satisfaction', *Journal of Sustainable Agriculture*, vol 26, no 2, pp61–79

Large, M. (2004) *From Community Supported Agriculture to Community Farm Trusteeship: Indian Line and Four Corners Farms*, Star and Furrow, Winter 2004, Stroud Common Wealth Community Farm Land Trust Project, Stroud

Large, M. and Pilley, G. (2007) 'Acquiring land for community supported agriculture (CSA) step by step guide and overview', Briefing paper, Stroud Common Wealth Community Farm Land Trust Project, Stroud

Lastarria-Cornhiel, S. and Melmed-Sanjak, J. (1999) 'Land tenancy in Asia, Africa and Latin America: A look at the past and a view to the future', Working Paper No. 27, Land Tenure Center, University of Wisconsin-Madison, Madison, Wisconsin, USA

Lea, E., Phillips, J., Ward, M. and Worsley, A. (2006) 'Farmers' and consumers' beliefs about community supported agriculture in Australia: A qualitative study', *Ecology of Food and Nutrition*, vol 45, no 2, pp61–86

Lefebvre, H. (1971) *The Production of Space* (translated by D. Nicholson-Smith), Blackwell, Oxford

Lockridge, D. (2005) 'Community supported agriculture', *Small Farm Today*, March, pp32–35

London Food Link (2007) *One Planet Dining*, Sustain, London

Lyson, T. and Guptill, A. (2004) 'Commodity agriculture, civic agriculture and the future of US farming', *Rural Sociology*, vol 69, pp370–385

McFadden, S. (2003a) 'The history of community supported agriculture, part 1: Community farms in the 21st Century: Poised for another wave of growth?', The New Farm, www.newfarm.org accessed 6 May 2008. (Note: the current web address is http://www.rodaleinstitute.org/new_farm accessed 30 July 2009)

McFadden, S. (2003b) 'The history of community supported agriculture, part 2: CSA's world of possibilities', The New Farm, www.newfarm.org accessed 6 May 2008.

(Note: the current web address is http://www.rodaleinstitute.org/new_farm accessed 30 July 2009)

Marsden, T., Milbourne, P., Kitchen, L. and Bishop, K. (2003) 'Communities in nature: The construction and understanding of forest natures', *Sociologia Ruralis*, vol 43, no 3, pp238–256

Natural England (2006) *England Leisure Visits, Report of the 2005 Survey*, Natural England, Cheltenham

O'Hara, S.U. and Stagal, S. (2001) 'Global food markets and their local alternatives: A socio-ecological economic perspective', *Population and Environment: A Journal of Interdisciplinary Studies*, vol 22, no 6, pp533–554

Ostrom, E. (2000) 'Crowding out citizenship', *Scandinavian Political Studies*, vol 23, no 1, pp3–16

Parker, G. (2002) *Citizenships, Contingency and the Countryside*, Routledge, London

Parker, G. (2005) 'Sustainable food? Teikei, co-operatives and food citizenship in Japan and the UK, Working Papers in Real Estate & Planning 11/05, University of Reading, Reading, UK

Parker, G. and Ravenscroft, N. (2001) 'Land, rights and the gift: The Countryside and Rights of Way Act 2000 and the negotiation of citizenship', *Sociologia Ruralis*, vol 41, no 4, pp381–398

Policy Commission on the Future of Food and Farming (2002) *Farming and Food: A Sustainable Future* (the Curry Report), Cabinet Office, London

Pretty, J. (2007) *The Earth Only Endures: On Reconnecting with Nature and Our Place in It*, Earthscan, London

Princen, T., Maniates, M.F. and Conca, K. (eds) (2002) *Confronting consumption*, Massachusetts Institute of Technology, Cambridge, MA

Ramblers' Association (1993) *Harmony in the hills*. Consultation document. Ramblers' Association, London

Ravenscroft, N. (1995) 'Recreational access to the countryside of England and Wales: popular leisure and the legitimation of private property', *Journal of Property Research*, vol 12, pp63–74

Ravenscroft, N. (1998) 'Rights, citizenship and access to the countryside', *Space & Polity*, vol 2, no 1, pp33–48

Ravenscroft, N. (1999) *Good Practice Guidelines for Agricultural Leasing Arrangements*, FAO Land Tenure Series, Food and Agricultural Organization of the United Nations, Rome

Rojek, C. (2001) 'Leisure and life politics', *Leisure Sciences*, vol 23, pp115–125

Shoard, M. (1987) *This Land is Our Land*, Paladin Books, London

Smith, S. (1997) 'Caretaker Farm', in Groh, T. and McFadden, S. (eds) *Farms of Tomorrow Revisited: Community Supported Farms – Farm Supported Communities*, Biodynamic Farming and Gardening Association, Kimberton, PA, pp 175–180

Soil Association (2005a) *Cultivating Communities: Farming At Your Fingertips*, Soil Association, Bristol

Soil Association (2005b) *Cultivating Co-Operatives: Organisational Structures for Local Food Enterprises*, Soil Association, Bristol

Stebbins, R.A. (1992) *Amateurs, Professionals and Serious Leisure*, McGill University Press, Montreal

Stebbins, R.A. (1997) 'Serious leisure and well-being', in Haworth, J.T. (ed) *Work, Leisure and Well-Being*, Routledge, London, pp117–130

Stevenson, G.W. (1998) 'Agrifood systems for competent, ordinary people', *Agriculture and Human Values*, vol 15, pp199–207

Stroud Common Wealth (2007) 'Community farm land trusts', Final report, Stroud Common Wealth, Stroud

Sumberg, J. (2008) *Public Land for the Public Good – A Concept Note*, New Economics Foundation, London

Sustain (2008) *Ethical Hijack: Why the Terms 'Local', 'Seasonal' and 'Farmers' Market' Should be Defended from Abuse by the Food Industry*, Sustain, London

Van En, R. (1988) *Basic Formula to Create Community Supported Agriculture*, Indian Line Farm, Great Barrington, MA

Vogt, G. (2007) 'The origins of organic farming', in Lockeretz, W. (ed) *Organic Farming: An International History*, CABI, Oxford, pp9–29

Winch, R. (2005) 'Community supported agriculture: A model for combating distance', Sustainable Development Final Research Paper ENVI 313, Center for Environmental Studies, Williams College, Williamstown, Massachusetts, USA

Winter, M. and Butler, A. (2008) 'Agricultural tenure 2007', Research Report No. 24, Centre for Rural Policy Research, University of Exeter, Exeter

Wodraska, W. (2008) 'Deep gardening: Soul lessons from 17 gardens, Chapter 9, New Hampshire 1989–1994', www.soulmedicinejourney.com/Book.html accessed 2 May 2008

10

A Story of Becoming: Landscape Creation Through an Art/Science Dynamic

Les Firbank, Helen Mayer Harrison, Newton Harrison, David Haley and Bruce Griffith

Introduction

Most research into landscape change projects forward from the recent past, through the present into the future, either using some form of statistical model or some form or scenario. In other words, landscape change is a passive outcome of other driving forces and processes. An alternative approach is to consider the kind of future landscape that we would like and work out how to get there. Such imagined landscapes can generate a much more proactive discourse about what we really need from our environment, and these needs are variously social, cultural, aesthetic, economic and ecological. They touch upon the major concerns of our time, about sustainability in the face of global environmental change, about the relationships between our human and other species, about the natural, the built and the social environments we wish to live in and that we need for survival. Arts and science (both natural and social) are needed to develop future landscapes that are rich enough to engage people's attention and commitment.

This is a story about a meeting between scientifically informed artists and aesthetically informed scientists, and about how their work is becoming increasingly interdependent in order to address this joint concern about future landscapes and how to bring these underlying questions to public attention and ultimately action. The story starts in 1997, when the Harrison Studio, at work on a large landscape-transformation arts proposal, was brought into contact with Les Firbank, who was looking at the British landscape from an ecological point of view.

Mapping a relationship: Beginnings

The Harrison Studio (composed of Helen Mayer Harrison, Newton Harrison, with David Haley as project manager), had been commissioned by the Henry Moore Foundation and the Liverpool Tate Gallery to do a visionary work as part of a larger exhibition entitled 'Art Transpennine98'. Basically, what was called for was a conceptual design to express a new multi-levelled vision for the Pennine region of Britain. With a team of students from an environmental design class at the Manchester Metropolitan University (MMU), the Harrisons were proposing a new design for revitalizing a green network of forest, meadow and hedgerow from Liverpool, on the Irish Sea, across the Pennines to the North Sea coast at Hull. This work proposed to rethink land use for an approximately 35-mile-wide swathe, bounded by Roman roads on the north and the south and including the National Parks of the Yorkshire Dales and Peak District.

The Harrisons had worked for many years as ecological artists, travelling to many diverse countries and exploring the potential for revitalizing landscape around the world although based in their studio at the University of California, San Diego. This new project was based at MMU, where David Haley, who was teaching a class in environmental art, was invited to be project manager and lead researcher. David had many years of experience in commercial and socially engaged arts direction, had studied the Harrisons' work and was keen to learn of their internationally acclaimed ecological arts practice and how their research engaged the sciences.

For Helen and Newton, the concept underlying this work was quite simple and one that they have employed repeatedly. It emerges from the question 'How big is here?', which defines the study area. The shape of this study area brought to mind a dragon-like creature with head and tail stretching out to the estuaries at Liverpool and Hull respectively, with the two National Parks suggesting wings. The work was entitled, 'Casting a Green Net: Can it Be We are Seeing a Dragon?' and was expressed in a complex narrative form of stories embedded in an array of large maps, photographs and hand-worked images. The Dragon work, in one of its images or 'ecological icons' envisioned the potential network of a 'green' connectivity of forest, meadow, town, field and hamlet between the two estuaries. Another map image showed how 900 square miles of farmland could be transformed to strengthen this network (Figure 10.1, facing page 236). In another image, a method of dispersal for 400,000 new houses in the area was generated that would not disrupt the existing landscape and village patterns within it. The final large-scale image showed how unfettered 'sprawl' development would choke the biodiversity, natural resources and cultural entityhood of this region. These issues seemed particularly poignant to the artists given that this region was to be a part of a European Economic Trade

Route, 'E20', promoting economic development on a line from stretching from Limerick on the west coast of Ireland, crossing Northern England, via the Baltic States to end at St Petersburg in Russia.

Overseen by the Harrisons, each of the large-scale (3.5 m × 2.5 m) images was assembled from 1:500,000 OS maps and then re-drawn by a group of Art As Environment and Textiles students. Using coloured pencils, first all the roads were whited-out, then the rivers, lakes and estuaries were intensified with blue and finally the particular features of that image were laboriously rendered. The process was very labour intensive and some of the students protested, but through an ongoing discourse with the artists, they gained an almost meditative state of learning and understanding of the ecology of the region in which they lived. They learned much about the detail of maps and many questions arose from the process of drawing and talking. Finally, these students realized that they could be in control of the data contained therein. This was a particularly meaningful experience as so many had previously believed that map content was a given and predetermined.

Les Firbank's involvement started with a phone call one Friday afternoon in 1997 from one of these students who wanted environmental data about the British countryside. This in itself was not unusual: Les worked in the Land Use group at the Centre for Ecology and Hydrology (CEH) at Merlewood, and they dealt with large data sets, not least Countryside Survey (Firbank et al, 2003). His own interests were increasingly focusing on land-use change and its consequences for biodiversity. What was less usual was that the student did not really know precisely what she wanted or why. But for some reason, despite this, Les was intrigued, and suggested that one of her colleagues call back. He was much more surprised to get a call back soon afterwards from this American artist working in Manchester looking at the future development of the British landscape between the Humber and Mersey. This Les had to see for himself.

Les visited the studio in Manchester to see young students with large boards of detailed maps of the whole area, around 6 × 2 m, adding detail: drawing flowers where there are nature reserves on one map, greying out areas due for future development on another. He could see that this was landscape design, very relevant to his developing interests in landscape ecology, and the way that landscapes could be designed to benefit biodiversity and flows of water and nutrients through the landscape (i.e. 'ecosystem function' in the current jargon). But he was intrigued to see that these artists (from California!) considered it their work. But then, whose work was it? Well at least he could try to help them get the ecology right. And he soon found out that there were other people from many other disciplines also contributing to try to get the hydrology right, the information on planning correct and so on. In fact, Les was witnessing a long-employed strategy that the artists had invented to generate a community of creativity around problems and issues of importance at a scale beyond their solitary capacities as creators.

The work developed as a discourse between the Harrisons, the many scientists and experts that came to contribute, and the maps themselves. Relevant new data would be generated at CEH and brought to the studio, where the artists digested and interpreted it for the students to apply to the images. To make this process more efficient, David Haley moved into the CEH station at Merlewood, Grange-over-Sands, for a few months. On his first Monday at Merlewood, David was invited to present a lunchtime lecture. He rose to the challenge with the opening gambit of 'ecology is not a science, but an art form'. He awaited the heckling, but was greeted with warm, keen interest and a convivial scholarly exchange ensued. The relationship that David built with Les and the staff of the Land Use Section became a two-way process. As David fielded questions from the Harrisons, he discovered new questions about scientific methodologies, and the scientists were intrigued by the creative process. At one point David was bemused by what appeared to be incorrect data for biodiversity across the designated region; the figures didn't add up! Les explained that the black shapes on some of the images represented 'Urban' and these were not included in the 'Countryside' data. David complained that he had learned some parts of central Manchester had higher biodiversity than rural Yorkshire and if these were ignored the current data would misrepresent what was actually on the ground. From this exchange 'Urban' was included in the 2000 update of the Countryside Information System. However, as one of the criteria for defining 'Urban' as a new landscape category had to be 'Humans', 'Humans' had to be introduced as a new criterion for defining all the other landscape categories. And so, humans finally entered the landscape in British research. This and many other anecdotal exchanges developed into an ongoing, reciprocal relationship of exchanging good-quality data for nonlinear-thinking and training.

From Les Firbank's perspective, the centrepiece of the exhibition held in Liverpool was the six large maps. It was fascinating to watch how people reacted; they went close to a map to find their home, moved around to see the different aspects of their locality. They'd chat, 'I never knew we had a nature reserve there', 'I never knew they are going to build houses here.' They would move back and forwards, hunting different places, interacting with the maps in a very dynamic way. As he said, 'We never managed this kind of involvement with our data at CEH.' But there was another aspect of the work that really impressed Les. Just as they had drawn him, a scientist, into the work, they had planners, hydrologists, art historians, architects, geographers, landscape architects and farmers all throwing ideas into the pot about the future. Normally these people would have been defensive, concerned about their roles, positions of their organizations and the limits of scientific data, in contrast to insight. But this was an *arts* project, run by *Americans*! So they didn't need to be defensive, they could imagine possibilities, rather than worry about making mistakes. They could play. The project was a safe area for people with divergent, even incompatible, interests to get together, bounce ideas around, listen and trust. Relationships were being formed that would outlast the project itself. Les could not imagine what else could have brought such a group of people together in such a creative way. Also, it made him realize that ideas about sustainability are not owned by or restricted to any discipline or group. He had thought this was the territory of ecologists: it was enlightening for him to realize that anyone could ask these questions, that anyone could come up with answers. From the Harrison's perspective, a trans-discipline group literally was forming around a large-scale interrelated array of issues where the scale, the issues and the visual format were the attractors. This *was* interesting.

Working practices: ideas

Since the 1970s, Helen and Newton have invoked research, image and story to create proposals for change. They believe that the cultural landscape in its entirety is formed by decisions made by people in power in the place (i.e. a town here, a forest cut there, a dam placed here, an industry placed there, etc.). The Harrison Studio focus is based on the perception that the cultural landscape as a whole is too often unconsciously generated by beliefs in the value of expansion and extraction and has taken an ecologically counter-productive position. They mean this literally, for when too many demands on an ecosystem are made, or too great an extraction of available resources practiced, the productivity of the whole system simplifies and becomes less effective and useful. In some cases this is catastrophic. Often, in the longer term, survival of adjacent systems and ultimately human well-being is at stake.

This position is now shared by the scientific and policy communities. The Millennium Ecosystem Assessment (2005) surveyed the world's capacity to deliver those goods and services that require functioning ecosystems, namely food, clean water, flood regulation, climate regulation and nutrient recycling, In addition the attractiveness of the landscape – visually and sensually – smell and sound, as well as habitats for wildlife and so on. The conclusions are stark. In many cases we are taking more from ecosystems than they can deliver and still sustain themselves, and that the production of these goods and services is likely to decline, no longer able to pace with growing human populations and expectations. Moreover, the Stern Report (Stern, 2007) concludes powerfully that the sooner we can reduce the pressure on ecosystems the more cost-effective any actions will be. Meanwhile both societies and ecosystems are in increasingly rapid transformation. We all agree that this is a time for creative and intensive rethinking about how we relate to our environment and natural resources.

As a scientist, Les sees his work increasingly addressing the biophysical processes that underpin human well-being as well as how this relates to human behaviour (especially to rural land use and farming). In other words, how can you tell when too many demands are being made and whether you are approaching some form of threshold? How does productivity simplify? Do people act in ways that increase or degrade the capacity for production? Can you place these questions in a social and economic framework that encourages sound decision-making? These questions involve the technical assessment of the capacity of landscapes to deliver particular outcomes. They tend to be addressed using the ideas of landscape ecology, which explores the spatial relationships between landscape, land management, biodiversity and ecosystem function (Burel et al, 2003), and (increasingly) using system models to forecast land use and productivity on a spatial basis according to different economic or climate scenarios (Rounsevell et al, 2006; Verburg et al, 2002). Both approaches tend to extrapolate from the present and so are better suited to evaluating problems than to generating creative solutions.

Moreover the solutions will involve change and creativity: we should be seeking to consciously invent future landscapes, rather than just awaiting landscape change through short-term economic benefit, conservation or abandonment. From a scientific point of view, these landscapes should include the building blocks of habitats, soils, water courses and species, managed in a manner to promote the flows of water, nutrients and energy through the system, to be productive in terms of food and energy, to sequester carbon and filter pollutants, and to be resilient under variable climate patterns. From the social point of view, the landscapes need to provide for the economic, physical and spiritual well-being of those that live in them and rely upon them. From both points of view, landscapes can be exploited beyond their limits, causing catastrophic decline in ecological function and in the quality of all life, including human, that can be supported.

Les, Helen, Newton and David all agree on the strategy to work with the existing cultural narrative in a place, to understand the state of the ecosystem, its potential and trends, and then veer that narrative into a more ecologically provident and more societally equitable direction. This approach resonates with the work of Lynn Hull, who also seeks to improve the environment directly through art, in her case by creating sculptures that also serve to enhance habitats for wildlife, providing roosts, nest sites and so on.[1] Their strategy is much more ambitious than much land-use planning and landscape architecture, though it harks back to the classic work 'Design with Nature' (McHarg, 1969). Here, McHarg looked at the planning of new urban developments near Washington according to spatial data of topography, biodiversity, geology and hydrology for example, using a set of transparent overlays that precisely anticipated the use of GIS. But McHarg's work was deterministic; he came up with an optimal way of reconciling human needs and environmental potential. For the larger challenges we face we need to accept that there is no optimal solution unless there is also an effective, transparent and plastic discourse about what we want and what our priorities are.

Time passes: proposals

The differences and commonalities of the scientific practice of Les Firbank and the artistic practices of the Harrisons and David Haley became the basis of a ten-year conversation that led to the work entitled *Looking Ahead: A Stability Domain for Dartmoor.*

David, Helen, Newton, and Les kept in touch, all the time seeing how the increasing concerns about sustainable development, climate change and (more recently) human migration, food security and water supplies were increasing concerns about future land use, both globally and within Britain. But the opportunity to work together didn't come round again until 2006, when Les was appointed as head of North Wyke Research, a grassland research station in Devon. North Wyke Research brings together a powerful group of experts on water, soils, biodiversity, carbon, pollution and how they interact in agriculturally productive ecosystems – the key science issues when conceiving of new sustainable landscapes. Furthermore, North Wyke had its own farmland that maybe, just maybe, could be used to create something new.

In 2006, Les invited the Harrison Studio to make a proposal for a quarter of the 300-ha farm, where the outputs might be compared with the outcome from other landscapes designed by the research community. The initial proposal the Harrisons made was to generate an open-canopy forest and complex multi-species meadow, where the task was to find out how much carbon could be sequestered from such a configuration. In the interim, Helen and Newton were working with David's invitation to initiate a global warming work that considered the effects of sea level rise on mainland Britain, developed using funding from Defra. The physical centrepiece of this work was a large model of the island of Britain, with an audio commentary and six projectors above it that demonstrated the rising of waters and storm-surges with successive loss of lands. The conceptual centre of this work posed the question: 'What would an ideal landscape look like if created in this new greenhouse world?' They believed issues raised by global warming and issues generated by the need for new habitat for the upward movement of people were at an intersection that would permit a new cultural landscape that could be first envisioned and then tested. A key part of this work involved a two-week charrette held at the University of Sheffield, put together by Dr Paul Selman, with Gabriel Harrison as project designer for the Harrison Studio, as well as Newton

Harrison, with a team of students from diverse disciplines to consider the following questions:

- Assuming ocean rise and the upward movement of people in the context of global warming, what would an ideal landscape look like and behave like in the Pennine moorlands?
- Assuming a four-ton/person local carbon footprint, what area of land, under what kind of forest/grassland balance would sequester enough carbon to bring the carbon footprint of 9000 people to neutral? That is: How much land is required to:
  - sequester 36,000 tons of carbon on a yearly basis?
  - to produce harvestable species both in terms of animal and forest products while avoiding mono-cultural practices?
  - produce food, generate fresh water and maintain biodiversity without conflict and so that ideally these activities are mutually enhancing?
  - include some wild areas and some harvestable areas in terms of bearing trees, with trees harvested only when they have stopped sequestering carbon (between 70 and 300 years for poplar and oak respectively)?

With these preconditions in mind the Harrisons invented a probable domain initially based upon the idea that properly harvested grasslands sequester one ton of carbon per year and temperate forests sequester two tons of carbon per year. From this simple calculation they concluded that 42 square miles, or approximately 27,000 acres that is two-thirds grassland and one-third forest would sequester 36,000 tons of carbon per year, sufficient to bring 9000 people's local carbon footprint to a state of neutrality. The artists knew well that this work was conceptual and useful as a framework for further analysis. But in the absence of well grounded and detailed assessment of how the landscape would operate in practice, the work could not in fact land on the ground. That is to say, serious experimental science would be required for the evolution of the work. Moreover, a complex economic argument would need to be designed to give the science and the conceptual work credibility. Further, the Harrisons understood that through their practice of embedding studio members in a community to transact with people in their everyday settings, that sociological study was an issue as well. Meanwhile, increasingly Les's own vision was to look at the redesign of Devon. The county has a lot of interesting features: it is rural, with agriculture a major part of the local economy: the land forms of uplands (Exmoor and Dartmoor) give rise to catchment basins that are well defined (the Exe and Tamar nearly split the centre of Devon off from the rest of England), and the landscape still retains many features that one would expect to enhance sustainability: riparian woodlands, blanket bog and wetlands. More importantly, there was an increasing appetite within the county to rethink how it should develop and function into the future. He could see a new arts–science project as a way of engaging people in the region and raising the profile of the research station, while exploring ideas that would also feed back into their own science agenda. And, who knows, they might even come up with plans that would actually get implemented. After all, this is a time of change and possibility.

Of course, it took a while to sort out pump-priming funding and organization. To coordinate the project and facilitate a Devon network of local arts and environment organizations, the Harrison Studio appointed the artist Janey Hunt as Project Manager. Janey had previously organized a Harrison lecture and a workshop at Dartington Art School for Transition Towns Totnes, so she already had knowledge of sustainability issues and represented a socially receptive community in South Devon.

The Harrisons circulated some tentative ideas to provoke creative responses and gain initial soundings:

- We assume there is extant grassland research that would indicate different types, species and productivity appropriate to the site.
- What might be the different forest types with some history and some probability of growth in this area? How might changing weather patterns modify selection?
- How would one find a balance between wildness and silviculture?
- Are the species recommended for the Pennines appropriate here; for example, the Wisant and several Highland cattle types as described in the earlier texts?
- If one site is chosen, for instance North Wyke, an array of elements need to be investigated and expressed as underlying factors to that which may grow there. They are: the average rainfall, typical earth types, the manner in which the watersheds are working, and where and what forest types would be most useful in creating the sponge effects necessary to reduce run-off and erosion.
- Should we find an economist, who might help design the economics of the system we are inventing? The reason for this is that on the 42-square-mile terrain (or about 11,000 ha) the harvesting system will be unique in modern agriculture. This is because many different plants and animals will be harvested at many different times, in contradistinction to the monocultural food production that is now the norm.
- The working theory here is that most growth and most harvesting is perennial in nature; that is to say, the harvest preserves the system. Does this concept need to be elaborated with each subset in the system?
- We have proposed a 9,000-person community, either in two or three small towns, or one larger town. Would it be wise at this point to begin some real

planning here, (i.e. town sites, etc.) or to put this aside as Phase 2 or Phase 3?

In early June 2008, Janey, Les, David, Helen and Newton assembled at North Wyke Research for a ten-day workshop, the first phase in what we hope will be a longer-term project. Much of the content of the research material came from the meetings Janey had programmed with many experts from different disciplines, from drop-in sessions and field trips with the North Wyke scientists, and from maps and spatial data of the region. Not everything worked: the planned fly-over of Devon had to be cancelled because of poor weather. Nevertheless, this extremely hectic and dynamic process generated valuable connections to evolve a data-rich picture of Dartmoor and surrounding region based on many perspectives.

As always, early discussions struggled to find focus. Bruce Griffith, the GIS expert and data manager, was finding the process of working with artists fascinating, but slow and frustrating: he was more used to science projects in which the objectives and requirements were decided up front. He also felt that the artists failed to appreciate the time it took to put images together, for them to be rejected because they didn't quite work somehow or were even the wrong colour. Nevertheless, it was Bruce's work that provided the catalyst: David asked Bruce to project the effects of limited sea level rise on the Devon/Cornwall peninsula. Then Newton followed up, asking to see the effects of a 100-m sea level rise (equivalent to the total polar ice melt).

This gave a radically different vision of the region (Figure 10.2, facing page 237), and a massive shift in reading and understanding of the landscape took place. The team saw new urgencies emerge and conventional ways of thinking ebb away. Preconceptions of the landscape and the nature of watersheds were radically changed within the team as Les interpreted the hydrological characteristics of the terrain. This led to the introduction of water in the landscape in terms of the sponge effect of forest, uplands and mature complex meadows as one of the principle themes in this work – particularly in the context of global warming and potential drought. We realized that carbon sequestration and water sequestration would have to work in tandem and would be sensitive to existing differences in soil type (the Harrisons, in their Sheffield charrette, had come to consider the soil differences, the sponge effect, the hydrology and the biodiversity with human involvement as a consciously invented complex system operating on the ground). That is to say, indeterminacy is an inevitable participant in the operation of complex systems and working with them.

The question arose, 'What if a certain percentage of people did not want to participate, even if they saw an economic advantage?' At this point the Harrisons introduced a conversation on redundancy. They felt that a complex system of this kind could work quite well with non-participants scattered among it. The question that had to be addressed was: 'What percentage could that be without causing systems breakdown?'. Another question emerged: 'What if people with land adjoining wanted to join and what if villages wanted to participate?'. The metaphor that the Harrisons argued for here was that the perimeter might be considered a permeable membrane and would contribute to systems sustainability. For instance, if a 3,000-person town wished to raise the numbers in the region up to 12,000 people, then the region would have to expand its territory to accommodate an increased carbon budget. In this way, carbon footprint might become a rough definition of carrying capacity. Moreover, the soil types in each site were, in the main, different. Further, the water-sequestering properties of certain of these lands differed. Thus the idea of creating a large-scale comparative analysis became available. Science would be well served by such a comparison as well as being served by an analysis of the internal operations of a particular site. The idea was to focus in from the original coast-to-coast, north–south Devon transect to three smaller domains in and near Dartmoor, permitting a finer selection of forests types and grassland configurations, and to allow us to explore how far common ideas of landscape evolution can be replicated across different topographies, soils and weather conditions. Three, approximately 42-square-mile sites were chosen, partly selected because the existing human settlements were restricted to a few villages (Figure 10.3, facing page 237, showing two of the three regions).

We were starting to look at the maps as future possibilities, rather than being restricted by retaining anthropogenic historic perceptions. These possibilities would be shaped by ecological responses to climate change. A heated discussion arose about the long-term fate of peat on Dartmoor. Peat is a major sink of carbon in Britain; might it dry out, oxidize and release that carbon to the atmosphere? If so, is it better to try to maintain water levels or to try to lock that carbon into woodlands?

We were not seeking to commit to solutions or plans in the working session, rather we wanted to start a fluid dialogue, whereby trajectories of dynamic processes could be considered as possible projections into probable futures. And so the emphasis was shifted from 'problem based learning' to 'question based learning', with the questions and propositions already posed in need of testing.

Is the scale of land suggested appropriate to:

- move from a carbon neutral to a carbon negative state
- and sequester the carbon for 9,000 people (a town)
- establish an experimental field
- find the right scale for redundancy?

As the creative/scientific process progressed, the team was aware that the project not only gathered strong supporters, but it had its doubters too. As the team prepared for its closing presentation, very big issues were brought to light that questioned the value of the whole approach. We decided to embrace them by dealing with them directly or identifying them and accepting that they are beyond our present remit:

- How would we generate the imagery that would determine whether public participation could happen at all?
- Are the ideas proposed viable in terms of: the carrying capacity of land, population, ownership?
- How is this permeable, semi-closed, malleable system to come into existence?
- How independent does it need to be from the world around it?

Other questions were perceived to be more easily approached, though no less difficult to resolve:

- What new architectural forms of dwelling and settlement are required?
- What new, non-monocultural economics are needed?
- What forms of management and governance will promote participation and respond to the needs of place?
- How might we teach literacy of place and vernacular skills?
- What is the new aesthetic?

Looking ahead: commitment

What was established out of the very fruitful closing presentation was a great enthusiasm for participation in the project, including a number of generous offers to carry the project forward, such as support to gain further funding for a next phase, planned for 2009. What has also emerged is a working, 'on the ground' collaboration between the preponderance of evidence provided by science and the prima facie understanding offered by art. In an esoteric sense, this may be likened to and represented by the *Taijitu*, yin and yang symbols of integral dynamic opposites, which can bring about change together.

There is no conclusion to this work: this story, as they say, will run and run, because the dialogues about our future landscapes are so important and have barely started. However, the issues before us are so imperative in terms of survival and wellbeing of large systems that new forms of creativity, collective in nature, are coming into being that require multi-tasking and interdisciplinarity. We believe that our experiment will generate a new complex system that blends the disciplines of the Arts and the Sciences and applies them to the generation of a new cultural landscape. It has little precedence and therefore pre-existing models are not available for comparative analysis. We perceive that it is vital to generate this new kind of community and the unexpected advantages to be found in creating this complex system. In fact, later discussions suggest that it would be advantageous to select one of the three areas and move away from comparative analysis towards a high focus on one of these three areas, North Wyke for instance. This approach has the value of permitting us to work on the ecosystems, economic and social systems simultaneously. We are undaunted by the questions raised because we expect the multi-layered community that our experiment calls for will, on the one hand, have unpredictable outcomes (true for any complex system), yet on the other will have self-organizing properties so that many of the questions thus far posed will be self-answering. The linkage between the Arts and the Sciences provides a space for these dialogues to progress powerfully, creatively and openly; it encourages playfulness and insight, and by its nature is open to all who wish to participate.

Finally understanding
that the news
is neither good nor bad
it is simply that great differences are upon us
that great changes are upon us as a culture
and great changes are
upon all planetary life systems
and the news is about how we meet these changes
and are transformed by them
or
in turn
transform them
(from *Greenhouse Britain*).

Acknowledgements

We thank Janey Hunt for project management, Lesley Ryan, Linda Ciesielski, Sue Blackburn, Jerry Tallowin and all those who took part in the workshops and events. This work was funded by the Biotechnology and Biological Sciences Research Council (BBSRC) through the Institute Strategic Programme Grant to North Wyke Research, 'Delivering multifunctional landscapes'. Figure 10.2 was constructed using the Landmap 25m DEM, © NOAA, and distributed by CHEST under licence from Infoterra International. http://repositories.cdlib.org/imbs/socdyn/sdeas/vol2/iss3/art3

Notes

1 http://www.eco-art.org/

References

Burel, F., Baudry, J. and Le Flem, Y. (2003) *Landscape Ecology: Concepts, Methods, and Applications*, Science Publishers, Enfield, UK

Firbank, L.G., Barr, C.J., Bunce, R.G.H., Furse, M.T., Haines-Young, R., Hornung, M., Howard, D.C., Sheail, J., Sier, A. and Smart, S.M. (2003) 'Assessing stock and change in land cover and biodiversity: Countryside Survey 2000', *Journal of Environmental Management*, vol 67, pp207–18

McHarg, I.L. (1969) *Design with Nature*, Natural History Press, New York

Millennium Ecosystem Assessment (2005) *Synthesis Report*, Island Press, Washington, DC

Rounsevell, M.D.A., Reginster, I., Araujo, M.B., Carter, T.R., Dendoncker, N., Ewert, F., House, J.I., Kankaanpaa, S., Leemans, R., Metzger, M.J., Schmit, C., Smith, P. and Tuck, G. (2006) 'A coherent set of future land use change scenarios for Europe', *Agriculture Ecosystems & Environment*, vol 114, pp57–68

Stern, N. (2007) *The Economics of Climate Change: The Stern Review*, Cambridge University Press, Cambridge

Verburg, P.H., Soepboer, W., Veldkamp, A., Limpiada, R., Espaldon, V. and Mastura, S.S.A. (2002) 'Modelling the spatial dynamics of regional land use: The CLUE-S model', *Environmental Management*, vol 30, no 3, pp391–405

11

Agricultural Stewardship, Climate Change and the Public Goods Debate

Clive Potter

Introduction

Policymakers have for decades presented public-good and social-equity justifications for shielding farmers from market forces, but it was not until the 1980s that an environmental public-good rationale was articulated with any great persuasiveness. Mounting evidence that agricultural intensification and the production of food on an industrial scale were depleting biodiversity, eroding soils and contaminating the wider environment throughout industrialized countries brought about a profound questioning of the idea that the rural environment could safely be left to the self-regulated farmer's care (Lobley and Potter, 2004; Fish et al, 2008). The result was a rethinking of policy models and approaches and arguably one of the largest scale policy experiments in what, in current parlance, would be called 'payments for ecosystem services' ever attempted in jurisdictions such as the EU or the USA (Potter, 1998). Since then we have seen the notion of agricultural stewardship established in Europe as a central policy idea, simultaneously justifying large expenditures of public money in order to support (chiefly extensive, grassland) farming as a beneficial land use in its own right but also to incentivize environmental management and a stream of public benefits on farms in a more narrowly conceived, 'fit for purpose' sense.

As the opening discussion to this chapter will show, the tension between these two ideas – what we have called a 'working lands' versus a 'public goods' approach to agri-environmental governance – has never been fully resolved at the level of the EU. Debates about the relative policy merits, cost-effectiveness and environmental sustainability of these two models have in recent years intensified in response to the threats and opportunities posed by the much larger policy agenda of agricultural trade liberalization and due to mounting budgetary pressures within the EU. The former has focused on the extent to which Europe's farmers could be vulnerable to increased market opening and has required policymakers, and those who lobby and advise them, to accept that there may be difficult trade-offs between rural environmental sustainability in the North and poverty reduction in the South through freer world trade (Losch,

2004; Potter and Tilzey, 2007). If more open markets are necessary in order that farmers in developing countries and newly emerging economies are to be able to export to markets in the North without prejudice, a central concern has been how best to ensure the continued provision of public environmental goods in the resulting, much more neoliberal context. As we have argued elsewhere (Potter, 2008, Dibden et al, 2009), agri-environmental governance has thus increasingly been drawn into an international debate about the contradictory nature of agricultural sustainability at different spatial scales.

Seen in these terms, climate change and the threat of global food insecurity have sprung relatively recently on to the international policy agenda. Compared with the transport and energy sectors, agriculture has come late to the climate change debate, and food security was rarely mentioned as a serious issue in agricultural policy circles until just two or three years ago. Nevertheless, their rapid emergence as policy narratives means that climate change and food security are already reframing domestic policy on rural land use in ways just as contentious as trade liberalization was doing a decade ago. The likelihood is that they will similarly test the validity of a range of public policy assumptions about what is the best and most sustainable relationship between agriculture, the environment and rural communities and what will be required in terms of a commitment of public funds. However, while the recent increased awareness of global food security can be seen as a return to a long-established (and, according to neoliberals, in the case of calls for greater self-sufficiency, largely misconceived) agricultural policy concern, the threat of climate change to agricultural production, biodiversity and landscapes takes the policy debate into altogether more uncharted territory.

Empirical uncertainty about the land use consequences is reflected in the still heavily science driven nature of the current land use debate surrounding climate change. As other contributors to this book have pointed out, the difficulty of arriving at any very coherent assessment of its likely impact across the range of rural interests means that little has yet been said about the long-range implications for farming livelihoods or patterns of countryside management. Rural social scientists are only now beginning to think about what climate change and the policy response to it will mean for the way land is managed, let alone whether it will favour particular policy models and the advocacy coalitions that support these.

As an initial contribution to this debate, this chapter argues that we need to think about what food security concerns and climate change adaptation will mean for *the way* government support to agriculture will be justified in the years ahead. In particular, how far, and in what manner, will this reframing intensify the battle that already rages between a working-lands and public-goods rationale for agricultural support? In relation to food security, we argue that an apparent resurgence of productivism is closely allied to working-lands thinking in the minds of many policy actors, albeit in ways which seem likely to complicate the environmental rationale for public support to farming for its own sake. This is roughly associated with an approach to climate change adaptation which emphasizes shifts in land use rather than improved stewardship and the promotion of essentially productivist measures such as an expansion of biofuel crops. Even if we leave aside the environmental impact of some of the huge projected increases in biofuels that have been advocated by renewable-energy interests, the difficulties of reconciling such an approach with neoliberal priorities means that a protectionist interpretation of food security will inevitably be heavily challenged in forums such as the World Trade Organization (WTO). Indeed the argument has been turned on its head, with the need for greater food security being used by neoliberals as a justification for more market opening, not less.

Climate change, by contrast, appears to map well on to the public-goods model. Indeed at a time when agri-environmental policy may be reaching the limits of expansion due to concerns about efficiency, effectiveness and public value for money, the need to encourage farmers to make climate change adjustments over long periods of time may encourage governments to renew their commitment to agricultural stewardship. Farmers are likely to be leading suppliers of mitigation and adaptation measures through better slurry and soil management, for instance, and the need to retain extensive grassland and peatland by dint of their carbon sequestration properties appears consistent with a long-standing agri-environmental concern with sustaining high nature value farming in the uplands. At the same time, however, there is evidence that policy-makers and those who advise them are anxious to draw agri-environmental management into a larger, more integrated system of thinking about ecosystem services. A new type of environmental policy discourse is emerging which is much more globally framed and anthropocentrically justified than the conservationist rationale that has held force in the past. The chapter concludes by raising questions about how best to develop the public-goods justification for continued agricultural support in the wake of this shift in thinking.

Contested models of agri-environmental stewardship

Environmentalist critiques of modern farming first emerged in Western Europe during the late 1970s. In the UK perhaps more than any other EU Member State, the resulting public debate about whether this required more regulation of farming activity was complicated by the contrary perception that farming and the countryside were not only synonymous but mutually constructed. It was thought that widely flung farming activities and the practice of good husbandry had largely fashioned the agricultural landscapes that are so highly valued by the public. A way out of this apparent contradiction was to argue that it was the policies which encouraged farmers to intensify production, remove hedgerows and drain ponds that were at fault and needed correction, rather than the farmers themselves (Potter, 1998).

The next step was to argue for a series of reforms which would rebalance the financial incentive in favour of conservation rather than against it. Over the next decade a series of agri-environmental programmes were therefore put in place which defined farmers as providers of first resort of environmental goods in the countryside and thus offered them payments in order to manage land in an environmentally sensitive manner. Importantly, whereas in the UK and a limited number of other Member States such as Denmark and Sweden, there was an attempt from the start to justify these measures strictly as payments for public environmental goods, elsewhere in Europe it was the activity of farming, qua farming, that was supported under more broad-brush schemes. Under this working lands model, broad swathes of grassland farming and traditional, extensive arable systems were enrolled in agri-environmental programmes wholesale, on the grounds that, without support, these systems of farming would no longer be viable. This is reflected in the very different patterns of enrolment that we have seen in agri-environmental programmes across the EU, with the 'broad and shallow' pattern favoured in Member States such as Germany, France, Finland and Greece contrasting with the more targeted and arguably more 'narrow and deep' enrolments achieved in the UK, Denmark and Sweden (Buller et al, 2000). A key assumption here is that nature conservation goals can best be achieved as a by-product of agricultural production. Buller et al (2000) comment that one of the most persuasive arguments in favour of working lands (*sic*) is the extent to which human and natural systems have indeed co-evolved in European rural settings. The maintenance of much of our semi-natural habitats and managed landscapes requires the continuation of farming practices such as grazing, mowing and burning in order to retain their biological and aesthetic value.

As time has gone on, this working-lands thinking has increasingly been extended as a justificatory rationale for other forms of direct support to agriculture under the CAP, notably in relation to the broad-brush support that has traditionally being on offer to hill and upland farmers in the 58 per cent of the EU's agricultural areas classified as Less Favoured Areas. A debate has raged for the last 25 years about the environmental and social justification of income-support measures which, in their original form, had an at best implicit conservation rationale (Bonn et al, 2009). Arguments about how far it is necessary for the state to continue offering farmers income support reached a height at the time of the Agenda 2000 CAP reforms. Member States such as France and lobby groups such as Committee of Professional Agricultural Organizations (COPA) drew on a working-lands rationale to justify the introduction of the controversial Single Payment Scheme (SPS) on the grounds that, in order to be able to adapt to reductions in price support agreed elsewhere in the reform package, and thus continue farming as viable operators, land managers had to be offered multi-annual income payments.

The EU thus entered the Doha trade negotiations, launched in early 2001 on the basis of an international commitment to reducing trade distortion in the agricultural sector, with a significant continuing reliance on income support and a CAP that was far from trade neutral in the substantial commodity supports and forms of market protection it still extended to its farmers. There was intense discussion among trading partners during the early stages of the round about the validity of 'agricultural multifunctionality' as a justification for state support, now recognized as code for a continuation of the working-lands approach. This stimulated a discussion within the academic and policy communities about the type and extent of restructuring of land use that might result from much greater market opening and the emergence of conflicting accounts of the vulnerability or adaptability of Europe's farmers as an occupational group (Potter and Lobley, 2004).

In more neoliberal Member States such as the UK, concerns about the wider trade effects of the CAP strengthened the hand of those within the Treasury, long critical of working lands and anxious to put in place a much more 'decoupled' (i.e. no longer linked to production) form of support to agriculture (MAFF, 1995). On the one hand, farmers were more adaptable than official information systems such as the EU's annual farm survey suggested, because an outdated construction of farmers as atomistic operators of farm businesses rather than members of agricultural households meant that the range of income streams typically flowing into their hands was often grossly underestimated. On the other, some state support to the sector can continue to be justified but where offered must be strictly justified in terms of public money for public goods.

By this time UK policymakers and those who advise them had learnt much from 20 years of agri-environmental policy implementation. This suggested that, while there had been some success from some of the more broad-brush schemes established early in the programme, the best-designed measures were those which matched management options to the environmental needs and biodiversity potential of particular sites (Boatman et al, 2008). Aware of the need to promote basic levels of environmental management throughout the countryside, however, the Policy Commission on the Future of Farming and Food had previously recommended an approach which combined both broad and shallow schemes with more demanding narrow and deep ones. This was enshrined in the overhaul of agri-environmental policy which followed, whereby a complex array of schemes was replaced with the more streamlined Environmental Stewardship Scheme. ELS is open to all farmers and landowners in England

and has meant that, for the first time, a majority of farmers will be involved in a scheme designed to encourage positive environmental management.

The HLS scheme is similar to previous schemes, with selective entry and individual payment rates for each management option. Practical experience in the difficulties of designing (and effectively monitoring) such schemes that, in standard Treasury phraseology, are 'fit for purpose', thus became a reference point for UK policymakers and their wider advocacy of a public-goods approach during the course of the CAP 'Health Check'. This was launched by the European Commission in November 2007 in order to take forward the reformist agenda for the CAP that had been set in motion with the 2003 reforms. Following statements of support for a much more explicit public-goods approach to agricultural support made early in the process by the Agriculture Commissioner, Marianne Fischer Boel, it was expected that policy-makers would agree to stricter cross-compliance (the environmental conditions attached to the direct payments farmers receive) and to a more rapid application of compulsory modulation (the rate of transfer of funds from commodity support under Pillar 1 of the CAP to more directly funded environmental and social payments under Pillar 2). With working lands advocates still resistant to the idea of full decoupling, however, the outcome fell well short of the public money for public goods principle that had by now become the common agenda of environmental NGOs. The debate though is far from concluded, and at the time of writing there are signs in Brussels and Whitehall that, with the Health Check behind them, policymakers are increasingly anxious to clarify the justification for continuing agricultural support (Cooper et al, 2009). In particular, attention is focusing on the terms of the EU Budget Review and the underlying rationales for public expenditure that will determine the level and pattern of agricultural support after 2013.

Food security and working lands

At first glance, it seems that the Review could not have come at a better time for public-good advocates. Spending on agriculture has declined from a peak of 70 per cent of the EU budget in the mid-1980s to around 40 per cent by 2008 (CEC, 2008), yet the call on resources to meet other, non-agricultural policy goals has continued to intensify in the wake of the Lisbon Treaty and new EU commitments on climate change and energy security.[1] The current economic recession is likely to sharpen the debate about competing shares further, with the comparatively still very large subventions going to agriculture (€55.1 billion for EU-27 in 2007) coming under growing public scrutiny. If agriculture is to continue to receive state support, it is argued, now is the moment to put in place an explicit 'public money for public goods' rationale. The debate about food security complicates this picture, however, for, somewhat to the surprise of even the most seasoned CAP analysts, it has been seized on by farming interests and some Member States to justify a continuation of production aids and a policy approach that some believe could return the CAP to its neo-mercantilist roots.

The reasons for food security concerns are complicated, not least because of a tendency of commentators to conflate the short-run factors which precipitated a sudden rise in the price of cereals and oilseed during 2006, with longer-run demand and supply reasons for market tightening and politically motivated reasons for disruptions to supply (Ambler-Edwards et al, 2009). What is important here is the rapidity with which some interests have managed to exploit a turn of events to their political advantage and the destabilizing effects this has had on the policy debates discussed above. Hence the calls early on from organizations such as the NFU and COPA-COGECA for a return to self-sufficiency targets under the CAP. Similarly, in 2006, the Chair of the European Parliament Agriculture Committee could warn of the dangers of continuing to rely on food imports to meet the needs of over 450 million Europeans. For agriculturally fundamentalist Member States such as France, the 'food crisis' has offered an opportunity to renew calls to place food security at the heart of a reinvigorated CAP. This is now impacting on the debate surrounding the Budget Review, with the French Presidency's Discussion Paper in autumn 2007, issued as the Health Check negotiations were nearing their conclusion, arguing for food security to be recognized once more as a primary policy objective of the CAP.

Public good advocates are wary of these developments and the return to a productivist CAP that they imply. While neo-classically inclined commentators have quickly dismissed the idea that governments should seek to resurrect self-sufficiency as a legitimate policy goal, it is not so easy to disregard the political appeal of a concept which appears so well to serve the interests of entrenched, if in recent times rather beleaguered, policy groupings. Significantly, some of these interests have begun to associate the need to secure food supplies with a working lands model of the countryside, the NFU in the UK in particular arguing for the countryside benefits of a more fully exploited land resource and, in evidence to a House of Commons Select Committee, of the need to 'exploit spare capacity in farming' in order to deliver more ambitious self-sufficiency goals (NFU/CLA/AIC, 2007). UK conservation NGOs and agencies such as Natural England have had particular difficulty deciding how to respond to these developments. This is illustrated by the recent story of set-aside in the UK.

Improved returns to arable farming, together with long-standing criticisms of the economic inefficiency and administrative burden of the policy, led to the abolition of set-aside as a production control measure under the Health Check review. While never designed as an environmental policy tool,[2] the measure had been broadly beneficial for farmland birds, invertebrates and plants, because it took land out of intensive cropping in arable landscapes that had been most resistant to enrolment and hardest to reach under agri-environmental schemes (Silcock and Lovegrove, 2007). In a classic example of a policy evolution towards multiple objectives, the scheme was steadily fine-tuned in favour of agri-environmental interests, starting out as a crudely designed measure which applied only to cereal, oilseed and protein (COP) crops but then being extended through a series of rule changes to enable farmers to enter into longer-term arrangements and/or to integrate their set-aside land with management regimes being paid for under arable stewardship. In arguing for the retention of set-aside, organizations such as the RSPB found themselves swimming directly against the productivist tide, having to make a case for keeping land out of a productive use in the very parts of the countryside where rising market prices seemed to be justifying an expansion of cropping. Interestingly, their success in persuading Defra to put in place a compensatory arable stewardship scheme turned strongly on public-goods arguments and the need to safeguard a public investment in conservation headlands and permanent set-aside that had been accumulated over many years.

The argument, however, does not end there because in an important cross-over between the food security and climate change agendas, the abolition of set-aside has created an opportunity to increase the area of land devoted to bioenergy crops.[3] The chief justification for growing more energy crops on farmland is that it increases the proportion of total energy coming from renewable sources, reducing dependency on fossil fuels and thus contributing to the reduction in GHG emissions. Unsurprisingly, the Directorate-General for Agriculture (DG Agriculture) was an early enthusiast for the expansion of bioenergy crops, seeing in the policy an opportunity to make a farming contribution to climate change mitigation while offering farmers an alternative income source.

The biofuels debate has coincided and to some extent been shaped by the decoupling of the CAP and the search for alternative land uses that this has brought about. In the event, a policy justified by climate change concern has impacted on food security and biodiversity protection in unexpected ways and has come for many to exemplify the law of unintended consequences in public policy. To begin with, Cooper and Arblaster (2008) observe that most of the incentives for the initial expansion of biofuels seen in the EU came initially from outside agricultural policy. The renewable energy sector and its sponsors in government in the UK played a key role in setting up the RTFO, which has fostered demand in the transport sector and created a lucrative domestic market for biofuels. Nevertheless, there was early enthusiasm within the farming industry for the policy and by 2008 there had been a significant conversion of cropland to first-generation biofuels.

Financial incentives to grow bioenergy crops on farmland have steadily increased and in the UK include the ECS, offering establishment grants and hectarage payments for energy crops grown on land other than set-aside. Here, it seemed, was a magic-bullet version of working lands, delivering an (albeit indirect) agricultural contribution to GHG reductions but also boosting farming incomes and thus keeping the countryside farmed in some fashion.

Doubts began to surface following publication of the UK's Biofuels Strategy in 2007 and its sweeping assertion that up to 1.6 million hectares were potentially available to be converted in the UK by 2030. Commentators such as Von Braun (2007) had previously estimated that some 4 million ha, equivalent to two-thirds of current arable land, would need to be converted in order to supply just 15 per cent of primary transport energy in the UK. Ambitious targets for biofuel consumption had been set worldwide, leading to changes in land use as countries respond to their own targets or see an export opportunity. Conservationists were thus at this moment digesting the evidence that up to 30 per cent of the increase in average world grain prices from 2000 to 2007 may have been due to first-generation biofuel conversion, particularly in places such as the American Midwest, where the 2005 Energy Policy Act had subsidized the conversion of cropland to bioethanol production on a massive scale (Von Braun, 2007). Questions were now asked about the biodiversity and landscape impact of such a large, targets-driven land-use shift.

This has been followed by further debate about the GHG savings reputed to be made from the increased transport use of biofuels. Again, conservation groups have found it hard to negotiate a policy which, while apparently very 'green' in its generic climate change justification, may actually be damaging to local biodiversity and landscape character in its policy effects. Putting more emphasis on biomass (the planting of perennial grasses and woody crops such as miscanthus) rather than biofuels has not resolved this debate, despite Natural England's apparent conviction that the extensive planting of miscanthus in the agriculturally more marginal areas where it is most viable will not impact severely on landscape character (Natural England, 2007).

Elsewhere, there are more encouraging signs that the productivist 'turn' in farm policy, if that is what it is, may yet be exploited in more nuanced ways by environmental and rural development interests. There have been attempts, for instance, to link growing policy interest in the *resilience* of our food supply with a longer-standing, if essentially consumerist agenda for food and farming reform. This speaks to an approach to rural policy situated somewhere between working lands and public goods in its advocacy of an expansion of more locally produced food through low-input systems. The presumption here is that state-centred debates about government intervention in rural areas are increasingly being eclipsed by consumer-driven concerns about food and where it comes from. In a recent contribution in this vein, the think-tank Chatham House, for instance, has argued that government should seek to protect British interests in a secure food supply in the UK, but in ways which promote other policy objectives such as biodiversity protection, healthy eating and sustainable production. This contrasts interestingly with the recent claim of the Country Land and Business Association that the promotion of niche consumption such as organics might actually increase vulnerability to supply-chain interruption due to the increase in imports that would be required to satisfy demand.

Climate change and public goods

These complications and displacements aside, the climate change agenda seems likely to strengthen the hand of those arguing for a public goods rationale, though it may also require a reinvention of the way government subsidies come to be justified in environmental and stewardship terms. We know that various agricultural practices have the potential to help mitigate GHG emissions, many of them already recognized as part of the suite of good agricultural practices that policymakers have attempted to incentivize through agri-environmental policy and regulation (for instance, improved slurry management, better soil conservation, more tree planting and active woodland management). At the same time, there is a case to be made on climate change mitigation and adaptation grounds for the maintenance of certain extensive farming systems, particularly in the uplands where the retention of a permanent grassland cover and the restoration of peat sinks may have a significant contribution to make (Smith et al, 2008). A long-term strategy might thus combine payments for the management of carbon sinks associated with land as well as payments that will bring about more general carbon-reducing and sequestering changes in the management of land. This seems to fit well with the existing system of agri-environmental subsidies and incentives designed to encourage active stewardship but also to prevent change and to maintain, possibly over large areas, certain systems of farming. Indeed, there are many mitigation options that could be pursued as part of a decoupled public money for public goods agenda under Pillar 2 of the CAP.

Rollett et al (2008) discuss the range of actions farmers could undertake as part of a climate mitigation programme, including the traditional one of ensuring that all cereal crop residues are returned to the land either as straw or farmyard manure. In the arable sector, many of the established elements of good arable stewardship such as wide field margins, reduced tillage and increased use of crop residues go with the grain of good carbon management (Rollett et al, 2008). For peat management it is suggested that carbon sequestration could become an explicit goal of hill farm support systems, payments being given to hill farmers to prevent the overgrazing, excessive burning and drainage which have damaged peat and reduced its carbon storage capabilities in the past (Hubacek et al, 2009).

These climate-driven extensions of agricultural stewardship come at a critical juncture, not only in the larger debate surrounding the EU budget and the need to find new ways to justify state support to farmers, but also, particularly within the UK, in terms of a still-unresolved debate about how to make agri-environmental policy schemes more fit for purpose in biodiversity protection terms. Recent years have seen mounting criticism of the effectiveness of agri-environmental measures following a hostile report from the European Court of Auditors which questioned the measurability and value for money of many programmes and a scientific study which concluded that there was insufficient evidence on which to assess the effectiveness of programmes. Subsequent evaluations have been more encouraging and in the UK there is now robust evidence to suggest that agri-environmental policy here has made a significant contribution towards meeting biodiversity objectives, particularly following improved incentives under Environmental Stewardship for arable habitats (Boatman et al, 2008). Part of the difficulty is that relationships between land management and improved biodiversity outcomes are complicated and still not fully understood. However, as Hodge and Reader (2007) comment, the gradual evolution of the UK strategy away from a more or less exclusive emphasis on the prevention of change on the extensive margin to payments on all agricultural land, some of which aim to restore biodiversity lost in the past to intensification, reflects the development of a more sophisticated ecosystem services understanding.

The pattern of stewardship likely to optimize climate change mitigation and adaptation, however, may differ from that required to deliver more traditional conservation goals. For example, while more extensively managed livestock is a central component of traditional conservation strategy and could be seen as an energy efficient source of protein, indoor systems may be better overall in terms of the regulation of GHG emissions such as methane. There might even be a case for reducing livestock numbers on climate change grounds. The 'new stewardship' may also need to operate at a systematically different spatial scale. Traditionally, agri-environmental policy has incentivized management at a field or farm level, largely ignoring landscape-level interactions. Ecosystems services thinking, the standard point of conceptual reference in recent debate (see further discussion below) emphasizes interactions between ecological structures, processes, services and benefits (Rollett et al, 2008).

The increasing dominance of ecosystems thinking in rural policy debate, exemplified in the most recent contribution from Woods (2009), reflects the extent to which neoclassical environmental economic concepts have colonized contemporary rural policy debate. First developed as part of the Millennium Assessment (MA), the ecosystem services concept has attracted widespread support as a way of thinking about the flow of 'provisioning, regulating and cultural' services from ecosystems, including those managed through agriculture. While a progressive idea to the extent that it emphasizes the dependence of human well-being on natural processes, it is fundamentally utilitarian in its assumptions about the sources of value and is closely associated with the payment for ecosystem services model that is now one of the most important growth areas for environmental economics research. There is evidence that rural policymakers are coming to regard it as a more scientifically refined and politically serviceable justification for future funding than the textbook version of public goods.[4] In climate change terms it brings referents such as ecosystem resilience into the heart of the policy process. According to Defra, for instance, ecosystem function thinking and the valuations associated with it need in future to be embedded within measures such as Rural Development Programmes and the revised strategies for upland areas and nutrient management (Defra, 2007). The scale of this ambition will have a number of implications for the future conduct of policy. In an obvious sense, the emphasis on the demand as well as the supply side of service provision means that it will be easier to justify the application of market-based instruments in an area of policy where the public good has often been largely defined through political process, NGO lobbying and stakeholder representation. Under this model, large-scale peat restoration could be financed as Certified Emission Reductions on the carbon-offset market. The voluntary carbon-offset market could be enrolled in paying land managers for the services they provide through carbon sequestration. This could signal a more general marketization of the environmental services farmers provide, though what this will mean for landscape and biodiversity outputs that are specific to place is unclear.

More fundamentally, an emphasis on ecosystem services may marginalize traditional justifications for government intervention rooted in the socially and culturally constructed values of landscapes and countryside that have long been at the heart of the UK's conservation debate. Although advocates of ecosystems services acknowledge the importance of landscape units, for instance, their understanding is essentially instrumental, landscape being merely a unit of measurement and analysis within which to embed the ecosystem approach (Swanwick, 2009). This is symptomatic of a growing reluctance on the part of policymakers to acknowledge the aesthetic, perceptual and place-specific values of landscape that are so important to people.

Conclusions

Growing international concern about the security of global food supplies and the impact of climate change is already reframing the land debate in the UK and throughout the EU in various ways. In this chapter we have compared this reframing to the rise of neoliberalism in the 1990s and the way in which this movement in thinking challenged long-cherished assumptions and destabilized public policy debates about government support to the agricultural sector. Policy analysts are only just beginning to consider what the implications of climate change imperatives might be for the future design of farm support, most analysis to date (including other chapter treatments in this book) focusing on the biophysical consequences of the requirements of mitigation and adaptation, for instance. The extent to which the insertion of these new policy concerns into the policy debate will see the emergence of new interest coalitions, and thus to profoundly different policy outcomes, is even more difficult to assess at this point. We have speculated here that while a new productivism associated with food security could cut against many of the agri-environmental achievements of the last 20 years, the contributions that agriculture as a sector will be required to make to climate change mitigation and adaptation will greatly strengthen the public good case for continued government support. The market-orientated nature of the policy discourse surrounding climate change nevertheless suggests that traditional, essentially conservationist justifications will be eclipsed by much more anthropocentric, environmentalist arguments for countryside management and habitat restoration.

Notes

1 A new EU Position Statement on climate change was published in Spring 2009

2 Set-aside was introduced in 1988 as a supply control measure in response to the overproduction of cereals and increased public expenditure on surpluses. Under 'obligatory' set-aside, introduced in 1993, farmers were required to set aside a given percentage of their land under COPs as a prerequisite for obtaining Arable Area Payments

3 Although, strictly speaking, set-aside was already subject to a derogation which allowed land managers to grow bioenergy crops on land taken out of production

4 Despite having its origins in the economic theory of public goods, the ecosystem services and payments for ecosystem services literature is more explicit in its emphasis on the need to assign a value to a service output. The dominance of this set of concepts is testimony to an increasingly integrated (and thus hard to disentangle) climate science and environmental economic research agenda

References

Ambler-Edwards, S., Bailey, K., Kiff, A., Lang, T., Lee, R., Marsden, T., Simons, D. and Tibbs, H. (2009) *Food Futures: Rethinking UK Strategy*, Chatham House, London

Boatman, N., Ramwell, C., Parry, H., Jones, N., Bishop, J., Gaskell, P., Short, C., Mills, J. and Dwyer, J. (2008) 'A review of environmental benefits supplied by agri-environmental schemes', Research Report for Land Use Policy Group, London

Bonn, A., Allott, T., Hubacek, K. and Stewart, J. (eds) (2009) *Drivers of Environmental Change in Uplands*, Routledge, London

Buller, H., Wilson, G. and Holl, A. (2000) *Agri-Environmental Policy in the European Union*, Ashgate, London

CEC (2003) Long range policy perspectives for sustainable agriculture, COM 2003/23, Brussels

Cooper, T. and Arblaster, K. (2008) 'Climate change and the rural environment in a European Union context: Implications for land use and land use policy', Research Report for Institute for European Environmental Policy, London

Cooper, T., Baldock, D., Hart, K. and Eaton, R. (2009) Analytical paper to inform LUPG's thinking on a vision for a future CAP, Unpublished Paper, IEEP, London

Defra (2007) *England Biodiversity Strategy – Towards Adaptation to Climate Change*, Final Report to Defra, Defra, London

Dibden, .J, Potter, C. and Cocklin, C. (2009) Contesting the neo-liberal project for agriculture: productivist and multifunctional trajectories in the European Union and Australia, *Journal of Rural Studies*, 25, pp299–308

Fish, R., Seymour, S., Watkins, C. and Steven, M. (eds) (2008) *Sustainable Farmland Management: Transdisciplinary Perspectives*, CABI, Wallingford

Hodge, I. and Reader, M. (2007) Maximising the provision of public goods from future agri-environmental schemes, Final Report to the Land Use Policy Group, Department of Land Economy, University of Cambridge

Hubacek, K., Dehnen-schmutz, K., Qasim, M. and Termansen, M. (2009) 'Description of the upland economy: areas of outstanding natural beauty and marginal economic performance', in Bonn, A., Allott, T., Hubacek, K. and Stewart, J. (eds) *Drivers of Environmental Change in Uplands*, Routledge, London

Lobley, M. and Potter, C. (2004) 'Agricultural change and restructuring: recent evidence from a survey of agricultural households in England', *Journal of Rural Studies*, vol 20, pp499–510

Losch, B. (2004) 'Debating the multifunctionality of agriculture: From trade negotiations to development policies by the South', *Journal of Agrarian Change*, vol 4, no 3, pp336–360

Ministry of Agriculture, Fisheries and Food (1995) *Agriculture: The Case For Reform*, MAFF, London

Natural England (2007) *Carbon Management by Land Managers*, Natural England, Peterborough

NFU/CLA/AIC (2007) *Part of the Solution: Climate Change, Agriculture and Land Management*, NFU, London

Potter, C. (1998) *Against the Grain: Agri-environmental Reform in the US and the EU*, CABI, Wallingford

Potter, C. and Lobley, M. (2004) 'Agricultural restructuring and state assistance: competing or complementary rural policy paradigms?' *Journal of Environmental Policy and Planning*, vol 7, pp34–50

Potter, C. and Tilzey, M. (2007) 'Agricultural multifunctionality, environmental sustainability and the WTO: Resistance or accommodation to the neoliberal project for agriculture?' *Geoforum*, vol 38, pp1290–1303

Rollett, A., Haines-Young, R., Potschin, M. and Kumar, P. (2008) 'Delivering environmental services through agri-environmental programmes: a scoping study', Research Report to Land Use Policy Group, London

Silcock, P. and Lovegrove, C. (2007) 'Retaining the environmental benefits of set aside: a policy options paper', Report prepared for LUPG, Cumulus Consultants, Broadway, UK

Smith, P., Martino, D. and Cai, Z. (2008) 'Greenhouse gas mitigation in agriculture', *Philosophical Transactions of the Royal Society, B*, vol 363, pp789–813

Swanwick, C. (2009) 'Landscape as an integrating framework for upland management', in Bonn, A., Allott, T., Hubacek, K. and Stewart, J. (eds) *Drivers of Environmental Change in Uplands*, Routledge, London

Von Braun, M. (2007) 'Agricultural market impacts of future growth in the production of biofuels', Report of a working party on agricultural policies and markets, OECD Committee for Agriculture, Paris

Woods, A. (2009) *Securing Integrated Land Management: Issues for Policy, Research and Rural Communities from the Relu Programme*, Relu, London

12

Regulating Land Use Technologies: How Does Government Juggle the Risks?

Claire A. Dunlop

Introduction

It is a fact of life that action to reduce one risk can lead to the unintended or unforeseen creation of another. When governments try to make things better, the consequences of their actions can conspire to make things worse; this is the law of unintended consequences. The complex technologies associated with new uses of land offer potentially huge benefits for the amelioration of the risks posed by climate change and energy insecurity. They also carry with them, however, their own significant social, environmental, economic, legal and human challenges. As a result, the interventions made by governments to harness these technical 'solutions' often create new sets of countervailing risks. By incentivizing the production of bioenergy crops for example, governments may reduce CO_2 emissions from transportation and diversify energy supplies. But such measures may shift risks elsewhere, for example increasing global food insecurity and deforestation. The extensive reach of government across many sectors and the fact that risks are experienced globally makes the central challenge faced by decision-makers double: they must choose the most appropriate policy instruments to manage the primary risks in question and also anticipate the secondary risks that may arise as the result of these choices.

Changes in land use are the result of decisions made by governments juggling multiple and often interdependent 'risk trade-offs' (Graham and Wiener, 1995)[1] that grow in number and change in form before their eyes. This chapter explores the interconnectedness of policy responses to risk (Graham and Wiener, 1995) and the risk trade-offs associated with land diversification by focusing upon two themes of government activity: how they *effect* policy action and how they can *detect* and resolve problems that arise from it (Hood, 1983). By outlining the pre-eminent policy measures and economic instruments deployed to harness new technologies associated with land-use diversification, some of the most significant risk trade-offs associated with these policy solutions to reduce CO_2 emissions and increase energy security are illuminated. Empirically a wide net is cast, with examples drawn from a variety of bioenergy technologies and countries – notably the UK, EU, USA, Canada, Brazil and Japan – and contextualized within the bewildering array of local, regional, supranational and international agreements and political pressures that are prevailing. The discussion is by no means exhaustive. Indeed, each of the examples could warrant a case study to themselves, and those readers looking for in-depth technical analysis should explore the individual chapters in this volume. Rather, the main aim is to provide some early reconnaissance of the risk trade-offs that are dominating emerging technologies and initial policy responses to climate change and energy security that implicate the land. What are the countervailing risks of policies that treat the land as the potential receptacle for a wide range of technological fixes? Who are the winners and losers in the trade-offs being made? Which risks are seen as more manageable and acceptable than others? How aware are decision-makers of the need to anticipate perverse effects of regulation and weigh up the risks associated with them? How can governments minimize overall risk?

The risk trade-offs that decision-makers encounter in attempting to exploit the potential of the land are pervasive and complex. This survey illustrates that, while technologies that draw upon the land carry their own risks for the environment and public, these are exacerbated by the variety of regulatory strategies deployed and have resulted in new risks to ways of life, some of which exert a disproportionate effect on vulnerable populations. Not all unintended risks are of equal gravity, and discussion here is limited to risks that are considered to be significant relative to the main aims of reducing CO_2 emissions and increasing energy security. We should be clear; the risks occasioned by particular policy measures involving the diversification of land use and new technologies designed to counter CO_2 emissions and increase energy self-sufficiency are not necessarily unhelpful in themselves but rather need to be analysed and balanced on a case-by-case basis. The impact of a countervailing risk is mediated by temporal and spatial context: what is a hazard in one particular time or place may not be seen in the same way in another. By learning from experts, citizens and experience, decision-makers are better placed to anticipate trade-offs coming down the line and identify so-called 'risk-superior' (Graham and Wiener, 1995) options that reduce overall risk. This is important because, while the propagation of complex countervailing risks appears to be a pervasive feature of regulations devised to exploit the potential of the land, we need not be entirely fatalistic about governments' ability to anticipate unintended consequences and knock-on effects of policy action. With this in mind, the second part of the chapter addresses the other side of the coin – how governments detect risk trade-offs and learn to resolve them. Again, these are illustrated using examples of practice from around the world.

The discussion is timely. The best way to exploit land resources and to mitigate the risks associated with climate change and insecure energy supplies are matters of global concern and contention. In particular, there is a growing awareness in both government and society that policy responses can be counterproductive and are prone to being overtaken by rapidly changing technology. Such policy limitations risk the original goals becoming displaced, and opportunities from new technologies may also be lost as a result of uncertainty and incomplete risk governance. Moreover, where society is not getting the level of protection that it expects, the credibility and legitimacy of decision-makers who are developing and implementing policy and of scientists and stakeholders advising government on future uses of land may be under threat, giving way to cynicism or apathy (Graham and Wiener, 1995).

Characterizing risk trade-offs

The intersection of two global risks – climate change and energy insecurity – sees environment and energy policy and interests intertwined at the top of political agendas around the world simultaneously, possibly for the first time ever. High stakes in ecological, economic and security terms have given rise to an ever increasing array of technological innovations and policy instruments with the potential to fundamentally change land use patterns across the world. Embedded within these solutions is a complex range of countervailing risks that pose technically and politically challenging risk trade-offs. A single-policy intervention may simultaneously give rise to multiple and often unrelated countervailing risks which leave decision-makers with risk trade-offs that are discrete and easy to discern if not easy to weigh. Countervailing risks may also form part of a domino effect, where one unintended consequence sets off chain reactions propagating new, and drawing in additional, countervailing risks. The complex web of risk trade-offs that results is far harder to negotiate. It is worth pointing out that while the idea of countervailing risks is associated often with negative side-effects, unanticipated but welcome consequences can also emerge from seemingly unrelated policy action. Analysis of such 'ancillary benefits' (Rascoff and Revesz, 2002) will also be included in the discussion that follows.

Coherent analysis of the challenges decision-makers face requires a clear characterization of the distinguishing features of these trade-offs. Risk analysts explore what is at stake through two questions (Graham and Wiener, 1995).[2] The first concerns whether or not the adverse outcomes are of a different type to those that the policy aims to eliminate. When compared with the risk being regulated or 'target risk', has the nature of the adverse outcome changed or remained the same? The second considers the population affected by the countervailing risk. When compared with the target risk, does the countervailing risk

Table 12.1 *Graham and Weiner's typology of risk trade-offs*

Effects of countervailing risk compared with target risk	Countervailing risk compared with target risk		
		Same type	Different type
	Same population	Risk offset	Risk substitution
	Different population	Risk transfer	Risk transformation

Source: Graham and Weiner, 1995, Table 1.2, p22

affect the same or a different population? The resulting four categories – offset, substitution, transfer and transformation – are not mutually exclusive. A single-policy intervention can give rise to multiple trade-off types.

What is the problem? How do governments define the target risks?

Governments have a variety of policy instruments or arrows from which to select to reach their policy targets (Black, 1995; Hood, 1983; Eliadis et al, 2005). To understand what governments are trying to achieve and why one arrow has been chosen to address a risk as opposed to another we must first understand how the cause of hazard is conceptualized. Problem framing determines who the key participants will be on an issue and sets the boundaries to conflict (Schön and Rein, 1994). The dual challenges of climate change and energy insecurity are recognized by governments across the world as externalities rooted in a single and global market failure: high carbon dependence (Stern, 2007). A low-carbon economy with a mix of energy types and sources has emerged as the pre-eminent paradigm informing the selection of policy arrows. Government interventions to address environmental, economic and geopolitical distortions and inefficiencies created by carbon dependence revolve around regulatory controls on CO_2 emissions and investment incentivizing low-carbon technologies that may implicate the land. We cannot assume, however, that actions work in the way decision-makers intend; just as markets fail so do governments (Graham and Wiener, 1995; Wolf, 1988). Harnessing the potential of the land to help correct the market is a complex task involving decision-makers from multiple sectors – most obviously energy, environment, agriculture, trade, development, R&D, economy and transport. The crowded nature of these policy arenas presents the considerable challenges of designing policies that are mutually supportive across sectors, anticipating the potential risk trade-offs that may occur and resolving them in ways which accommodate different geographical contexts and balance competing interests.

Effecting risk reduction and the countervailing risks of policy instruments driving new uses of land

Regardless of the specific policy instrument selected by decision-makers, there are countervailing risks associated with changing old technologies and encouraging new ones. Such iatrogenic potential – where a remedy threatens to be worse than the malady – is rooted in the inherent uncertainty of moving innovations from small-scale, controlled environments to large-scale, where they must function technically, economically and environmentally. That said, decision-makers still exercise a good deal of influence over how risks play out. It is policy instruments that determine the ways in which and extent to which iatrogenesis occurs. Where markets are encouraged or even created by regulatory measures, the action of governments help to determine levels of certainty and confidence in industry. The same is true of institutional structures. Innovations often raise legal as well as technical dilemmas. The manner in which decision-makers negotiate these will bear heavily upon the character of risk associated with innovation.

When it comes to complex policy problems, silver bullets and single global approaches are rarely found. Policy instruments are rarely used in isolation. Rather, the overall picture of governance reflects a mix of measures (Howlett, 2005). Indeed, since the Kyoto Protocol's agreement in 1997, more than 1000 carbon abatement policies have been implemented worldwide (Simeonova, 2007). Changes in behaviour are sought through a cocktail of incentives and capacities. This chapter discusses the risk shifts and trade-offs implicating the land that have been facilitated by four policy interventions that dominate governments' responses: target setting, price and rights-based trading schemes, subsidies to encourage market penetration, and direct intervention through grants for research and development (R&D).

Setting targets

Setting quantitative targets on emission reductions and energy self-sufficiency, and achieving regional and international agreement on how to reach them have emerged as a central policy agenda driving global changes in land use. Such targets provide the overarching framework within which more detailed strategies and diverse policy tools are developed. The use of fiscal tools to create a carbon price, deploy subsidies and give R&D grants are all driven by the initial targets set at local, national and international levels. Before exploring the countervailing risks that arise from these three policy choices that flow from targets, we explore the risk trade-offs that can arise as the result of how targets themselves are designed. Specifically, the issue of goal displacement is scrutinized,; where the targets that are set under conditions of uncertainty fail to anticipate future developments perverting the overall policy intention.

To explain how goals become displaced and the original risks of climate change and energy security are offset, transferred and transformed, the framing power of targets must be understood. Targets set the tone for policy action; they embody the specific behaviour governments want to address and the level of commitment that a government attaches to a problem. The more ambitious the targets are, the more effective institutional arrangements are required to ensure both the successful achievement of the targets themselves and the link-up with the delivery of overall policy goals. Where the problems are both pressing and characterized by uncertainty and technical novelty, as is the case here, the targets set may be overly ambitious and lacking in appraisal as to how they are to be achieved. In such situations, the technological solutions favoured by those who must comply may meet the target but ultimately pervert the overall goal.

The targets set around the world for biofuel use encapsulate the problems decision-makers face in trying to establish regulatory regimes in areas marked by technical uncertainty, and the rapid change in land used for energy crops represents an archetypal example of goal displacement. Nearly 20 countries have mandates for blending biofuels into vehicle fuels (Monfort, 2008). The EU took the lead in 2003 by adopting its first biofuels directive setting targets on biofuel use in petrol and diesel at 5.75 per cent by 2010 (Council of Ministers, 2003). Recent proposals extend this to 10 per cent by 2020 (European Commission, 2008). In its 2007 Energy Bill, the USA announced plans to cut gasoline use by 20 per cent over the coming decade. Three-quarters of this is to be achieved by using 136 billion litres of renewable and alternative fuels by 2022. Similarly, Japan and China have set targets of 6 billion litres per year by 2030 and 13 billion litres of ethanol by 2020 respectively.

Why have such ambitious targets formed around biofuels in particular? When faced with challenges as complex as climate change and energy insecurity, the search for 'like-for-like' replacements is a common default position adopted by governments to address risk reduction. The aim here is to select a substitute compatible with current technology that, on the face of it, will cause least disruption to the status quo (Gray and Graham, 1995). As one of the first technologies to emerge as a viable alternative to fossil fuels that could be produced on a large-scale, first-generation biofuels have been embraced as a sustainable improvement on oil. The absence of any fundamental re-thinking of policy problem – in this case to focus on the demand side as opposed to supply side of the contribution of transportation to the target risks – means that such substitute technologies are liable to produce risk offsets. Encapsulated in the idea of a 'rebound effect' (Schipper, 2000; UKERC, 2007),[3] the environmental savings made by biofuels may be undermined by the systemic responses and inefficiencies that accompany their production. For example, millions of hectares of peatland in Southeast Asia have been drained for oil palm crops. As carbon sinks, it is estimated that 40–50 billion tons of carbon is stored in this watery land and its clearance is calculated to be equivalent to 10 per cent of global emissions from fossil fuels (Wetlands International, 2007).

Similarly, the need for fertilizers, increased water quantities, the importance of fossil fuels in harvesting energy crops, and release of carbon in clearing the land for crops reduce the benefits of biofuels in reducing CO_2 emissions. Biofuel crops also represent a major risk offset where changes in land use have been facilitated by deforestation that increases CO_2 emissions. For example, the UN estimates that if current rates of production of palm oil – the world's main biofuel – continue, virtually all Indonesian and Malaysian rainforests will be destroyed by 2022 (Uryu et al, 2008). This offsetting of risk – the same risk of increased CO_2 emissions from a different source – has been enough for some experts, such as Dr Richard Pike (2008), Chief Executive of the Royal Society of Chemistry, to reject first-generation biofuels as a technological dead end. There are those who take the more benign view, arguing that while offsetting risk in these ways is clearly not desirable in the long term, in the early phases of policy when the goal is to engender behavioural change, such targets alter mind-sets about dependencies on fossil fuels in transport. In this way, first-generation biofuels may represent a necessary but interim 'transitional technology' whose rebound effects will be mitigated later, as argued by bioenergy researchers at Imperial College (Templer and Woods, 2008).

The unintended and indirect consequences of mandatory targets for biofuels in environmental terms are also now becoming clear. Making the adoption of a technology obligatory distorts markets by creating certainty for biofuel producers but transferring risk to other groups. Biofuel targets have stunted other forms of land diversification, with biofuel technology and the well-organized industry behind it becoming 'locked in' at the expense of other, potentially superior innovations. Most notably, an argument often championed by the carbon capture and storage (CCS) and anaerobic digestion industries is that 'what gets measures gets done' and setting targets that support like-for-like technologies does little to encourage investment in technologies that are less well known and arguably riskier but address emissions more directly.

In addition to these risk offsets, the external impact of biofuels targets also leads to risk being transformed and transferred. Forests being felled and replaced by palm oil crops increase CO_2 emissions but also displace forest animals (most notably the already endangered orang-utan; UNEP and UNESCO, 2007) and take energy away from ecosystems through soil erosion and nutrient depletion (Haberl et al, 2007). Here both the adverse outcome and population affected are changed and the nature of risk transformed from the reduction of CO_2 emissions produced by citizens in the industrialized North to threatening future land use of those in the rural South. This human dimension has been brought into sharp focus by the UN. Victoria Tauli-Corpuz (2007), chair of The UN Permanent Forum on Indigenous Issues, warns that land displacement as the result of deforestation could create up to 60 million 'biofuel refugees' – 5 million in Kalimantan, Indonesia, alone – who will either migrate to the mega-cities of the global South or grind out a subsistence living in the rural communities that remain.

Driving innovation through carbon pricing and trading in pollution rights

Lasting innovations are rarely realized by a simple investment by government in R&D) into new technologies developed in universities or industry (Freeman and Soete, 1997; OECD, 2002).[4] Rather, governments aiming to achieve socially defined objectives encourage market opportunities for new technologies by combining *rights-based measures* that give actors the right to emit CO_2 up to a certain limit, *price mechanisms* that allow these rights to be traded and *taxes* to persuade heavy polluters to reduce their emissions (Beder, 2001; Hahn and Hester, 1989). Setting a price for carbon has become established as one of the main ways to energize these innovation networks by incentivizing efficient energy use and stimulating investment in technologies that aim to manage CO_2 or lower it. Specifically, in the past two decades, industrialized countries have introduced emissions-pricing systems – notably CO_2 taxes and tradable allowances – as the main policy mechanisms through which to meet their emissions targets. Commercial energy users and suppliers are levied for the amount of fossil fuels they use and produce through a cocktail of schemes administered at local, regional, national, transnational and international levels. For example, since 1991 eight European countries have introduced CO_2-based taxes – Denmark, Finland, Germany, Italy, the Netherlands, Norway, Sweden and the UK, and since 2005 all energy-intensive firms in the EU have been able to buy and sell CO_2 permits in the most ambitious attempt to date to establish an international carbon market – the EU Emissions Trading Scheme (ETS) (Anderson et al, 2005).

Similar initiatives are under way in North America: one joining eight US states with three Canadian provinces and another of ten northeastern and mid-Atlantic States.[5] Indeed, according to a recent survey for the Point Carbon consultancy, the expectation among the majority of traders (73 per cent of a survey of 3700) is that a global price for carbon will have emerged by 2020 (Point Carbon, 2008).[6]

The design of these markets underlines the view of climate change and energy insecurity as negative externalities, unwelcome costs that have been imposed on societies by industry users and energy suppliers. Through emissions rights and carbon pricing these costs are reflected in the decisions taken by these firms; with the rate of CO_2-based taxes usually dependent upon the CO_2 content of fossil fuels being used and/or produced. The primary intention of these levies is to change behaviour rather than raise revenues and as such the success of these schemes is measured in the extent to which revenue falls. In the most basic schemes, firms and suppliers that increase their energy efficiency reduce their tax burden. Additional incentives are common where companies receive rebates for reaching their efficiency targets or can trade unused allowances in carbon markets. For example, the Climate Change Agreements (CCA) introduced to complement the UK Climate Change Levy (CCL) reward success with up to 80 per cent rebates.

The logic of these price mechanisms is to link abatement with innovation. Behavioural change in energy consumption among firms alters the signal communicated to energy providers and potential investors, stimulating investment in new technologies that among other things implicate the land. Market approaches such as this are associated with the idea of an 'option value' (Gross and Foxon, 2003) – they create choices between different technologies from which investors can select. For this value to be realized, however, a threshold effect must be achieved whereby innovations enjoy enough investment to become commercially viable.

While increased investment in alternative energy sources mean that the dual targets of reducing CO_2 emissions and mixing the balance of nation states' energy supplies are addressed by combinations of rights and price-based measures, there is significant variation in the type of land-based technologies that are beginning to benefit. While energy crops for biofuels are the recipient of significant levels of private investment, this is not the case for technologies whose commercial viability is yet to be established (UNEP, 2008). The key issue here is the strength of the price signal sent out to firms. By setting carbon prices through fixed-period schemes such as the CCL and ETS, the stable incentives required to encourage large-scale investment in risky technologies are not generated. A type of risk offset may result where lack of investment in emerging technologies turns them into missed opportunities and limits future energy-generation options. Such restriction on future solutions could jeopardize countries' abilities to ameliorate CO_2 emissions and create an optimal energy supply mix. Nowhere is this scenario more marked than in relation to CCS. Though this is an 'end of pipe' technology that promises a smart way of dealing with CO_2, CCS is a costly technology, is yet to be proven and will push up the cost of electricity.[7] It also requires significant re-thinking of key parts of the production and regulatory processes. Questions concerning what types of land are suited to CCS and what are acceptable leakage rates can only be answered through trial-and-error research.

Slow progress in securing investment has also encouraged the same inertia in terms of setting the framework for regulation. Filling the considerable legal vacuum that exists around CCS is pressing, however. In particular, who should be liable for leakage is arguably the thorniest issue yet to be addressed by governments. As carbon will be sequestered over decades by multiple actors, it will not always be possible to identify a single actor responsible for injecting CO_2 into a site. And pinpointing the cause of a leak – for example injection method or site stewardship – is unlikely to be straightforward. These concerns have led to suggestions from industry, for example Dr Mike Farley (2008), Director of Technology Policy Liaison for Doosan Babcock, that long-term financial liability should be assumed by national governments, but a consensus on this solution is yet to emerge (International Risk Governance Council, 2008). Thus while it is expected to take more than a decade to assess the efficiency of a CCS plant, it may take just as long for the required investment to be secured and for the institutional rules of the game to be established.

For commercial viability, a self-reinforcing innovation system needs to exist around high-risk technologies such as CCS. This requires significant long-term financial support. For example, it has been estimated that commercial CCS projects require a sustained price of around €30 t^{-1} of CO_2 emissions avoided. However, the EU phase one of trading has seen market prices peaking at €30 for only one month and falling as low as €0.10 t^{-1} (Kema, 2007, in IRGC, 2008). Even if the costs of emitting carbon stabilizes at the level analysts estimate is required to incentivize investment in emerging technologies such as CCS, this will do little for those who want to invest early (Forward, 2008), and the fact that such a price is set for a fixed and relatively short period of time makes forecasting long-term revenues impossible. Thus, while the emissions trading scheme has been welcomed across Europe's energy and industry sectors, the absence of stronger long-term pricing signals or commitments by governments to underwrite a price jeopardizes large-scale investment.

Experts from the worlds of both science and industry agree that CCS could become a missed opportunity, with the uncertainty generated by short-term price mechanisms creating an impermeable blockage in the innovation system (Stromberg and Metz, 2008).[8] The absence of legislation requiring new coal-fired power stations to be CCS compatible in the world's largest energy producers[9] and recent withdrawals of governments from CCS projects will do little to embolden the private sector to move beyond R&D ventures.[10] In one of the most recent high-profile cases, Saskatchewan's planned development of CCS has been characterized by protracted uncertainty, with the public electricity company SaskPower undermined by the new federal government's commitment to nuclear power, which is viewed as a better investment than CCS (Salaff, 2008). The surge of interest in nuclear power underlines the competitive market within which emerging technologies must attract investment. It also suggests that societies are not taking into account the risks of missing the CCS opportunity and are reducing technology options for tackling climate change and energy security in the future.

The lack of full commercial exploitation of land-based technologies such as CCS may, however, also transfer risks to developing countries whose ability to harness land resources to address climate change and energy insecurity is heavily reliant upon the transfer of knowledge from the industrialized world. As climate change economist Sir Nicholas Stern observes: 'unless the rich world demonstrates, and quickly, that CCS works, developing countries cannot be expected to commit to this technology' (Stern, 2007).

Subsidies for land diversification

For innovation systems to operate successfully, actors must be incentivized to alter their behaviour in a way that facilitates adoption. Subsidies are a key way in which governments attempt to remove informational, financial and cultural barriers to behavioural change and defray the costs of mandatory standards. Those who manage and tend the land are encouraged to respond to the changing energy market and meet emissions targets through direct economic support. In particular, subsidies to encourage the growth of energy crops for biofuels and combined heat and power generation, and incentives for anaerobic digestion schemes and on-shore wind farms are commonplace across both industrialized and developing worlds. Early experiences confirm what policy analysts have long known: the success of subsidies is a function of the context within which they are applied and the technology to which they are applied.

The operation of the UK's RO scheme illustrates the importance of the legal and administrative context for the success of subsidies. The RO scheme obliges electricity providers to source an increasing proportion of their electricity from land-based renewables. Policy tools are not deployed in a vacuum, however, and must compliment the existing rules and regulations on the statute books. There is concern that problems in obtaining planning permission (Upreti 2004) and access to the grid for projects involving technologies with which regulators and citizens are unfamiliar – notably anaerobic digestion – have led to a risk offset (Defra, 2007). Consumers are effectively paying for projects regardless of whether or not they help mitigate CO_2 emissions (Ofgem, 2007).

The suitability of a particular technology for subsidies also has implications for the intensity of the countervailing risks that arise. This is perhaps exemplified by the contrasting impacts of first-generation biofuel crops in Brazil and the USA (Mol, 2007). From the 1970s–1990s, the production of sugar cane for the production of ethanol was subsidized. This transformed Brazil's agricultural economy, boosting employment and reducing both CO_2 emissions (with more than 40 per cent light vehicles' fuel consumption from ethanol) and oil imports. In spite of these benefits, there is famously another side to the biofuels story, where the subsidized displacement of traditional agricultural products by biofuel crops has produced only modest reductions in CO_2 emissions. Keen to emulate Brazil's success, the USA subsidizes maize production for ethanol at the cost of US$5.5 billion to US$7.3 billion a year through over 200 support measures (World, Bank, 2008).[11] Significant direct and indirect side-effects have resulted from this subsidy regime.

In terms of direct trade-offs, the cost-effectiveness of such subsidies in both cases is low and use of land inefficient – when the costs of motorists and taxpayers are taken together, biofuels costs are almost double those of fossil fuels (Doornbosch and Steenblik, 2007; Global Subsidies Institute, 2007; IEA, 2006). Moreover, the conversion of maize to ethanol is also significantly less energy-efficient than using sugar cane. Biofuels from sugar in Brazil are estimated to have 86 per cent CO_2 emissions reduction as compared to 10–30 per cent of maize-based biofuels produced in the USA (Macedo et al, 2004; Searchinger et al, 2008). Significant research published in *Science* estimates that once land-use changes are factored in, instead of producing a 20 per cent saving, maize-based ethanol nearly doubles CO_2 over 30 years (Searchinger et al, 2008).

The USA's drive for ethanol production combined with income growth in industrializing economies, widespread drought (attributed by many to climate change), an increased population and changing patterns of food consumption, has resulted in emissions and energy insecurity effectively traded for increased food prices globally and food insecurity in world's poorest regions (Evans, 2008; FAO, 2008; World Bank, 2008). In 2006 the US Department of Agriculture (USDA) estimated that around 20 per cent of the total corn harvest in the USA was used by the biofuel industry – an increase of 63 per cent since 2004. This figure is expected to increase to 30 per cent by 2009. The result was an increase in wheat prices by 130 per cent between March 2007 and 2008 (IAASTD, 2008) and 60 per cent in maize prices from 2005 to 2007. The prices of feedstock and fertilizers have increased accordingly. The magnitude of the competition faced by dairy and cattle farmers in the feed markets and food producers and consumers across the world is only just becoming clear. While these risk substitutes affect the world population, voice was given to these concerns most forcibly by those most adversely affected – the poor of the developing world – when in 2008 demonstrations in Egypt, Mexico and Yemen and almost a dozen other countries became increasingly violent.

Purist economists would argue that such market distortions are temporary – higher prices result in more planting and increased production. Indeed, this thesis is borne out with agricultural production projections for 2008/2009, indicating that farmers are cutting cotton and sugar production in favour of wheat, maize and soya. Notably, wheat production is expected to increase by 16 per cent in the USA, 17 per cent in the EU and 25 per cent in Canada (Robinson, 2008). The argument that such risks could be neutralized by increased agricultural productivity may stack up in industrialized and industrializing countries. In those locations most at risk of food shortages, however, it requires greater qualification. For example, with 63 per cent of all plant production in south Asia used or destroyed by humans, it is unclear how developing countries can boost productivity and re-calibrate the market (Haberl et al, 2007). And so, while bioenergy crops have the potential to reduce the price of oil, which brings obvious benefits to poorer countries, if these crops occupy the lion's share of the best land in these countries the resulting increase in global food prices can be expected to induce a similar increase in the cost of food aid (UN-Energy, 2007).

There are also countervailing risks associated with subsidy-induced land-use change that target farmers more specifically. The costs of changing land use and generating abatement are straightforward enough – these involve farmers calculating the opportunity costs of using their land to sequester carbon or reduce emissions in some way rather than producing more traditional commodities. The transaction costs involved in identifying a market for their goods and enforcing contracts are more problematic, however (Lipper and Cavatassi, 2004). The markets for most energy crops and biogas are emerging ones and often highly regional, meaning that opting to diversify in this way involves a higher level of risk than more conventional changes in land use, thereby compounding transactions costs further. This increased burden is rarely reflected in the subsidy levels (Lipper and Cavatassi, 2004). Where it is not, risk transformation is possible where the adverse outcome of increased CO_2 emissions is replaced by significant economic risks that directly affect small-scale landowners and farmers who are already financially vulnerable. Early indications of the new bio-economy that has begun to emerge as a result of the diversification of crop investment confirm that subsidies have not prevented smaller farmers in both the developed North and developing South suffering as large companies determine and control the prices that are paid for energy crops (UN-Energy, 2007).

With more than 70 per cent of the world's poor living in rural areas, policy action to address the challenges of climate-change abatement must also alleviate poverty (IFAD, 2001). International agencies use subsidies to unlock financial and social barriers that inhibit land-use change (Cacho et al, 2002; Olsson and Ärdo, 2002; Smith and Scherr, 2002) and enable the creation of synergies between the diversification of land to increase income and carbon sequestration, principally through cropland management, reforestation and revegetation (UNFCC, 2008; World Bank, 2002). Such double returns make subsidies relatively cost-effective (Smith and Scherr, 2002). Research illustrates that the key to success is to match subsidies to technologies that fit the local context and on-the-ground support provided by NGOs and aid agencies (Scott, 2006; del Rio and Burguillo, 2008). For example, developing CCS units may be problematic in areas that rely on off-grid energy. In such contexts, subsidizing the growth of biomass for heat and power generation would represent a more efficient low-emissions boost to rural communities and economies.

Subsidizing land diversification has a further dimension. Addressing climate change and energy insecurity by encouraging farmers to focus on the carbon output of their land may also imply an additional form of risk transformation where farmers lose a sense of intimacy with the land. For some, the contrast between economic and personal relationships is a stark one; these represent 'hostile worlds' governed by incompatible logics (Zelizer, 2005). In this context, encouraging farmers to react to market signals and engage in carbon management is detrimental to sustainable land stewardship, destabilizes biodiversity and erodes a sense of intimacy and kinship with the land.[12]

R&D grants

Large-scale technological shifts cannot be stimulated by indirect regulation through subsidies or market competition alone. Such innovation requires direct public intervention. The uncertainty that surrounds scientific knowledge of inventions and the high degree of financial commitment and risk involved in piloting and commercializing large-scale technological innovations means private investment in R&D is often far lower than social objectives require (Arrow, 1962; Nelson, 1959; Socorro, 2003). This gap between investor confidence and socio-political targets is particularly marked where innovations seek to combine environmental and economic goals (Klassen and Whybark, 1999). Direct public investment in the form of R&D grants is essential to avoid this potential market failure and the best way to 'de-risk' technology. Neoclassical theory tells us that governments aim to secure the optimal allocation of resources required for an innovation to succeed. However, economics textbooks tell us little about how well-equipped government may be to manage highly specialized projects involving a network of actors with their own agendas. Large-scale R&D projects enable government to pool finances with multiple actors with the aim of creating something greater than the sum of their parts. Innovation by consortia however, implies a whole range of organizational and logistical challenges. The most pressing is to ensure that all parties possess a shared understanding of what constitutes a successful outcome in the initiative and that government monitors the allocation and use of funds.

How 'success' is defined in a large-scale technological initiative depends upon why a project has been funded in the first place. With R&D monies, governments generally aim to achieve one of two goals: either to try out a risky technology or to showcase a technology to secure further investment and achieve market penetration. Where an unreliable technology is simply to be tried out any failure of that innovation does not equate with the failure of the policy project itself. Indeed knowledge of what does not work can be seen as an ancillary benefit. Where the main goal is to prove to the market that an innovation works technically and is financially viable on a large scale, however, the failure of an initiative can give rise to significant and negative countervailing risks. Most notably, not only may the target risk remain undiminished, it may be compounded by reduced confidence within government, industry, stakeholders and the wider public, offsetting the risk further still.

Whether a countervailing risk is seen as an acceptable cost or an unacceptable burden often lies in the eye of the beholder and has significant implications for grant-giving activities. Where there is no shared vision of the ends to be achieved, the acceptability of risk trade-offs and side effects are subjective and may become sites of conflict between partners. An early partnership between the EU Commission, UK government and a biomass investment consortium is illustrative. Despite a 20 per cent reduction in energy-related R&D in the last two decades (Stern and HM Treasury, 2006), significant funding opportunities exist for biomass projects. While they involve higher capital costs than fossil fuel technologies (in some cases by a power of five), they also carry the promise of almost net carbon neutral heat and power generation (Thornley, 2006) and are potentially perhaps less socially controversial than other land-based renewables (see Upreti, 2004; Upreti and van der Horst, 2004; Upsham and Shackley, 2006a,b). In 1994 it was decided in the EU that all non-fossil fuel-funded projects involving energy crops should use the emerging and costly new conversion technologies of gasification and pyrolysis.[13] The main flagship demonstration plant – the £30 million project Arable Biomass Renewable Energy (ARBRE) in Yorkshire – generated electricity for only eight days before succumbing to technical problems associated with gasification and bankruptcy of the contractor managing the project (Piterou et al, 2008). In an effort to ensure that those farmers already contracted to grow willow were not left without a market for their crops and to boost confidence in biomass, the Government relaxed its policy, allowing the co-firing of coal and wood dust allowed in conventional power stations,[14] and left the conventional energy industry importing wood sourced from deforestation in countries such as Latvia and Indonesia, the transport of which produces carbon emissions that are estimated to be three times that of UK biomass sources (Lindegaard, 2005).

The EU's selection of a risky technology indicates that its goal was for ARBRE to be a test bed for gasification. The UK Government, however, provided the project with a 15-year contract from the NFFO scheme deployed for innovations that were expected to make it to market (Piterou et al, 2008). This clash of goals was never resolved and resulted in the project being judged by standards, in the UK at least, which it could never have hoped to satisfy. The ARBRE case also underlines the problems governments may have in keeping a grip on their R&D initiatives. Grants enable government, to increase the stock of knowledge in an area and draw on specialist capacity that they do not have. But the highly specialized nature of R&D projects means that, after providing the funds, governments often takes an arm's-length approach to management, and progress monitoring is often cursory. The result can be a pronounced asymmetry of information, where decision-makers find themselves out of touch with events on the ground and are often not alerted about problems until it is too late. Certainly in the ARBRE case, the main protagonists involved agree that the lack of structured oversight obscured the most fundamental problems that led to the project's collapse (Piterou et al, 2008).

High-profile cases such as ARBRE represent important learning experiences for those involved and the trade-offs can be rationalized as part of the process of creative destruction that fuels successful innovation (Schumpeter, 1975). But where negative perceptions become attached to a technology, the ripples often extend well beyond any specific case and time, giving rise to non-technical barriers to future innovation which may offset the original risk and 'lock out' emerging technologies (IEA, 2005). High-profile failures do little to encourage farmers to change their land use or to convince citizens and local government, already prone to 'nimbyism', when faced with applications to host large-scale plants, that such projects will be successful in the long term. In the ARBRE case the efforts of government and industry to demonstrate the benefits of a technology were undermined by this lack of confidence. Specifically, there has been concern that the development of biomass for heat generation may have been slowed as a result of the lack of obvious success with electricity-focused initiatives (UK Biomass Task Force, 2005).

Detecting 'risk-superior' policy options: learning from experts, citizens and experience

The propagation of complex countervailing risks is a pervasive feature of decision-making in general and, as this account demonstrates, of the policy tools deployed to exploit the potential of the land in particular. The emergence of high-profile risk trade-offs however, should not be mistaken as an indication of policy failure. Not all side-effects can be anticipated prior to policy action, particularly where the risks being regulated are complex and pressing and the technological solutions novel. Though the preceding discussion has highlighted the fallibility of policy action, this is no tract against Government intervention. The risks occasioned by particular policy measures involving the diversification of land use and new technologies designed to counter CO_2 emissions are suggestive of incomplete risk governance (Graham and Wiener, 1995). However, it is through learning that corrective action can be taken and trade-offs coming down the line anticipated more readily.

How can decision-makers know which trade-offs deserve close attention? How might they decide which ones to ignore? In short, from what and whom can decision-makers learn? Examples of how governments are beginning to meet the countervailing risks that swirl around land-diversification technologies and fashion risk-superior options that reduce overall risks brings into relief three learning types that help complete risk governance: epistemic learning, social learning and institutional learning (Dunlop and James, 2007). These are discussed in turn.

Epistemic learning

Unintended side-effects often occur because, like all humans, decision-makers have limited cognitive capacity and difficulty in attending to more than one thing at a time (Simon, 1957; Jones, 1994). And so, while the choice of policy instrument is motivated by self-interest, decision-makers' bounded rationality means that their sense of what is politically rational and efficient is not always accurate. As a result they may resort to received wisdom, ideology and inaccurate 'heuristics' to guide policy design (Wiener, 1995, p233; Kahneman and Tversky, 1979). If they are to anticipate the consequences of policy action more readily and provide information to reduce uncertainty in the marketplace, decision-makers must find ways to develop more accurate policy heuristics. This is where evidence and experts come in. By engaging in analysis and with analysts – epistemic learning – decision-makers are better placed to develop the peripheral vision they require.

The information production at the heart of epistemic learning carries significant material costs. There are also temporal barriers. The time constraints that decision-makers must negotiate and those that govern knowledge creation are very different. While decision-makers face political pressure to act now, research into the different scenarios that may flow from policy action can stretch out indefinitely – there is always something more to be learned and taken into account about a policy problem and the proposed solutions. Mining the analytical capacity already established by industrialized governments for which the costs have already been sunk – domestic research services and international specialist agencies for example – is a way to mitigate these barriers. International players such as the International Energy Agency (IEA) already show early signs of becoming an epistemic hub around which national regulators can converge for substantive advice as well as proactive lobbyists for emerging technologies.

Governments are also making significant new investment into learning about the pros and cons of technological responses to climate change and energy insecurity. Much of this takes the form of specialist domestic and transnational agencies and institutions that have been established to commission research. Of course having the knowledge is one thing; using it is quite another. The example of biofuels and the UK's RTFO illustrates that taking an evidence-based approach to weighing the pros and cons of competing trade-offs is a matter of political will as opposed to lack of information. The defence that such risk trade-offs were inadvertent at the time initial targets were set is undermined by the fact that decision-makers who are now armed with a growing body of evidence of the additional adverse outcomes that are arising have been slow to reform their approaches or build headroom into policy targets to allow for the rebound effects. Most notably, the formulation of sustainability clauses to limit the cultivation of 'bad' biofuels and require producers to prove that their biofuels are not the product of damaging agricultural practices (Council of Ministers, 2008) is not expected to come into force until 2011 at the earliest. Environmentalists and scientists argue that in the meantime irreversible damage may be done (Monbiot, 2005; Barlow, 2007). The implication of transferring and transforming risks in the biofuels case is that in the short term it is acceptable for some groups to bear more risk than others. Such choices are inherently political but to be credible and legitimate they require evidence that the benefits of pressing ahead without sustainability criteria justify the costs implied by the use of unsustainable biofuels. Despite the growing evidence based on the subject, such cost-benefit analyses are yet to be conducted by Government.

Thus substantive knowledge creation is not an end in itself. For real epistemic learning to take place and for evidence to have policy relevance, decision-makers must ensure that it is used to inform the analysis of the policy choices under consideration. Industrialized countries are certainly well placed to make this happen using the analytical tools – notably impact assessment and risk analysis – that have become commonplace in the last two decades. Thus far, however, very little commitment has been shown to risk-based philosophy in relation to the countervailing risks that flow from policy-tool selection that aims to harness the potential of the land.

Social learning

Innovation requires invention and capital (Schumpeter, 1975). As the parties who fund and bear the consequences of policy action and, by their reaction to a policy intervention, are the source of many countervailing risks, citizens are often best placed to remind governments that innovation cannot be at any cost and that social capital is as essential to making inventions work as financial. While decision-makers acknowledge that 'public acceptance is an issue that is almost as dire as financing and in relation to technologies such as CCS, will become more and more critical as we try to implement the projects' (Panek, 2008, p21),[15] it is marked, however, how rarely public voices have featured in policy tools selected by governments. Lay people and experts process risk in very different ways (Slovic, 2000). In order to anticipate how citizens will respond to regulation, it is essential that decision-makers understand that the political environment within which they operate has a strong 'public' dimension

(May, 1991) and accept the need to engage in reflexive dialogue or social learning (Sanderson, 2002).

A key challenge of social learning concerns timing – when should dialogue begin and when should it end? Consultation processes that invite citizens to comment on policy proposals are the pre-eminent way in which decision-makers draw lessons from society. Such an approach can only go so far however. Often occurring 'downstream' in the policy process – after policy goals have been settled and technologies' trajectories been 'black boxed' (Collingridge, 1980) – consultation exercises can limit citizens' contributions to commenting on the strategies that have been selected rather than the fundamental policy objectives. The developmental nature of many of the technologies that implicate the land offer decision-makers important opportunities to make citizens a core part of the innovation process by going beyond the orthodox approach and engaging citizens in the 'upstream' before policy choices are actually made (Latour, 1997) and in the evaluative processes after pilots have been implemented.

Social learning can also be generated through the selection of policy tools that directly implicate citizens. By adapting market-based policy instruments to also empower citizens, some governments have begun to engage in a reflexive dialogue. The most high profile example of such a policy mechanism with implications for land use that links the market to society are feed-in tariffs (FiTs). Widely used across Europe (18 countries at present), FiTs are the main policy alternative to subsidies and certificated incentive mechanisms. By requiring energy companies to buy renewable electricity which is exported to the grid at rates above market value, this system gives energy suppliers the long-term certainty they need to invest in renewables and gives citizens, farmers and small businesses the confidence to diversify their own land by installing renewable energy systems to sell back any excess power to the grid. Research tells us that citizens often readily accept the need for renewable technologies but are opposed to land use changes that impact upon their local community. Yet where such changes do take place local communities often change their minds (Upreti, 2004). By democratizing the economic potential of the land, policy instruments such as FiT allow citizens to control directly how they engage with renewables, boosting buy-in and helping to engender confidence in technologies. Much of the success of anaerobic digestion and wind turbines in continental Europe and mainstreaming of land diversification strategies in new house design – in particular in Austria and Germany – has been attributed to the security and civic involvement provided by the guaranteed premium energy prices provided by FiTs (Broer and Worthington, 2008; Needham, 2007).[16]

The 'omitted voices' (Wiener, 1995) of electorates are not the only ones to which decision-makers aiming to develop their peripheral vision would find it useful to attend. The potential for risk transfer and transformation illustrates the importance of how policy action is experienced by citizens beyond decision-makers' immediate political radar. The importance of this is probably best understood in relation to knowledge transfer projects where consortia of countries, industry and universities work in partnership to site large R&D initiatives in the industrializing and developing world. For example, the UK CCS initiative, Near Zero Emissions Coal (NZEC),[17] is currently exploring the feasibility of building coal-fired power plants in sedimentary basins in China. To address public acceptance issues, the initiative plans to dovetail with the EU's Co-operation Action within CCS China–EU programme (COACH). The greater challenge will be for governments to engage with communities whose co-operation they do not need and whose bargaining power is weaker.

Institutional learning

As well as learning from experts and citizens, governments can learn from their own experiences and from those of other governments. When faced with a pressing problem, ad hoc policymaking is always a danger. The risk of countervailing risks increases where decision-makers select policy instruments without consideration of the context within which they are to be applied, or of lessons that might be drawn from previous experience or other jurisdictions, and how efforts might be coordinated internationally.

Given that there are very few new policy instruments for governments to try, to exploit the land in ways that achieve the so-called risk superior options, decision-makers increase their chances of success by combining instruments to create bespoke ways of delivering technologies that address both the target risks and potentially significant trade-offs. For example, to move CCS beyond pilot stages to commercial operation, emissions schemes can be supplemented with other policy tools such as mandatory compatibility on future plants, tax breaks or targets. In relation to the latter, Japan has become a significant contributor to the global development of CCS by setting the target of 200 million $tyear^{-1}$ cut (half in Japan, half overseas) in CO_2 emissions using CCS technologies and underpinning these with government funding of the demonstration phase of the technology until the cost of reducing emissions reaches US\$26 t^{-1} (Innovation Norway, 2008).

Part of the move to more sophisticated cocktails of policy instruments has, of course, been stimulated by the direct experience of countervailing risks thrown up by initial policy choices. For example, decision-makers now acknowledge that the leap of faith required for companies to invest in risky technologies is perhaps too wide (Panek, 2008). In the case of CCS in the EU, it has been proposed that plants with CCS are regarded as not having emitted – thus imposing costs only on those power plants that are not CCS compatible and reducing the likelihood that old equipment will be kept in use (European

Commission, 2008).[18] Similar lessons are being drawn from the EU experience elsewhere.[19]

Identifying the optimum instrument mix or innovative policy solutions is not necessarily straightforward. For example, the market distortion and countervailing risks, which have been the result of biofuels targets coupled with subsidies, highlight the difficulty in achieving a palatable mix. Bolting on sustainability criteria and setting quotas for second generation biofuels not made from food crops is expected to militate against some of the most obvious countervailing risks. But these correctives will not address the loss of revenues for other innovations. Thus, the mix here will be far from perfect but is superior to the current situation. Similarly, reverse auctions – where industry bidders requiring the lowest amount of public money are subsidized – have been suggested as a cost-effective way to inject market discipline into subsidizing the development of second-generation, cellulosic biofuels. Convincing the private sector that it should bear the development risks, accept low incentives and technically develop the auctions are challenges which cannot be met overnight (Koplow, 2006).[20]

An awareness of the work being done by colleagues in other jurisdictions and sharing of data is an essential part of institutional learning. This is especially true for technologies that carry the greatest knowledge deficits, where wide-scale deployment can only be achieved once a broad base of knowledge of what works and what does not is established (Wilson et al, 2007). Decision-makers are beginning to recognize the importance of pursuing complementary strategies. For example, the decision by the UK's Department for Business, Enterprise and Regulatory Reforms (BERR) to restrict its CCS competition to post-combustion on coal was based on the fact that a gap in this particular form of carbon sequestration existed on the international research landscape. Moreover, a key criterion for success of a bid will be the inclusion of initiatives for information sharing (Northmore, 2008[21]).

Conclusions

The choice of policy strategies is not dispassionate; it is political and reflects the prevailing value hierarchies of a polity and the roles these ascribe to markets, citizens and rules. The aim of this chapter has been to analyse the impact of the choices that have been made so far in the regulation of climate change and energy diversification with respect to technologies that implicate land-use change. The question of why certain policies have been selected over others has not been addressed (though this is important of course). The wide-ranging reconnaissance of government interventions presented here underlines the embryonic state of policy responses and, given that, it would be unwise to draw any specific conclusions on future land use patterns. This survey does, however, highlight two fundamental challenges with which decision-makers must get to grips if they are to produce more complete risk governance.

The discussion of the pre-eminent policy interventions, and their often ad hoc and opportunistic nature, suggests that, in their early policy responses, decision-makers did not fully recognize that when they act they affect two levels of risk: target risks at the first level and the countervailing risks that flow from government intervention at the second. Reality has intruded, however. The quick wins promised by first-generation biofuels have been spectacularly displaced by excessive technological lock-in and undesirable trade-offs with food, the environment, vulnerable populations and climate change itself. The sheer scope and range of the side-effects created by government policies encouraging land-use change for fuel crops has the potential to become a sort of cognitive tipping point that forces decision-makers to confront the interconnected reality which they must negotiate.

Recognition of the complexity of the task they face and considering the implications of this in policy design are not the same thing of course. To analyse these trade-offs and reduce uncertainty about policy action, decision-makers must be willing to engage with a variety of sources of policy learning. This palliative to countervailing risks can itself be seen as a risk trade-off. Learning from experts and society and reflecting on experience are not cost-free exercises; they can delay policy action and increase the range of conflict around decision-making.

Learning can go much further of course. Organizational analysts exhort decision-makers to engage in 'double loop' learning (Argyris and Schön, 1978), where the fundamental assumptions of how problems are approached and solutions exploited are explicated and, where necessary, revised. Given the complexity of getting mainstream policy programmes off the ground (even where there is political will), double-loop learning encourages thinking the unthinkable and, in so doing, threatens to undermine accepted ways of addressing policy problems. It is the political equivalent of the circus high wire. Adopting the spirit of double-loop learning, however, where goals are capable of being adjusted and the selection of policy tools problematized, may serve as an antidote to the wishful thinking that underpins some of the more ad hoc policy selection and improve decision-makers' peripheral vision.

Notes

1 The economic impacts of externalities have long been analysed by economists. However, risk trade-offs are a wider phenomena – encapsulating all the possible side effects of regulations designed to mitigate adverse outcomes. The idea of risk versus

risk was first outlined by Lave (1981) and an analytical framework was first articulated by Graham and Wiener (1995) and Viscusi (1992)

2 For a detailed view of risk transfer and substitution see Whipple (1985)

3 This concept is sometimes referred to as the 'Jevons paradox', which was formulated in mid-19th century Britain in relation to the impact of changes in coal consumption (Jevons (1906) [1865]). In their report *2007 The Rebound Effect*, the UK Energy Research Centre (UKERC) offers one of the most in-depth reviews of the literature and concludes that while the evidence base is weak, rebound effects should not be discounted by decision-makers but rather more empirical studies and economic modelling of the impact of policy targets is required (http://www.ukerc.ac.uk/Downloads/PDF/07/0710ReboundEffect/0710ReboundEffectReport.pdf)

4 See Foxon (2003) for a review of the innovations systems literature

5 For more on the Western Climate Initiative (WCI) and regional Greenhouse Gas Initiative (RGGI) see http://www.westernclimateinitiative.org/WCI_Documents.cfm and http://www.rggi.org/. Both these schemes could be overtaken by a federal cap-and-trade scheme proposed by Senators Lieberman and Warner

6 This may also precipitate the establishment of an independent international energy agency making the value of carbon more predictable and investment more attractive

7 A comprehensive assessment from the IPCC (Metz et al, 2005) Working Group III estimated this at around 25 per cent and a 2007 UN report estimated the increase in costs as lying anywhere between 40 per cent and 90 per cent (cited by Harris, 2007)

8 Lars Stromberg is Director of Vattenfall and Bert Metz was co-chair of IPCC Working Group III

9 The construction of a regulatory regime to ensure that new coal-fired power stations are CCS compatible is under way in the EU; however, the USA plans no such demand. States such as Nevada have introduced a compromise whereby firms must sign Memorandums of Understanding (MOUs) with the Nevada Division of Environmen-tal Protection (NDEP) that once CCS becomes commercially viable they will engage in carbon capture (*Las Vegas Sun*, 2008, in *Carbon Capture Journal*, 2008 vol 1, no 1 p11)

10 For example, in January 2008, the USA announced that it was cancelling its involvement in TransGen and restructured its financing of FutureGen Alliance's coal-fired plant

11 Renewable Fuels Association (US) reports 134 ethanol plants currently in operation in the USA as compared to 50 in 1999

12 See MacKenzie (2007) 'The Political Economy of Carbon Trading' in *London Review of Books* for a more general discussion of Zelizer's hostile world's thesis and carbon trading

13 For more on pyrolysis see Van Loo and Koppejan, 2008

14 The SRC planted for that project is now being used in co-firing at the Drax Power Station in North Yorkshire, UK. Drax Power is now advertising for producers of SRC, miscanthus and rape grown within 50 miles of Drax power station to supply them through long-term contracts

15 Jan Panek is Head of Unit, Coal and Oil, for the European Commission

16 For an in-depth appraisal of FiTs, see Menanteau et al, 2003

17 NZEC is funded by the UK Government through Defra and BERR and is coordinated by AEA Energy & Environment (UK) and ACCA21 (China)

18 There are those within the CCS industry who want these proposals to go further and operate a 'double credit' whereby CCS power plants are allocated an additional allowance that can then be sold on
19 For example, both Australia and New Zealand plan similar clauses for their nascent trading schemes http://www.med.govt.nz/templates/ContentTopicSummary_ 38284.aspx
20 The 2005 United States Energy Policy Act made $250 million available to establish a reverse auctions programme (Section 942)
21 Bronwen Northmore is Director of the Cleaner Fossil Fuels Unit, BERR

References

Anderson, D., Barker, T., Ekins, P., Green, K., Köhler, J., Warren, R., Agnolucci, P., Dewick, P., Foxon, T., Pan, H. and Winne, S. (2005) 'Technology policy and technical change: a dynamic global and UK approach', Tyndall Centre Technical Report No. 23, Norwich

Arrow, K. (1962) 'Economic welfare and the allocation of resources for invention', in *The Rate and Direction of Inventive Activity: Economic and Social Factors*, Conference of the Universities National Bureau Committee for Economic Research and the Committee on Economic Growth of the Social Science Research Council, Princeton University Press, Princeton, NJ

Arygris, C. and Schön, D.A. (1978) *Organizational Learning: A Theory of Action Perspective*, Addison-Wesley, Reading, MA

Barlow, P. (2007) 'What's in your tank', *The Daily Telegraph*, 20 April

Beder, S. (2001) 'Trading the Earth: The politics behind tradeable pollution rights', *Environmental Liability*, vol 9, no 2, pp152–160

Black, J. (1995) 'Which arrow?', *Public Law*, vol 165, June, pp94–117

Broer, S. and Worthington, D. (2008) 'Renewables energy lessons from Austria', *Building Services Journal*, March, http://www.bsjonline.co.uk/story.asp?storycode=3108302

Cacho, O., Marshall, G.R. and Milne, M. (2002) *Smallholder Agroforestry Projects: Potential for Carbon Sequestration and Poverty Alleviation*, FAO ES Technical Series Working Paper, FAO, Rome

Collingridge, D. (1980) *The Social Control of Technology*, Pinter, London

Council of Ministers (2003) 'Promotion of the use of biofuels and other renewable fuels for transport', Directive 2003/30/EC of the European Parliament and of the Council of 8 May 2003 on the promotion of the use of biofuels and other renewable fuels for transport (OJEU L123 of 17 May 2003), EC, Brussels

Council of Ministers (2008) 'Council decision 8847/08 on Sustainability criteria for biofuels', http://register.consilium.europa.eu/pdf/en/08/st09/st09048-re01.en08.pdf accessed 10 June 2008

Defra (2007) 'Increasing the uptake of anaerobic digestion', Workshop at the University of Exeter, 3–4 September, session 2, p5, Defra, London

Doornbosch, R. and Steenblik, R. (2007) 'Biofuels: Is the cure worse than the disease?', Report for the Chair of the Round Table on Sustainable Development at the OECD, Paris, 11–12 September, OECD, Paris

Dunlop, C.A. and James, O. (2007) 'Principal-agent modeling and learning', *Public Policy and Administration*, vol 22, no 4, pp403–422

Eliadis, P., Hill, M. and Howlett, M. (eds) (2005) *Designing Government*, McGill-Queen's University Press, Montreal

European Commission (2008) Proposal for a Directive of the European Parliament and of the Council on the promotion of the use of energy from renewable sources, 23 January, EC, Brussels

Evans, A. (2008) *Rising Food Prices*, Chatham House Briefing Papers, London

FAO (2008) 'Climate change: Implications for food safety' http://www.fao.org/ag/agn/agns/files/HLC1_Climate_Change_and_Food_Safety.pdf accessed 12 February 2008

Farley, M. (2008) 'Clean coal technology', *Carbon Capture Journal*, vol 1, no 1, p16

Forward, K. (2008) 'Editorial', *Carbon Capture Journal*, vol 1, no 3, p2

Foxon, T.J. (2003) *Inducing Innovation for a Low-Carbon Future: Drivers, Barriers and Policies*, The Carbon Trust, London

Freeman, C. and Soete, L. (1997) *The Economics of Industrial Innovation*, 3rd edn, Pinter, London

Global Subsidies Institute (2007) *Biofuels at what Cost? Government Support for Ethanol and Biodiesel in Selected OECD Countries*, GSI, Geneva

Graham, J.D. and Wiener, J.B. (eds) (1995) *Risk Vs. Risk*, Harvard University Press, Cambridge, MA

Gray, G.M. and Graham, J.D. (1995) 'Regulating pesticides' in Graham, J.D. and Wiener J.B. (eds) *Risk Vs. Risk*, Harvard University Press, Cambridge, MA

Gross, R. and Foxon, T.J. (2003) 'Policy support for innovation to secure improvements in resource efficiency', *International Journal of Environmental Technology and Management*, vol 3, no 2, pp118–130

Haberl, H., Erb, K.H., Krausmann, F., Gaube, V., Bondeau, A., Plutzar, C., Gingrich, S., Lucht, W. and Fischer-Kowalski, M. (2007) 'Quantifying and mapping the human appropriation of net primary production in earth's terrestrial ecosystems', *Proceedings of the National Academy of Sciences*, vol 104, 31 July, pp12942–12947

Hahn, R. and Hester, G. (1989) 'Where did all the markets go? An analysis of EPA's emissions trading program', *Yale Journal of Regulation*, vol 6, pp109–153

Harris, J. (2007) 'The great global coal rush puts us on the fast track to irreversible disaster', *The Guardian*, 30 August

Hood, C. (1983) *The Tools of Government*, Macmillan, London

Howlett, M. (2005) 'What is a policy instrument?' in Eliadis, P., Hill, P.M. and Howlett, M. (eds) *Designing Government*, McGill-Queen's University Press, Montreal

Innovation Norway (2008) International CCS Technology Survey Issue 3, July, Oslo

International Assessment of Agricultural Knowledge, Science and Technology for Development/UN Educational, Scientific and Cultural Organization (2008) Synthesis report IAAST, Washington, DC

International Energy Agency (IEA) (2005) Bioenergy Newsletter No 6 July, Paris

International Energy Agency (2006) *IEA World Energy Outlook*, IEA, Paris

International Energy Agency (2008) *CO_2 Capture and Storage: A Key Carbon Abatement Option*, International Energy Agency, Paris

International Fund for Agricultural Development (2001) *Rural Poverty Report*, Oxford University Press, New York

International Risk Governance Council (2008) *Regulation of Carbon Capture and Storage*, IRGC, Geneva

Jevons, W.S. (1906) [1865] *The Coal Question: Can Britain Survive?*, Macmillan, London

Jones, B.D. (1994) *Reconceiving Decision-Making in Democratic Politics*, University of Chicago Press, Chicago.

Kahneman, D. and Tversky, A. (1979) 'Prospect theory: an analysis of decision under risk', *Econometrica*, vol 47, pp263–291

Klassen, R.D. and Whybark, D.C. (1999) 'The impact of environmental technologies on manufacturing performance', *Academy of Management Journal*, vol 42, no 6, pp599–615

Koplow, D. (2006) 'Biofuels – At What Cost? Government Support for Ethanol and Biodiesel in the United States', Global Subsidies Initiative of the International Institute for Sustainable Development, Geneva, http://www1.eere.energy.gov/biomass/pdfs/obp_roadmapv2_web.pdf accessed 12 December 2007

Latour, B. (1997) *Science in Action*, Harvard University Press, Cambridge, MA

Lave, L.B. (1981) *The Strategy of Social Regulation*, Brookings, Washington, DC

Lindegaard, K. (2005) 'The chips are down', *The Guardian*, 19 January

Lipper, L. and Cavatassi, R. (2004) 'Land-use change, carbon sequestration and poverty alleviation', *Environmental Management*, vol 33, Supplement 1, ppS374–S387

Macedo, I., Lima, M.R., Leal, V. and Azevedo Ramos da Silva, J.E. (2004) *Assessment of Greenhouse Gas Emissions in the Production and Use of Fuel Ethanol in Brazil*, Government of the State of São Paulo, Brazil

MacKenzie, D. (2007) 'The political economy of carbon trading', *London Review of Books*, vol 29, no 7

May, P. (1991) 'Reconsidering policy design: Policies and publics', *Journal of Public Policy*, vol 11, no 2, pp187–206

Menanteau, P., Finon, D. and Lamy, M-L. (2003) 'Prices versus quantities: Choosing policies for promoting the development of renewable energy', *Energy Policy*, vol 31, pp799–812

Metz, B., Davidson, O., Coninck, H. de, Loos, M. and Meyer, L. (eds) (2005) *Carbon Capture and Storage*, IPCC Working Group III, Cambridge University Press, Cambridge, UK, http://www.ipcc.ch/ipccreports/special-reports.htm accessed 4 January 2007

Mol, A.P.J. (2007) 'Boundless biofuel?', *Sociologica Ruralis*, vol 47, no 4, pp297–315

Monbiot, G. (2005) 'Biodiesel enthusiasts have accidentally invented the most carbon-intensive fuel on earth', *The Guardian*, 6 December

Monfort, J. (2008) 'Despite obstacles, biofuels continue surge', Worldwatch Institute, 23 April, http://www.worldwatch.org/node/5450 accessed 24 April 2008

Needham, A. (2007) 'Presentation to the anaerobic digestion stakeholder workshop', 4 September, University of Exeter, Exeter

Nelson, R.R. (1959) 'The simple economics of basic scientific change', *Journal of Political Economy*, vol 67, no 3, pp297–306

Northmore, B. (2008) 'The UK CCS demonstration project', *Carbon Capture Journal*, vol 1, no 1, p17

OECD (2002) *Dynamizing National Innovation Systems*, OECD, Paris

Ofgem (2007) Ofgem's Response to BERR Consultation on Reform of the Renewables Obligation, Ref 222/07, 13 September, Ofgem, London

Olsson, L. and Ärdo, J. (2002) 'Soil carbon sequestration in degraded semiarid agro-ecosystems – perils and potentials', *Ambio*, vol 31, no 6, pp471–477

Panek, J. (2008) 'Inaugural talk at the European carbon capture and storage summit', *Carbon Capture Journal*, vol 1, no 1, p21

Pike, R. (2008) 'Letters: Fuel for a new agricultural revolution', *The Guardian*, 29 March

Piterou, A., Shackley, S. and Upham, P. (2008) 'Project ARBRE: Lessons for bio-energy developers and policy-makers', *Energy Policy*, vol 36, pp2044–2050

Point Carbon (2008) 'Largest survey ever conducted into the world's carbon market released today' http://www.pointcarbon.com/aboutus/pressroom/pressreleases/1.266583 accessed 12 October 2008

Rascoff, S.J. and Revesz, R.L. (2002) 'The biases of risk trade-off analysis', *The University of Chicago Law Review*, vol 69, no 4, pp1763–1836

Rio, P. del and Burguillo, M. (2008) 'Assessing the impact of renewable energy deployment on local sustainability: Towards a theoretical framework', *Renewable and Sustainable Energy Reviews*, vol 12, no 5, pp1325–1344

Robinson, E. (2008) 'World wheat production — expected higher for 2008-2009', *Western Farm Press* June 25 http://westernfarmpress.com/ news/wheat-production-0625/ accessed 12 October 2008

Salaff, S. (2008) 'Saskatchewan moves from CCS to nuclear', *Carbon Capture and Storage*, vol 1, no 1, p12

Sanderson, I. (2002) 'Evaluation, policy learning and evidence-based policy-making', *Public Administration*, vol 80, no 1, pp1–22

Schipper, L. (2000) 'On the rebound: The interaction of energy efficiency, energy use and economic activity: an introduction', *Energy Policy*, vol 28, pp351–353

Schön, D.A. and Rein, M. (1994) *Frame Reflection*, Basic Books, New York

Schumpeter, J.A. (1975) [1942] *Capitalism, Socialism and Democracy*, Harper, New York

Scott, A. (2006) 'Small is sustainable', *New Statesman*, 15 May

Searchinger, T., Heimlich, R., Houghton, R.A, Dong, F., Elobeid, A., Fabiosa, J., Tokgoz, S., Hayes, D. and Yu, T-H. (2008) 'Use of US croplands for biofuels increases greenhouse gases through emissions from land-use change', *Science*, vol 319, no 5867, pp1238–1240

Simeonova, K. (2007) *Policy Developments and Projected GHG Emissions*, UNFCC Secretariat, Bonn

Simon, H. (1957) *Models of Man*, Wiley, New York

Slovic, P. (2000) *The Perception of Risk*, Earthscan, London

Smith, J. and Scherr, S. (2002) *Forest Carbon and Local Livelihoods: Assessment of Opportunities and Policy Recommendations*, Centre for International Forestry Research, Occasional Paper No. 37, Bogor, Indonesia

Socorro, M.P. (2003) *Optimal Technology Policy: Subsidies and Monitoring*, Unitat de Fonaments de l'Anàlisi Econòmica (UAB) and Institut d'Anàlisi Econòmica (CSIC) Working paper 570/03

Stern, N. (2007) 'Climate change, ethics and the economics of the global deal', Royal Economic Society (RES) lecture, 28 November, Manchester, http://www.res.org.uk/society/lecture.asp accessed 10 January 2008

Stern, N.H. and HM Treasury (2006) 'The economics of climate change: The Stern Review', London, http://www.hm-treasury.gov.uk/independent_reviews/stern_ review_ economics_climate_change/stern_review_Report.cfm accessed 12 December 2007
Stromberg, L. and Metz, B. (2008) 'The Economics of exploiting CCS on an industrial scale', *Carbon Capture Journal*, vol 1, no 1, p18
Tauli-Corpuz, V. (2007) 'Oil palm and other commercial tree plantations, monocropping: Impacts on indigenous peoples' land tenure and resource management systems and livelihood', The UN Permanent Forum on Indigenous Issues, 6th session, 14–25 May 2007, http://www.un.org/esa/socdev/unpfii/documents/6session_crp6.doc accessed 4 January 2008
Templer, R. and Woods, J. (2008) 'Letters: Fuel for a new agricultural revolution', *The Guardian*, 29 March
Thornley, P. (2006) 'Increasing biomass based power generation in the UK', *Energy Policy*, vol 34, no 15, pp2087–2099
UK Biomass Task Force (2005) Final Report, October, Defra, London
UK Energy Research Centre (2007) 'The rebound effect', London, http://www.ukerc.ac.uk/Downloads/PDF/07/0710ReboundEffect/0710ReboundEffectReport.pdf accessed 10 December 2007
UNFCC (2008) *Land Use, Land-Use Change and Forestry*, UN, New York
United Nations-Energy (2007) *Sustainable Bioenergy: A Framework for Decision Makers*, UN, Geneva
United Nations Environmental Programme (2007) 'The last stand of the Orang-utan', http://www.unep-wcmc.org/resources/publications/LastStand.htm accessed 12 December 2007
United Nations Environment Programme (2008) *Global Trends in Sustainable Economic Investment*, UNEP and New Energy Finance Ltd, Nairobi, Kenya
UNEP and UNESCO (2007) 'The last stand of the Orangutan. State of emergency: Illegal logging, fire and palm oil in Indonesia's 25 National Parks', http://www.unep-wcmc.org/resources/PDFs/LastStand/full_orangutanreport.pdf accessed 12 December 2007
Upreti, B.R. (2004) 'Conflict over biomass energy development in the UK', *Energy Policy*, vol 32, pp785–800
Upreti, B.R. and Horst, D. van der (2004) 'National renewable energy policy and local opposition in the UK', *Biomass and Bioenergy*, vol 26, pp61–69
Upsham, P. and Shackley, S. (2006a) 'Stakeholder opinion on a proposed 21.5MWe biomass gasifier in Winkleigh, Devon', *Journal of Environmental Policy and Planning*, vol 8, no 1, pp45–66
Upsham, P. and Shackley, S. (2006b) 'The case of a proposed 21.5MWe biomass gasifier in Winkleigh, Devon', *Energy Policy*, vol 34, no 15, pp2161–2172
Uryu, Y., Mott, C. and Foead, N. (2008) 'Deforestation, Forest Degradation, Biodiversity Loss and CO_2 Emissions in Riau, Sumatra, Indonesia', WWF Technical Report, Jakarta, http://assets.panda.org/downloads/riau_CO2_report__wwf_id_27feb08_en_lr_.pdf accessed 10 October 2008
Van Loo, S. and Koppejan, J. (2008) *The Handbook of Biomass Combustion and Co-Firing*, Earthscan, London
Viscusi, K.W. (1992) *Fatal Trade-Offs*, Oxford University Press, Oxford
Wetlands International (2007) *Global Peatland Assessment*, Wageningen, Netherlands

Whipple, C. (1985) 'Redistributing risk', *Regulation*, May/June, pp37–44

Wiener, J.B. (1995) 'Protecting the global environment' in Graham, J.D. and Wiener, J.B. (eds) *Risk Vs. Risk*, Harvard University Press, Cambridge, MA

Wilson, E.J., Friedmann, S.J. and Pollak, M.F. (2007) 'Risk, regulation and liability for carbon capture and storage', *Environmental Science and Technology*, vol 41, no 17, pp5945–5952

Wolf, C. (1988) *Markets or Governments: Choosing Between Imperfect Alternatives*, MIT Press, Cambridge, MA

World Bank (2002) 'Biocarbon Fund', www.biocarbonfund.org accessed 12 December 2007

World Bank (2008) *Biofuels: The Promise and the Risks*, Washington, DC

Zelizer, V.A. (2005) *The Purchase of Intimacy*, Princeton University Press, Princeton, NJ

13

The Land Debate – 'Doing the Right Thing': Ethical Approaches to Land-Use Decision-Making

Peter Carruthers

Introduction

This book addresses the question 'What is land for?' It is a crucial question, but a difficult one. It is difficult, first, because there are many different things we can, wish to and need to do with land, as this book demonstrates. It is relatively easy to identify the options; it is much less easy to decide which are the 'right' ones and to reconcile competing demands.

The question is difficult, also, because the core purpose of land is no longer self-evident. In the past, farming provided the essential rationale for land use, with other functions spinning off the core business of producing food. But farming is no longer regarded as the mainstay of rural economies, and its role as the 'rural metanarrative' has been systematically de-constructed. Instead, there is a multiplicity of perceived roles for land and a plurality of ruralities (Murdoch et al, 2003). 'Sustainability' and 'multifunctionality' have helped fill the gap and provided some basis for engaging the debate and shaping policy. But these concepts, themselves, are being challenged by the 'new environmentalism' of climate change, prompting what some regard as a 'new productivism'.

A further difficulty arises from the nature of the question itself, which is suggestive of a particular philosophical stance – an instrumental and consequentialist one. Although this may represent the dominant modern (Western) worldview, there are other ways of looking at things. Alternative voices became increasingly audible and articulate in the latter half of the 20th century, strongly influencing the environmental debate, pressing the agenda beyond the optimization of options to a more fundamental consideration of humanity's relationship to land and nature.

These are ethical questions, and this chapter aims to bring an ethical perspective to the 'new land debate', to deciding 'What land is for?'. No attempt is made to address specific dilemmas, though it is hoped that the ideas set out here will provide some basis to do so. Rather, reflecting the overall purpose of this volume, what is offered is simply an initial reconnaissance of the territory – a mapping of the contours and establishment of a baseline of ideas.

In the quest for ethical guidance for land use, the account below considers what might be gleaned from the different meanings attached to land, from moral philosophy and, especially, its application to the environment and agriculture, and from metaphors and movements that express or configure the relationship between people and land. Finally, some conclusions and pointers forward are offered.

Land

'Land' means different things to different people. It is seldom just land, however, but is overlaid with social and cultural meanings that reflect the breadth of human identity, experience and aspiration (Breugemann, 1977; Woods et al, 2008).

Land is, first, a resource for subsistence and production, and for dwellings and living space. Land meets human physical needs and provides the basis for economic development. Land is the essential basis of the rural economy (though not necessarily of the economy of rural areas as now 'officially' defined in the UK). Land is also property, conferring wealth and status on its owners, and capable of being bought and sold, and inherited and bequeathed.

'Resource' and 'property' are the predominant constructions of land, at least in Western, capitalist economies, and the basis for land's perceived value. As a 'resource employed to produce goods and services', a factor of production, land acquires a 'value' related to the income (or 'rent') it can generate. As a tradable asset, a commodity, land is the subject of speculation and accumulation, with a value indicated by the price it commands in the market. Production and trade have become, therefore, the central guiding principles for land use, the essential 'land ethic'.

Land, at least rural land, also connotes 'environment', 'landscape' and 'nature' (i.e. 'wilderness'); and environmental responsibility, landscape preservation and connecting to nature are essential elements of any land ethic, as we shall see below. Although landscape can be regarded just as another resource, a scene to be consumed visually, it has deeper meanings, with often complex historical, cultural, political, aesthetic and symbolic connotations (Woods et al, 2008). Similarly, although closeness to nature might be just to do with aesthetic appreciation, there is growing recognition that it goes deeper than that, influencing people's physical, mental and spiritual health and wellbeing (Pretty, 2007, 2008; ACORA, 1990).

Closeness to nature may partly explain why land and landscapes play a central role in national identity in many countries. For the English, it is the rural landscape that epitomizes the quintessential character of England. 'Country' in English means both 'rural land' and 'native land', and it was in the English countryside that the 18th- and 19th-century romantics, for example, found intimations of 'wilderness', which they saw as 'an antidote to the corrupted industrial city' (Murdoch et al, 2003, p1).

However, the idealization of rural land, of countryside, as a virtuous counterpoint to the 'brutality of the market or the anonymity of the city' (Short, 1991, p3) was also associated with agriculture, which was seen as more praiseworthy than industry and commerce. Land is not only a place to encounter nature; it is also a place of work. For the British, especially, rural land is both wilderness and workplace: 'a working (and not merely scenic) countryside is important to the British imagination' (N. Wirzba, 2007, personal communication).

Identification with specific areas of land, with places, may also extend to the individual and local: 'people do form bonds with place ... and territory is vitally important to people and may serve as an integral component of self-identity' (Storey, 2001, p17). Indeed, meanings of land are not primarily expressions of abstract thought, but arise from encounter and experience of real places (Breugemann, 1977; Teo and Huang, 1996). The significance of place for community identity and development is increasingly recognized, as evidenced, for example, by 'sense-of-place projects' and 'ecomuseums' (Convery and Dutson, 2006; ICU, 2008; Woods et al, 2008).

Underlying many of the above constructions is a human hunger for rootedness and belonging, not only in community, but also in the land itself. Human life is embedded in the land (Wirzba, 2002a) and land is a vital dimension of human identity: 'our humanness is always about historical placement in the earth' (Breugemann, 1977, p3); land is '*the* context of our earthly life and the plinth or foundation of all civilized life' (H.J. Massingham, 1942, in Abelson, 1988, p109).

This placement of (particular) people and communities on land in the context of history is especially evident in pre-industrial traditions, as the following examples demonstrate:

- African land-tenure systems affirm the social structure rather than seeking to change it (Rowland, 1990). The land-owning community is perceived in both place and time. Land is received from ancestors and passed on to children, all of whom are (in contemporary jargon) 'stakeholders'. In words attributed to Ghanaian Chief Nana Sir Ofori, 'land belongs to a vast family of whom many are dead, a few are living and countless hosts are still unborn' (Rowland, 1990, p20).
- Within the ideology of 'peasantry', no person had individual or exclusive property rights, and individuals could not sell off their share of the family property (Schluter and Ashcroft, 1990). 'The central feature of 'peasantry'

is the absence of absolute ownership of land, vested in a specific individual. The property-holding unit is a 'corporation' which never dies. Into this an individual is born or adopted, and to it he gives his labour' (Macfarlane, 1978, quoted in Schluter and Ashcroft, 1990, p6).

- In the Biblical economy, land is, again, not a possession but an inheritance and a gift (Berry, 2002a), with an inalienable connection to the social unit (Breugemann, 2002). Land was not really 'owned' in our modern economic sense (i.e. property to be used and disposed of at the owner's pleasure), but rather held in trust, received from one generation and passed on to another. Ultimately land is not received from ancestors, but from God. And the gratitude due for, and the obligations and responsibilities that come with, the gift of land are owed to the Divine giver himself.

The necessity of land for human existence, the deep human desire and need to connect to land, the role of land and place in the history and identity of people and communities, and the richness of meanings and human aspirations invested in land, combined with the fact that the majority of the world's land is controlled by a minority of people, makes it inevitable that land is also the focus of rights and conflicts, and essential that a land ethic not only recognizes the above constructions, but also accommodates land rights and provides a basis for resolving conflicts.

Ethics

'Ethics' refers to principles of right and wrong (i.e. morals) and to the philosophical study of these principles (i.e. moral philosophy). Within the latter, a distinction is made between 'metaethics', 'normative theories' and 'applied ethics'.

Metaethics investigates the sources, meaning, application and justification of moral principles (Armstrong and Botzler, 1993; Newell, 2005; Sayre-McCord, 2008, Vardy and Grosch, 1999). Aspects of metaethics of interest here are nature as a source of ethics, the extension of ethical consideration beyond people and the social contract as a justification for moral behaviour.

Normative theories provide principles to guide behaviour in order that such behaviour becomes 'normal' (Vardy and Grosch, 1999, p109). Contemporary moral philosophy recognizes three groups of theories: consequentialism, deontological theories and virtue ethics (Hursthouse, 2008).

Applied ethics uses the (seemingly abstract) insights of metaethics and normative ethics to inform often complex and controversial real-life decisions (Fieser, 2006); our particular interest here is the application of ethics to the environment and agriculture.

Metaethics

The wisdom of nature

A leading candidate for the source of ethics is 'nature'. The 'natural law tradition' argues that, like physical laws, moral laws are embedded in nature – inherent in the nature of things – and can be discovered by reason (Anderson, 1998; Murphy, 2008; Thompson, 2003). Although 'nature' in this context means nature and human nature and experience, the belief that 'our deepest moral guidance comes from understanding nature and our "natural" place in it' (Armstrong and Botzler, 1993, p54), is a recurrent theme in the material considered here (despite nature's 'mixed messages').

To whom or what do ethics apply?

It is widely accepted that all people, including future generations, merit moral concern. But should moral consideration be extended beyond people to other living things, ecosystems or to the earth as whole? Addressing this question is a central concern of environmental ethics, as will be seen below, and central to formulating an ethic for land.

A related question asks whether moral judgements are universal or restricted in application. One response is to argue that different moral frameworks are needed for current stakeholders, future generations, other sentient creatures, inanimate objects and so on (i.e. 'pluralism'). Another response argues that moral judgements are tied to the place and/or time in which they occur (i.e. 'contextualism'; Armstrong and Botzler, 1993, p54). Both of these have special relevance to land-use issues.

The social contract

One answer to the question 'Why be moral?' is provided by the idea of the 'social contract', an implied agreement among members of society to accept a limited set of rules to enable social co-operation (Clark, 2003). Predicated on self-interest (i.e. 'a good society is good for me') or duty (to serve the 'common good'; Cudd, 2008, p1), the social contract provides a rationale for individuals to act morally and for governments to create and maintain a just and ordered society.

Thompson et al (1994) and Thompson (2002, 2007) identified four ways in which the social contract is specified, noting how each applied to land/farming issues, as follows:

- *Libertarianism* restricts government's powers to protecting 'non-interference' rights (e.g. life, liberty and property), arguing that were there no government (a 'state of nature'), rational individuals would agree not to harm others if they could be assured that others would not harm them.

In relation to land, libertarians are concerned with ownership, control and property rights.

- *Egalitarianism* extends government protection to 'opportunity' rights, which entitle citizens to primary goods (e.g. food, shelter) or which specify social goals, such as the elimination of poverty – using the argument that if people were ignorant of their own social position (behind a 'veil of ignorance') they would wish to limit the inequality of access to the primary goods that is allowed by libertarian theory. Egalitarians see land as a means of subsistence and advocate redistribution and the 'right to farm'.
- *Utilitarianism* (of which more below) argues that because individual actions, each rational, can lead collectively to undesirable results ('prisoners' dilemmas'), government must act to achieve the 'greatest good for the greatest number'. Utilitarians focus on land's asset value and use efficiency and advocate a regulated market. Land and environmental issues can present prisoners' dilemmas, thus justifying state intervention or privatization (i.e. 'enclosure'), as Hardin (1968) argued in his seminal paper, 'Tragedy of the Commons'.
- *Procedural theory* focuses solely on the rules used to make decisions. Policy choices are right if arrived at in an agreed way using principles of fair procedure.

The above offers at least two guidelines for ethical land-use decision-making. First, because recognition of the rights of individuals and groups defines the acceptable limits of government, in any specific situation possible interventions could be identified a priori and put through a 'rights proofing' process before proceeding further. Second, existing decision-making procedures should be evaluated against principles of fair procedure (e.g. participation, consent, fairness, impartiality, participation, consent, free choice) and new approaches developed.

Normative theories

Consequentialism

The most important form of consequentialism (Sinnott-Armstrong, 2008) is utilitarianism (as above), which evaluates consequences in terms of the sum of pleasure and pain – hence, the common slogan 'the greatest happiness for the greatest number'.

Utilitarianism is the popular 'common-sense' ethic', 'dominant moral paradigm' and 'the standard form of moral argument deployed in government' in the Western world (Northcott, 1996, pp 92 and 90). And it provides an ethical justification for policy-informing procedures such as opinion polls, stakeholder consultation, contingent valuation and cost-benefit analysis.

However, utilitarianism reduces the human experience to a game of pleasure and pain and disregards nobler virtues and transcendent values. By regarding people as essentially self-centred, utilitarianism favours individual- and consumer-focused institutions (e.g. as above) rather than community- and citizen-based approaches, and constructs people's preferences and values accordingly (Jacobs, 1997). Its focus on aggregate human utility, gauged by subjective sensations, means that it fails to protect the moral welfare of particular individuals, minority groups and communities, non-human and non-sentient life, ecosystems and the biosphere (Northcott, 1996).

Efforts have been made to extend utilitarian arguments to non-human life (Singer, 1990; Attfield, 1991), but applying utilitarianism to land-use decisions is, essentially, a conventional exercise in establishing who is affected and how – the process that has largely, though not always effectively, been employed hitherto.

Deontological theories

Deontological theories judge the morality of an action by its conformity to a moral norm or rule, independent of human inclination, divine volition or consequences (Northcott, 1996). Modern deontological theories may be duty-based, rights-based or contractarian (Alexander and Moore, 2008). The latter two include the terms of the social contract described above.

Deontological approaches to land/environment centre on duties owed to the natural world (Hargrove, 1989) or the rights of non-human life forms (Regan, 1988; Rolston, 1988; Taylor, 1993) or natural objects (Stone, 1993). It could be argued also that ecocentrism (see below), which assigns a place for humanity alongside other citizens or constituents of the earth, is a form of deontological contractarianism.

Virtue ethics

Virtue ethics focus not on the act, but the agent. The central question in virtue ethics is not 'What ought I to do?', but 'What sort of person should I be?'. Virtues are 'deep' dispositions, which can be cultivated and developed. It is not primarily right action, but right attitudes that make a person. But people with good attitudes, virtuous people, do good things and advance the flourishing of both themselves and society.

Of special interest here are virtue ethics' emphasis on 'community', 'narrative' and 'flourishing':

- Moral life is only meaningful in the context of social relationships, of *community*. The virtues both sustain and are sustained by 'socially established co-operative human activities' (practices) (MacIntyre, 1985, p187), which are themselves embedded in, and made intelligible by 'traditions' (MacIntyre, 1985; Vardy & Grosch, 1999).

- *Narrative* history explains human actions and provides the essential context for moral decisions. 'I can only answer the question "What am I to do?" if I can answer the question "Of what story or stories do I find myself a part?"' (MacIntyre, 1985, p216). And, as we saw above, people's stories are situated, contextualized, in places and on the land. Narrative myth or fiction also provides ethical 'texts' – witnesses to interrogate in the search for ethical evidence and stages on which to rehearse and perform moral values and principles.
- Connection to land/nature, it is claimed, is part of human *flourishing*, is inherently virtuous and promotes virtue (an idea at the core of agrarianism – see below). More tentatively, 'flourishing' could be applied beyond the human to land and nature – 'one's community in a larger sense' (Cheney, 1993, p89) (as in Leopold's 'land ethic').

Environmental ethics

Contemporary environmental ethics emerged as an academic discipline in the 1970s, in response to a growing environmental awareness and sense of environmental responsibility. There is now a considerable body of literature, including several journals in the field.

A key debate within environmental ethics is over the distinction between instrumental and intrinsic value (Brennan and Lo, 2008). It is commonly agreed all people have intrinsic value and merit moral concern. But can intrinsic value be extended to animals, ecosystems and the earth itself? Three positions, anthropocentrism, individualism and ecocentrism, are summarized below. Theocentrism and environmental virtue are then offered as alternative approaches.

Anthropocentrism

Traditional Western ethical systems are 'anthropocentric': only human beings have intrinsic value. Other entities may have instrumental value in serving human interests. Human impacts on the environment are an issue in so far as they affect the health, welfare and well-being of present, and future, generations of people. Concern for the environment is self-interest, indirect concern for one's fellow or, as Fraser Darling (1903–1979) argued, a duty, as 'biological aristocrats, to serve the lesser creation, to keep our world clean and pass on to posterity a record of which we shall not feel shame' (Darling, 1993, p301).

Individualism

Individualism extends moral value to non-human individuals (variously sensate animals, animals, all living organisms), but not to collections of individuals.

Proponents include Singer (1979, 1990), who used a utilitarian argument based on the fact that animals feel pain; Regan (1988), who argued for the rights of animals; and Albert Schweitzer, who developed a philosophy of 'reverence for life': 'it is good to maintain and to encourage life; it is bad to destroy life or to obstruct it' (Schweitzer, 1993, p343). Taylor (1993) developed Schweitzer's reverence for life into a philosophy of 'respect for nature', seeing human beings as members, along with other living things, of 'earth's community of life', a position suggestive of ecocentrism, as below.

Ecocentrism

Ecocentrism assigns intrinsic value to the natural world in a holistic rather than an individualistic way. The 'classic' expression of ecocentrism is Leopold's (1949) 'land ethic'. Leopold argued that moral community should be extended beyond people to the land (meaning the biosphere as a whole). Human beings should regard themselves as plain members and citizens of the biotic community. He inveighed against the tendency to invent devices to assign economic value to non-economic categories (such as song birds or salt-marshes) when they are threatened, summarizing his core principle as follows: 'a thing is right when it tends to preserve the integrity, stability and beauty of the biotic community. It is wrong when it tends otherwise' (Thompson, 1995, pp 43–44).

James Lovelock's Gaia hypothesis extends the ecocentrism of Leopold's land ethic to the chemical and physical structures and processes of the earth as a whole. The hypothesis conceives the planet as self-organizing and self-regulating, manifesting qualities of intelligence and apparently purposive (though Lovelock denies that his model makes the earth a teleological system). Gaia's unconscious goal is a 'planet fit for life'. Human beings are partners with or even subjects of natural and planetary forces in an egalitarian 'planetary democracy', rather than owners, tenants, guardians or managers. As such, we need to reconnect with nature and limit our impacts on it, or 'Gaia may throw off the human race' (Northcott, 1996, p110–112).

'Deep ecology' was developed initially by Arne Naess to counter or challenge the ecological movement of the 1960s and 1970s with its 'shallow' and short-term concern with solving the problems of resource depletion and environmental pollution. Not strictly a rational ethical theory, deep ecology advocates transformation of the basic principles guiding a long-term relationship with the environment and incorporates ideas of 'living a life that is simple in means, but rich in ends; honouring the right of all forms of life to live and flourish; empathizing with other forms; maximizing the diversity of human and non-human life; and maximizing the range of universal self-realization' (Armstrong and Botzler, 1993, pp369–370).

Theocentrism

Theocentrism shifts the focus from created things to the Creator. In the Biblical view humanity, 'made in God's image', is to exercise dominion and practise stewardship in relation to other creatures. Dominion is not, however, domination, but the wise and benevolent rule of a viceroy accountable to the ruler. Stewardship encapsulates ideas of tending and protecting. Both mandates are tempered by the Mosaic Law, which includes concern for animals and the land (Carruthers, 2002). Biblical theocentrism assigns value to non-human life, to the land and the earth, primarily on the basis that 'the earth is the Lord's' (Psalm 24).

Environmental virtue

Sandler (2005) has argued that the environment needs not only an ethic of action, but also an ethic of character, proposing four ways to specify environmental virtue:

- *Extensionism.* An environmentally virtuous person extends the standard interpersonal virtues such as gratitude, compassion or justice in an appropriate way to non-human entities.
- *Agent benefit.* The many benefits the environment offers individuals justify not only a disposition to preserve and protect these goods and opportunities, but also cultivation of the kind of character that allows one to enjoy them.
- *Human excellence.* A virtue enables its possessor to flourish as a human being. Human beings are social beings, so excellence includes a disposition to promote good social relationships. If human beings are also 'ecological beings', then excellence includes a disposition to promote the well-being of the wider ecological community.
- *Role models.* Environmental virtue can be discerned from studying exemplars of environmental excellence (Sandler mentions John Muir, Rachel Carson and Aldo Leopold, but is anxious to point out that the role models do not need to be public figures).

Northcott (1996) combines insights from anthropological studies of tribal cultures, feminist philosophy (Gilligan, 1981; Plumwood, 1993) and virtue ethics (MacIntyre, 1985) to present a compelling alternative to the abstract and individuated approach characteristic of much modern moral philosophy. Environmental virtue flourishes in the context of relationships, intimacy, connectivity and care, in the particular and local, in historical traditions and cultural identities, in small communities. 'The reversal of the environmental crisis' will come about 'when we recover a deep sense for the relationality of human life to particular ecosystems and parts of the biosphere, and where communities of place foster those virtues of justice and compassion, of care and respect for life, and prudence in our appetites and desires' (Northcott, 1996, pp122–123).

Agricultural ethics

Agriculture and environment

Agriculture uses more than one-third of the world's land, about half of the USA, and more than two-thirds of the UK, plus the largest share of fresh water. It might be expected, therefore, that environmental ethics would have a primary focus on agriculture, while agricultural ethics would emphasize environment.

In fact, as Thompson (2001) lamented, agriculture has been an infrequent topic within environmental ethics, while agricultural ethics has focused on issues related to technology and biotechnology, human and animal health, and animal welfare. This neglect of agriculture arises from the belief that substantial land exists where human beings have no significant effect and that environment ethics is solely about environmental impacts. The first assigns moral priority to wilderness, the second automatically vilifies agriculture, which by its nature involves impact on nature (i.e. 'the less agriculture the better'). 'The rhetoric of the environmental movement engendered critique of production rather than an ethic of production' (Thompson, 1995, p11) and environmental ethics became focused on conservation and preservation.

This marginalization of agriculture, Thompson (2001) argues, prevents us from asking the right philosophical questions and, hence, developing an adequate environmental ethic. What is needed are ethics for land and water that are 'first and foremost ethics for agricultural production' (Thompson, 2002, p172).

Equally, however, agriculture also needs to put its house in order. Agriculture has been driven by 'technocentric productionism, the headlong and unreflective application of industrial technology for increasing production' (Thompson, 1995, p70). This is anti-environmental as evidenced by agriculture's environmental record and the trenchant environmental critique of agriculture over the last decades. 'What we are after is an ethic of farming, a philosophy of agriculture, with particular attention to agriculture's impact upon and integration with the wider natural world' (Thompson, 1995, p19).

A further category of interest to agricultural ethics is the 'cultural, historical and social significance of agriculture as a way of life and a system of connected institutions' (Thompson, 2001, p42). And it is here that Thompson finds the way forward for an agricultural environmental ethic, 'drawn from the very practices of farming itself' (Thompson, 1995, p46) and from the insights of traditional agrarianism, yet informed by the science of sustainability (Thompson, 2001). Important aspects include the reconnection of food producers and consumers, and the development of an ethic of community, identity and solidarity (including with nature). Citing Wes Jackson (1996), Thompson concluded with a now familiar theme: 'human communities must concentrate on becoming native to the places they inhabit, on being attentive to landscape, and organizing life's rituals in a manner that will make consumers and producers alike more aware of ecosystem health' (Thompson, 2001, p470).

It might be argued that the above reflects a primarily US perspective and one of a few years ago. In the UK, there is no wilderness to fixate on and farming has 'greened up' appreciably. However, I would argue, the core diagnosis applies equally and the challenge remains to develop an ethic that embraces, rather than ring fences, production, people and environment. In fact, an ethic that legitimizes land for production may be particularly needed in the face of a shift in favour of non-product public goods.

Biofuels

One area that has interested agricultural ethicists recently is the production of biofuels to help mitigate climate change. Thompson (2008) has examined the 'push towards biofuels' (p185) in the North American context in terms of two philosophical perspectives on the purpose of agriculture, industrial and agrarian, arguing that the 'current trajectory of biofuels places it squarely in the domain of industrial agriculture' (p197) and the power structures associated with it. As such, biofuels are disconnected from, and antithetical to, the diverse moral goods associated with agrarianism, characterized by sustainability, alternative food networks and so on. Thompson (2008) calls for a 'democratic' construction of the future of biofuels in open fora in which everyone's ideas are taken seriously.

Jordaan (2007) has come to a similar conclusion. She argues that biofuels have the potential to transform many of the earth's natural landscapes into monocultures. This land transformation leads to ethical trade-offs and competing social institutions with different values. These issues need to be addressed prior to enacting policy. A scientific approach is too reductionist to achieve a fair outcome, and she advocates the use of a wide reflective equilibrium (WRE) process – an interactive dialogue between all interests, subjective and objective, scientific and value-laden, working back and forth, revising judgements, until a just equilibrium is reached.

Traditional land ethics

Traditional ethics for farming and land offer a challenging perspective on our present situation and echo some of the themes already explored above. Wibberley (1996) has outlined the traditional African approach to land as follows: landlessness is bad, but so are land speculation and expansionism; land is joint property and the right to use it derives from community membership (i.e.

family, clan, tribe, village); the Chief is the custodian of land on behalf of all; land should be allocated according to need; failure to cultivate forfeits the right to use land; collective responsibility is good and necessary for security.

Several of these ideas are reflected in the land economy of the Old Testament, with its lynch pins of Sabbath and Jubilee. The sabbath day was a day of rest for people *and* livestock (Exodus 20:8–11). It anticipated the sabbath year, when debts were cancelled (Deuteronomy 15:1–11) and the land itself rested (Leviticus 25:1–7). In the Jubilee year, every 50 years (Leviticus 25:8–55), 'each was to return to his property and each to his family'. The Jubilee emphasized the inalienability of family land and, in effect, strictly limited the growth of private wealth. Sabbath and Jubilee placed a radical constraint on relentless production (and, by implication, consumption) and protected those without a voice and without power – the poor, livestock and the land itself (Carruthers, 2002).

Metaphors and movements

Wilderness

'Wilderness' is a powerful metaphor configuring the relationship between humanity and land. 'Wilderness' can be understood as referring to places where the original ecology is intact and unchanged by human intervention (Lamb et al, 2006) (though such places are largely an ideal rather than a reality), or to unsown, waste or wild land, a social definition that emerged with the development of settled agriculture (Short, 1991). But it is the idealized wilderness, even of places that do not qualify under either of these two definitions, that is the most significant.

Throughout history, wilderness has been both feared and revered. On the one hand, it is dangerous, fearsome, the abode of evil things. On the other hand it is sacred space, a place of enhanced spiritual awareness, a 'symbol of earthly paradise' (Short, 1991, p6). Short (1991) explains this dichotomy in terms of classical and romantic perspectives. For the classicist, who looks *forward* to a Golden Age, taming the wilderness is a mark of progress and the imperative is to turn wilderness into 'garden' (see below). For the romantic, who looks *back* to a Golden Age, the conquest of the wilderness is a 'fall from grace' and the imperative is to regain paradise by preserving it.

The romantic wilderness ideal has done much to shape environmentalism and environmental ethics, particularly in the USA, through figures such as Henry David Thoreau (1817–1862) and John Muir (1838–1914), whom Schama (1996, p7) describes as the 'founding fathers of modern environmentalism'. Muir's thought was profoundly shaped by his walks in the mountains of

California, and he went on to found the Sierra Club, which remains one of the most influential environmental organizations in North America.

The roots of the wilderness movement can, however, be traced to the English romantics, especially the Lake Poets, especially William Wordsworth (Bunce, 1994). Wordsworth's *Lines written a few miles above Tintern Abbey* (1798) encapsulates many aspects of the wilderness ideal:

- The poem is prompted not by abstract reflection, but by a concrete encounter with nature in a particular place and time, a place which had become part of the poet's own history.
- Many lines in the poem are suggestive of the poet's deep longing to connect with nature, which stirs emotions of peace, joy, love and even worship.
- People are present and human artefacts evident, but they blend into the natural scenery ('among the woods and copses lose themselves') and do not 'disturb the wild green landscape'.
- Nature is a source of moral value and spiritual awareness ('the anchor of my purest thoughts, the nurse, the guide, the guardian of my heart').
- Nature has an existence and value in itself, symbolizing the 'spiritual unity of all things in which humans are united with the cosmos' (Northcott, 1996, p87) ('a presence that disturbs me with joy ... a sense sublime of something more deeply interfused').

The poem itself is indicative of the creative impetus of wilderness, a creative impetus that also issued in the 'establishment of voluntary societies and trusts for the preservation of areas of great natural beauty, such as the Lake District in North West England, by Romantic exponents such as Ruskin and Wordsworth' (Northcott, 1996, p88).

The wilderness metaphor, therefore, focuses many of the themes already explored here: connecting to land and nature is part of being human and this connection is made in particular places in the context of one's own story; land and nature are self-existent, have intrinsic value and virtue; land and nature are a source of virtue, wisdom and a moral guide; human beings belong to the land, are part of nature and, by implication, part of a greater unity along with the other members of the Earth community.

The garden

The converse of wilderness is the garden. As suggested above, the classical view of progress can be symbolized by the transforming of wilderness to garden. But there remains a tension between the impetus to tame the wilderness and a deep resonance with wild nature. This wilderness versus garden dialectic, it can be argued, is fundamental to the relationship between people and land.

Further, the garden itself is something of a two-edged metaphor. On the one hand, it conveys the idea of peaceful and constructive co-operation between humanity and nature or God (Carruthers, 2005), a place of safety and security, of self-realization and fulfilment. The Persian word translated 'paradise' referred originally to an enclosed garden. 'In the agrarian setting God's country is often a garden, tended lovingly and faithfully by the farm families that are God's faithful stewards' (Thompson, 1995, p56). Perhaps, most importantly, the garden emphasizes simply that people need to work and eat; that what land is for is, first and foremost, to meet human needs and maintain human life.

On the other hand, the garden is suggestive of conquest and control. Gardens are, by nature, places where people are at work transforming nature, reshaping, ordering, modifying it to achieve human goals. The formal gardens of the 18th century were in some senses part of a political landscape and expressions of status and power. Further, if all of nature is a garden, then all of nature is to be managed. As Thompson (1995) has argued, the garden metaphor, along with the ethic of hard work, has been used to reinforce the productionist ethic and has pitted agriculturalists against environmentalists, in a conflict that is ultimately philosophical and ethical.

Agrarian stewardship

Agrarianism perhaps expresses the garden metaphor in its best sense. Essentially a North American phenomenon, its leading 'prophet' is Wendell Berry (Wirzba, 2002b). Berry is a poet, novelist and essayist, but it is his life as a farmer in Kentucky that has most significantly shaped his agrarian thought. Contemporary agrarianism is variously a movement, a culture, a way of life and a philosophy. The latter is of special interest here. The main themes of agrarian thought are as follows:

- *Virtue.* Agriculture, working the land, is both virtuous and imparts virtue, not only to individuals, but also to society as a whole (Hilde and Thompson, 2000; Thompson, 2001, 2008).
- *Knowledge.* Cultivating the land imparts knowledge and moral understanding – 'a privileged outlook upon questions of human conduct and, sometimes, the nature of reality itself' (Hilde and Thompson, 2000, p1). Wirzba (2003, p1) has suggested that 'a culture loses its indispensable moorings, and thus potentially distorts its overall aims, when it foregoes the sympathy and knowledge that grows out of cultivating (*cultura*) the land (*ager*)'.
- *Humanness.* Human life is 'embedded in the land' (Wirzba, 2002a, pxiii) and in working the land: 'I farm, therefore I am'! If manipulating nature is essential to human ontology and identity, then agriculture, 'the intimate

nurturing, giving, taking, and renewing between humans and the land, brings us closest to nature in its fullest sense' (Hilde and Thompson, 2000, p2).

- *Community.* The social context and contribution of farming is fundamental to the agrarian vision. Individuals live and work in dynamic networks of specific people (family, neighbours, suppliers, buyers and so on). Diversified, smaller family farms make for healthy rural communities (Thompson, 2008; Pretty, 2002).
- *Work.* Work is also fundamental to being human, and working the land is the most authentic form of work. Work also connects us to the land – 'people are joined to the land by work' (Berry, 2002b, p189) (and, by implication, working the land is a more meaningful way of connecting to nature than reflecting and using it for recreation, for example).
- *Place.* Agrarianism stresses the particular and local. Relationships among people and with the land are focused around the 'immovable geography of the farm' (Thompson, 2001, p468). Farmers have a sense of place and connectedness to place, which is established over time and advanced by community. 'A healthy culture holds preserving knowledge in *place* for a *long* time' (Berry, 2002b, p189).
- *Stewardship.* Stewardship of the land and nature is fundamental, and a 'virtue reinforced by the feedback mechanisms of nature itself' (Thompson, 2001, p469). Closeness to the land, continuity in time, permanence in place and connection to community all promote long-term commitment to maintaining soil fertility and environmental health and prevent the pursuit of short-term goals and the 'tragedy of the commons'.

Agrarianism's proponents believe we face a choice between industrialism, the way of the machine, of violence, of war (Berry, 2003; Shiva, 2003), and agrarianism, a 'compelling alternative to the modern industrial/technological/economic paradigm ... concerned with the health and vitality of a region's entire human and nonhuman neighbourhood' (Wirzba, 2003, p4) and, as evident above, peaceful, permanent and virtuous. The agrarian vision may seem idealistic, and its advocates are quick to point out that a return to traditional agrarian society is unfeasible. But agrarianism provides some profound challenges and some clear ethical guidelines for our future use of land and our relationships with one another and with nature.

Organic farming

The closest UK counterpart to the USA agrarian movement is, I would suggest, organic farming, in particular organic farming as practised and described by the movement's founders.

Organic farming originated from the pioneering work, in the 1930s and 1940s, of Sir Albert Howard and his associates, notably Lady Eve Balfour, author of the now classic text *The Living Soil* (Balfour, 1943), and instrumental in the founding of the Soil Association in 1946.

The core of the approach was a concern for the health of the soil. Fundamental is the 'Rule of Return, which decrees that the soil's health and fertility is maintained by encouraging the presence of humus' (Conford, 2001, p17) – 'feed the soil and the soil will feed the plant'. A major factor prompting this focus on soils was the American Dust Bowl, which served as a powerful warning of the reality and risks of soil erosion (Jacks and Whyte, 1939). Man's 'contract' with the soil came to be understood not only as the basis of production, but also as the foundation on which civilizations stood or fell (Hyams, 1952; Carter and Dale, 1955).

However, the pioneers' agenda was much wider than simply an alternative way of farming. Their vision was of a new society. Organic proponents argued that a 'revivified rural life, based on the principles of husbandry, could provide an answer to the problems which beset Britain in the 1920s and 1930s. Land work would reduce unemployment; humus-grown food and open-air tasks would improve standards of health; industry would be smaller-scale and geared to the needs of rural activities, and this would counteract centralization and alienation' (Conford, 1988, p14).

Spiritual values and the link between spiritual and social were also important. Balfour (1948, p188) wrote that 'when a new generation has arisen, taught to have a living faith in the Christian ideals, to value and conserve its soil, and to put service before comfort, then not only will our land have citizens worthy of it, but it will also be a land of happy contented people'. Conford (1988, p12) argued that organic farming could be seen as 'inherently religious in a general sense, since it is based on reverence for the laws of nature – literally on humility – rather than the arrogant assumption that the earth can be indefinitely persuaded or forced to do Man's will'.

Organic farming shares many ideas and ideals with agrarianism. While it may in its present form place less explicit emphasis on the virtues of working the land and the moral value of the farming community, it perhaps places greater emphasis on harmony with, caring for and learning from nature. Together, it could be argued, they provide a potent paradigm for an environmentally and socially sustainable agriculture and a holistic land ethic.

Conclusions

The above leaves many stones unturned, many depths unplumbed, as I am acutely aware. I have not done full justice, for example, to natural law ethics, religious perspectives or postmodernism, nor dealt in any depth with issues of power, rights and the resolution of land-use conflicts. No attempt has been made to address specific dilemmas or to review the ways in which specific problems have been tackled. And there are many important thinkers and activists who have not had a look in. This is 'work in progress', though, and these are all tasks for the future.

Nevertheless, this brief survey does offer some tools to employ and some pointers to direct us in deciding 'what land is for'. These are summarized below, along with some critical comments and personal perspectives.

Encounter

A central theme above is the role of encounter in shaping thinking. Philosophy and ethics are informed by real-life experience. We may not necessarily agree with the conclusions of Wordsworth, Muir or Berry, but we should take note of their methodology, of the way in which their thought is deeply rooted in experiential reality and authentic encounters.

We need, then, to 'enhance our land experience' and to create opportunities for people (perhaps especially young people and city-dwellers) to encounter land, nature and farming, both directly and through art, literature and music. These encounters can stimulate imagination, create vision and help build a kind of 'moral credit'.

I am less convinced that encountering land, whether as 'wilderness' or 'garden', whether through walking or farming, whether in the John Muir or Wendell Berry way, is automatically virtuous or imparting of virtue. Few would argue that ramblers or farmers are somehow by definition moral exemplars (although these pursuits may be more admirable than many others). But encountering land can be virtuous and impart virtue. Also, the agrarian vision is as much about vibrant human relationships as it is about embeddedness in the land – virtue is encased in the whole land/place/community/continuity package.

Philosophy and ethics are also informed by reading and reflection. The 'role models' above have established a dialogue between observation and study, between encounter and expression, rehearsing, refining and communicating their thought through their writing. The 'concrete' and the 'abstract' speak to each other, and scholarship, in science, arts and humanities, including moral philosophy, has an essential role to play in shaping our land ethics.

Finally, our philosophy of land, environment and agriculture, and of what it means to be human, shapes our decisions and action; how we think determines what we do. Such a formula may not be of great interest to practitioners and policy makers, but it nevertheless reflects their and all of our conscious or unconscious modus operandi, and the philosophical and ethical task remains vital.

Procedure

A second important theme concerns the way we make ethical decisions. Ethical beliefs might differ quite fundamentally, yet a significant measure of agreement as to what needs to be done can often be reached. Equally, ethical pluralism means there may be no moral consensus, no 'received values'. In both cases, 'procedure' plays a central role. If people with diverse views can agree on a procedure, then some progress can be made. While it would be absurd to argue that ethics is only about procedure, procedure's role in practice is undeniable.

Some of the material considered above rightly challenges the reductionist, anthropocentric, utilitarian mindset, and the tendency to submit or off-load responsibility to the 'experts', which characterizes so much contemporary decision-making, and policy formulation and evaluation. Rights-based approaches help define the nature and limits of acceptable intervention. Ecocentrism and theocentrism challenge an exclusive focus on human beings and human welfare. Virtue ethics shift our focus from the things we do to the sort of people we are, emphasizing relationships, community and connectedness among people and between people and land. In developing procedures, I propose the following as particularly important or promising:

- The way 'people' and, by association, preferences and values (Jacobs, 1997) are constructed in decision-making procedures is fundamental. Here, deliberative approaches (which treat people as citizens, making informed decisions for the good of society) are to be preferred to consultative methods (which treat people as self-centred consumers or representatives of 'public opinion', presumably on the basis that Adam Smith's 'invisible hand' is still operating!).
- Procedures that allow all sorts of voices to be heard and heeded, and avoid a focus just on the experts (i.e. those 'in the know'), the pressure groups (i.e. 'those who shout the loudest') or the 'movers and shakers' (i.e. 'men of power' and 'power groups') are also desirable. This is not an easy target to hit, and the tendency to resort to the 'usual methods' is partly because doing things differently is difficult. The WRE approach mentioned above may be one way forward.
- 'Charters' provide a basis for expressing shared values and agenda and the outcomes of agreed procedures. Charters have been effectively employed, for example, in setting standards for animal welfare, focusing a measure of agreement between people with sometimes quite diverse ethical views.

In relation to land, these techniques could be especially usefully applied to complex, multi-faceted, multiple-stakeholder land/people-based issues such as the current debate about the future of the English uplands. Greater emphasis could be placed on deliberative approaches (e.g. centred on proposals or scenarios) rather than consultative approaches (i.e. listening to stakeholders). A dynamic interactive and purposeful dialogue could be established, incorporating all sorts of perspectives, including ethical ones. And a 'charter for the uplands' could be developed.

Relationships

Relationships are my third theme. Relationships and community are the essential context of ethics. Indeed, ethics are only really intelligible in a relational context, and, as MacIntyre argued, in a historical and geographical setting. The dynamic interplay between ethics and encounter applies as much to community as it does to land. Land ethics are developed and practised in the context of the 'land community'. I would suggest there are three particularly important and related issues for us here:

1 It matters how specific human communities position themselves in relation to land. A holistic focus on land–place–community–history–work–time (as per the agrarian vision) and narrative and sense-of-place approaches are to be preferred to disaggregated, individuated, reductionist viewpoints.
2 It matters how many people farm the land, how diverse farms are and who owns them. Smaller family farms are not necessarily more 'ethical' or 'environmental' than larger farms (indeed, in the UK, most farmers of whatever size operate out of an industrial mindset), but the retention of people and families on the land seems the most promising way to secure the land-related values and virtues described above.
3 It matters how we as human beings position ourselves in relation to the rest of nature. We may not wish to embrace a view that values people no more highly than pigs or potatoes, but there is a compelling case, on the basis of many of the ethical systems considered here, both to accept our place as citizens of the Earth community and, at the same time, to recognize our special responsibility to its other members.

The future

Finally, the overriding imperative regarding land is to do something and do it now! The question is: 'What?'. Our present agenda is profoundly influenced by intense concern about the future, especially the future that, it is believed, will result from climate change. Our concern about climate change is, however, based on predictions; scientific, strongly evidenced and widely accepted predictions admittedly, but, nevertheless, predictions. And predictions can be wrong. The future does not send us any signals; we only have signals from the present about what we think might happen and, 'beyond the short term, such signals are inherently unreliable' (Foster, 2008, p15).

This is not to say that we can afford to ignore climate change, which, as this volume demonstrates, presents both problems and opportunities for land use (and which, it would seem, is already kicking-in in many parts of the world). The need, however, seems to be to build a deeper and wider robustness and resilience into land use and society that is capable of weathering not only the storm of climate change, but also other, as yet unforeseen, challenges. And the ethical viewpoints and visions reviewed here may help to reset some of the balance, to allow our decisions and actions to be shaped ultimately by an understanding of human identity and purpose and of our relationships with one another and with land and nature.

As good as any expressions of this sentiment are the thoughts on the future of Wendell Berry, to whom I give the last word:

> We do not need to plan or devise a 'world of the future'; if we take care of the world of the present, the future will have received full justice from us. A good future is implicit in the soils, forests, grasslands, marshes, deserts, mountains, rivers, and oceans that we have now, and in the good things of human culture that we have now; the only valid 'futurology' available to us is to take care of those things. We have no need to contrive and dabble at 'the future of the human race'; we have the same pressing need that we have always had – to love, care for, and teach our children (Berry, 2002c, p73).

References

Abelson, E. (1988) (ed) *A Mirror of England: An Anthology of the Writings of H.J. Massingham (1988–1952)*, Green Books, Bideford, UK

Alexander, L. and Moore, M. (2008) 'Deontological ethics', in Zalta, E.N. (ed) *The Stanford Encyclopaedia of Philosophy*, Fall 2008 edn, plato.stanford.edu/archives/fall2008/entries/ethics-deontological/ accessed in October 2008

Archbishops' Commission on Rural Areas (1990) *Faith in the Countryside*, Churchman Publishing, Worthing, UK

Anderson, J.N.D. (1998) 'Law', in Ferguson, S.B. and Wright, D.F. (eds) *New Dictionary of Theology*, Inter-Varsity Press, Leicester

Armstrong, S.J. and Botzler, R.G. (1993) *Environmental Ethics, Divergence and Convergence*, McGraw-Hill, New York

Attfield, R. (1991) *The Ethics of Environmental Concern*, University of Georgia Press, London

Balfour, E. (1943) *The Living Soil*, Faber, London

Balfour, E. (1948) *The Living Soil*, Revised Edition, Faber, London

Berry, W. (2002a) 'The gift of good land', in Wirzba, N. (ed) *The Art of the Commonplace: The Agrarian Essays of Wendell Berry*, Shoemaker and Hoard, Washington, DC

Berry, W. (2002b) 'People, land and community', in Wirzba, N. (ed) *The Art of the Commonplace: The Agrarian Essays of Wendell Berry*, Shoemaker and Hoard, Washington, DC

Berry, W. (2002c) 'Feminism, the body, and the machine', in Wirzba, N. (ed) *The Art of the Commonplace: The Agrarian Essays of Wendell Berry*, Shoemaker and Hoard, Washington, DC

Berry, W. (2003) 'The agrarian standard', in Wirzba, N. (ed) *The Essential Agrarian Reader*, The University Press of Kentucky, Lexington, Kentucky

Brennan, A. and Lo, Y.-S. (2008) 'Environmental ethics', in Zalta, E.N. (ed) *The Stanford Encyclopedia of Philosophy*, Fall 2008 edn, plato.stanford.edu/archives/fall2008/entries/ethics-environmental/ accessed in October 2008

Breugemann, W. (1977) *The Land*, Fortress Press, Philadelphia, PA

Breugemann, W. (2002) *Place as Gift, Promise, and Challenge in Biblical Faith*, Augsburg Fortress, Minneapolis, MN

Bunce, M. (1994) *The Countryside Ideal. Anglo-American Images of Landscape*, Routledge, London

Carruthers, S.P. (2002) 'Farming in crisis and the voice of silence', *Ethics in Science and Environmental Politics*, vol 2, pp59–64

Carruthers, S.P. (2005) 'Creation and the Gospels', in Tillett, S. (ed) *Caring for Creation*, The Bible Reading Fellowship, Oxford

Carter, V.G. and Dale, T. (1955) *Topsoil and Civilisation*, University of Oklahoma, Oklahoma

Cheney, J. (1993) 'Postmodern environmental ethics: Ethics as bioregional narrative', in Armstrong, S.J. and Botzler, R.G. (eds) *Environmental Ethics, Divergence and Convergence*, McGraw-Hill, New York, pp86–96

Clark, K.J. (2003) 'Why be moral? Social contract theory versus Kantian-Christian morality', *Markets and Morality*, vol 6, no 1, www.acton.org/publications/mandm/mandm_article_60.php/ accessed in October 2008

Conford, P. (1988) 'Introduction', in Conford, P. (ed) *The Organic Tradition*, Green Books, Bideford, Devon

Conford, P. (2001) *The Origins of the Organic Movement*, Floris Books, Edinburgh

Convery, I. and Dutson, T. (2006) 'Sense of place project report', University of Central Lancashire, Newton Rigg, UK, www.theuplandcentre.org.uk/pubs.htm

Cudd, A. (2008) 'Contractarianism', in Zalta, E.N. (ed) *The Stanford Encyclopedia of Philosophy*, Fall 2008 edn, plato.stanford.edu/archives/fall2008/entries/contractarianism/ accessed in January 2009

Darling, F. (1993) 'Man's responsibility for the environment', in Armstrong, S.J. and Botzler, R.G. (eds) *Environmental Ethics, Divergence and Convergence*, McGraw-Hill, New York, pp278–301

Fieser, J. (2006) 'Ethics', in Fieser, J. and Dowden, B. (eds) *The Internet Encyclopaedia of Philosophy*, www.iep.utm.edu/e/ethics.htm#SSH2c.i./ accessed in June 2008

Foster, J. (2008) *The Sustainability Mirage, Illusion and Reality in the Coming War on Climate Change*, Earthscan, London

Gilligan, C. (1981) *In a Different Voice: Psychological Theory and Women's Development*, Harvard University Press, Cambridge, MA

Hardin, G. (1968) 'Tragedy of the commons', *Science*, vol 162, pp1243–1248

Hargrove, E.C. (1989) *Foundations of Environmental Ethics*, Prentice Hall, Englewood Cliffs, NJ

Hilde, T.C. and Thompson, P.B. (2000) 'Agrarianism and pragmatism', in Thompson, P.B. and Hilde, T.C. (eds) *The Agrarian Roots of Pragmatism*, Vanderbilt University Press, Nashville, TN, pp1–21

Hursthouse, R. (2008) 'Virtue ethics', in Zalta, E.N. (ed) *The Stanford Encyclopaedia of Philosophy*, Fall 2008 edn, plato.stanford.edu/archives/fall2008/entries/ethics-virtue/ accessed in October 2008

Hyams, E. (1952) *Soil and Civilisation*, Thames and Hudson, London

International Centre for the Uplands (2008) 'English northern uplands sense of place project', www.theuplandcentre.org.uk/sense.htm/ accessed in September 2008

Jacobs, M. (1997) Environmental valuation, deliberative democracy and public decision-making institutions, in Foster, J. (ed) *Valuing Nature*, Routledge, London

Jacks, G.V. and Whyte, R.O. (1939) *The Rape of the Earth: A World Survey of Soil Erosion*, Faber, London

Jackson, W. (1996) *Becoming Native to this Place*, Counterpoint, San Francisco

Jordaan, S.M. (2007) 'Ethical risks of attenuating climate change through new energy systems: The case of a biofuels system', *Ethics in Science and Environmental Politics*, vol 7, pp23–29

Lamb, J., Goodrich, G., Brame, S.C. and Henderson, C. (2006) *NOLS Wilderness Ethics: Valuing and Managing Wild Places*, Stackpole Books, Mechanisburg, PA

Leopold, A. (1949, republished 1977) *The Land Ethic. A Sand County Almanac: and Sketches Here and There*, Oxford University Press, Oxford

MacIntyre, A. (1985) *After Virtue*, 2nd edn: Duckworth, London

Murdoch, J., Lowe, P., Ward, N. and Marsden, T. (2003) *The Differentiated Countryside*, Routledge, London

Murphy, M. (2008) 'The Natural Law Tradition in Ethics', in Zalta, E.N. (ed) *The Stanford Encyclopaedia of Philosophy*, Fall 2008 edn, plato.stanford.edu/archives/fall2008/ entries natural-law-ethics/ accessed in October 2008

Newall, P. (2005) 'Introducing philosophy 11: Ethics', www.galilean-library.org/manuscript.php?postid=43789/ accessed in October 2008

Northcott, M.S. (1996) *The Environment and Christian Ethics*, Cambridge University Press, Cambridge

Plumwood, V. (1993) *Feminism and the Mastery of Nature*, Routledge, London

Pretty, J. (2002) *Agri-culture, Reconnecting People, Land and Nature*, Earthscan, London

Pretty, J. (2007) *The Earth Only Endures*: Earthscan, London

Pretty, J. (2008) 'Green exercise', www.essex.ac.uk/bs/staff/pretty/green_ex.shtm/ accessed in June 2008

Regan, T. (1988) *The Case for Animal Rights*, University of California Press, Berkeley

Rolston, H. (1988) *Environmental Ethics: Duties to and Values in the Natural Environment*, Temple University Press, Philadelphia

Rowland, C. (1990) 'The goals and values of traditional African land tenure systems', in Schluter, M. and Ashcroft, J. (eds) *Christian Principles for the Ownership and Distribution of Land*, Jubilee Centre Publications, Cambridge

Sandler, R. (2005) 'Introduction: Environmental virtue ethics', in Sandler, R. and Cafaro, P. (eds) *Environmental Virtue Ethics*, Lanham, Maryland: Rowman and Littlefield Publishers, Lanham, MD

Sayre-McCord, G. (2008) 'Metaethics', in Zalta, E.N. (ed) *The Stanford Encyclopaedia of Philosophy*, Fall 2008 edn, plato.stanford.edu/archives/fall2008entries/metaethics/ accessed in November 2008

Schama, S. (1996) *Landscape and Memory*, Fontana Press, London

Schluter, M. and Ashcroft, J. (1990) *Christian Principles for the Ownership and Distribution of Land*, Jubilee Centre Publications, Cambridge

Schweitzer, A. (1993) 'The ethics of reverence for life', in Armstrong, S.J. and Botzler, R.G. (eds) *Environmental Ethics, Divergence and Convergence*, McGraw-Hill, New York, pp342–346

Shiva, V. (2003) 'Globalization and the war against farmers and the land', in Wirzba, N. (ed) *The Essential Agrarian Reader*, The University Press of Kentucky, Lexington, Kentucky

Short, J.R. (1991) *Imagined Country*, Routledge, London

Singer, P. (1979) *Practical Ethics*, Cambridge University Press, Cambridge

Singer, P. (1990) *Animal Liberation*, Avon, New York

Sinnott-Armstrong, W. (2008) 'Consequentialism', in Zalta, E.N. (ed) *The Stanford Encyclopaedia of Philosophy*, Fall 2008 edn, plato.stanford.edu/archives/fall2008/entries/consequentialism/ accessed in October 2008

Stone, C. (1993) 'Should trees have standing?' in Armstrong, S.J. and Botzler, R.G. (eds) *Environmental Ethics, Divergence and Convergence*, McGraw-Hill, New York pp255–264

Storey, D. (2001) *Territory: The Claiming of Space*, Prentice Hall, Harlow, UK

Taylor, P. (1993) 'Respect for nature', in Armstrong, S.J. and Botzler, R.G. (eds) *Environmental Ethics, Divergence and Convergence*, McGraw-Hill, New York pp353–367

Teo, P. and Huang, S. (1996) 'A sense of place in public housing. A case study of Pasir Ris, Singapore', *Habitat International*, vol 20, pp307–325

Thompson, M. (2003) *Teach Yourself Ethics*, Hodder Arnold, London

Thompson, P.B. (1995) *The Spirit of the Soil, Agriculture and Environmental Ethics*, Routledge, London

Thompson, P.B. (2001) 'Land and water', in Jamieson, D. (ed) *A Companion to Environmental Philosophy*, Basil Blackwell, Oxford and Malden, MA, pp460–472

Thompson, P.B. (2002) 'Land', in Comstock, G.L. (ed) *Life Science Ethics*, Iwoa State Press, Iowa, pp169–190

Thompson, P.B. (2007) 'Agriculture and working-class political culture: A Lesson from *The Grapes of Wrath*', *Agriculture and Human Values*, vol 24, no 2, pp165–177

Thompson, P.B. (2008) 'The agricultural ethics of biofuels: A first look'. *Journal of Agricultural and Environmental Ethics*, vol 21, pp183–198

Thompson, R.B., Matthews, R.J. and Ravenswaay, E O. van (1994) *Ethics, Public Policy and Agriculture*, Macmillan, New York

Vardy, P. and Grosch, P. (1999) *The Puzzle of Ethics*, Revised Edition, Fount, London

Wibberley, E.J. (1996) 'Agricultural ethics and ethical agriculture', in Carruthers, S.P. and Miller, F.A. (eds) *Crisis on the Family Farm: Ethics Or Economics*, CAS Paper 28, Centre for Agricultural Strategy, Reading

Wirzba, N. (2002a) (ed) *The Art of the Commonplace: The Agrarian Essays of Wendell Berry*, Shoemaker and Hoard, Washington DC

Wirzba, N. (2002b) 'Introduction', in Wirzba, N. (ed) *The Art of the Commonplace: The Agrarian Essays of Wendell Berry*, Shoemaker and Hoard, Washington DC

Wirzba, B. (2003) 'Introduction', in Wirzba, N. (ed) *The Essential Agrarian Reader*, The University Press of Kentucky, Lexington, Kentucky

Woods, M., Richards, C., Watkin, S., Heley, J., Convery, I., Dutson, T., Rogers, J. and Storey, D. (2008) *Rural People and the Land – the Case for Connections*, Report to the Commission for Rural Communities, Commission for Rural Communities, Cheltenham

14

Conclusions: The Emerging Contours of the New Land Debate

Michael Winter and Matt Lobley

Introduction: the end of the beginning

We start our conclusions with a Churchillian turn of phrase for two reasons: first because it implies crisis and secondly because it implies possibility. Churchill, for much of his life not the most successful of politicians, discovered his true worth when he confronted crisis. His actual words, in a speech at the Mansion House on 10 November 1942, marking the defeat of Rommel at El Alamein, were as follows: 'this is not the end. It is not even the beginning of the end. But it is, perhaps, the end of the beginning.' The crisis of our time is presented by the twin challenges of climate change and resource depletion. We are certainly not claiming that this modest volume contributes greatly to the solution of such a crisis! Far from it. What we are suggesting is that this volume and many, many other scientific publications in the last decade have been a beginning. No longer are questions of resource depletion dismissed by serious commentators as merely neo-Malthusian scaremongering. No longer is climate change denial a serious scientific position. The battle to put these issues on the global policy and science agendas has been won. The entry into public discourse of carbon trading, low carbon activity, one-planet living, food security, energy security, water security, biosecurity, and ecosystem services, are all evidence of that victory. And the significance of this beginning should not be underestimated. For those of us who have been students of environmental issues for several decades, the transformation of discourse is dramatic, as discussed in the second section of this chapter.

But this is only the end of the beginning for some very obvious reasons. Winning the hearts and minds of many scientists, policymakers and citizens is not the same as real actions that will reduce both resource use and carbon emissions. The translation of political rhetoric and goodwill into regulatory regimes is fraught with difficulty; so too is ensuring that the green branding ('ecological modernization'; Buttel, 2000) developed by manufacturers and retailers will deliver genuine environmental gains. The new environmental citizenship (Barr, 2008) from waste recycling to the transition town movement, competes with the continuing pull of consumerism and the lure of affluence. Not surprisingly, for some, these manifest failures to act on knowledge long held suggest an apocalyptic scenario. James Lovelock (2009) in his most recent book issues a 'final warning' that the hard-wired desire in humans to continue 'business as usual' will probably 'prevent us from saving ourselves'.

Notwithstanding these chilling words, the task of social and natural scientists, in this context, is to provide warnings, but more than that, to offer opportunities and options. Our view, which we return to in the final section of this chapter, is that the UK scientific community faces some formidable challenges in contributing to the debate around the specific issue of land use but great opportunities too. Our focus in this book has been on rural land use primarily in a UK context. We realize, of course, that in a highly urbanized society and for such a small island this may seem to some an extravagant use of our time. Why not the larger challenge of sustainable urban systems? Why not a global perspective? Apart from the obvious limitations of our own research interests and expertise, our answer lies in our passionate belief that all land and all places matter. It is one of the delusions of the age of mobility and of speed that the world is smaller and the appropriate scale for investigation is necessarily always global. Mass communications, both material and electronic, may alter perceptions and perspectives but they do not alter the localness and specificity of underlying ecological processes. Scale is all in seeking to understand the land, its opportunities and challenges. We can only understand how 'the land' might be healed and also contribute to solutions to our crisis if we understand soil microbes, soil chemistry, field processes, local landforms, natural areas and catchments. We can only appreciate and rejuvenate the relationship between people and land if we understand people in place, what it means to be a citizen and a family within a locality.

So, for us, the end of the beginning has been to set out in this book at least some of the relevant ideas and thinking from both social and natural scientists which we hope will contribute to the continuing land debate. When we commenced our thinking on this volume in the autumn of 2007 and even when we held a seminar for contributors later, in the Spring of 2008, we only slowly became aware of the speed with which this debate was gathering pace. Critical to this developing debate have been the emerging findings of projects funded under the Relu Programme, some of which are represented in this volume, and the numerous knowledge-exchange and publishing endeavours which characterize Relu (see Phillipson et al, 2009). Equally important has been the decision by the UK Government's Foresight Programme to conduct separate programmes on Land Use Futures and Global Food and Farming Futures. There is a self-deprecatory tendency in UK science to lament the weakness of the relationship between scientists and policymakers and the lack of jointly forged strategies to better manage the future. Foresight with its avowed purpose of seeking to use 'the best evidence from science and other areas to provide visions of the future,' offers one important way to improve the relationship and bring science into the heart of policy consideration. Nor should we underestimate the importance of think tanks, select committees and commissions, all of which in recent years have focused on relevant land issues in ways that bring scientists and scientific evidence into dialogue with policy formulation.

The collection in this volume has not covered everything. There is a natural history to any edited collection and alongside those possible and putative chapters that blossomed and flourished are those that decayed and failed; some for reasons of competing and pressing demands on potential contributors' time. We rejected two draft chapters because we felt they did not quite fit the purposes of the volume, and our own overtures were rejected by some too busy to make a contribution. But the editors are culpable too – some issues we just did not address in our own minds until it was too late to find a contribution. No doubt there are other issues that even now we have yet to contemplate. Reviewers and commentators on this volume will no doubt draw these to our attention. It is also the case that space limitations forced us to make choices, and therefore some topics we chose not to cover. In that category lie various technological developments that may have an impact on land use. For example, the fibre and plastics industries are both heavily dependent on petrochemicals, and R&D to reduce dependence on oil-based products almost invariably implies moves towards crop-based products. We are aware of research on this but it is not represented in this volume. We have chapters on energy crops and AD, but acknowledge that we could have included work on woodfuel and indeed forestry more generally in a climate change context, and the land use and landscape implications of wind power. We consider biodiversity and ecosystem services in some detail but chose not to include conventional work on landscape and recreation planning (although the chapter by Firbank et al takes an innovative and sideways look at the landscape approach).

We feel more exposed in our failing to consider in any greater detail the implications of the new land agenda for agriculture itself! We admit that in part this was because we are working together on agriculture in a separate project, which can loosely be labelled as a study of the fitness of purpose of contemporary farming systems for sustainably delivering food in the era of climate change. But that is no real excuse for not presenting in this volume a fuller picture of the current state of UK agriculture, in short the rather obvious issue of what it means to use land to produce food. Our defence is our full recognition that food production remains the overwhelmingly most important use to which our land is put and that to do this topic justice would require a separate volume.

The evolving rural land discourse

This is not the place for a detailed discourse analysis of the changing ways in which land is conceptualized within the policy world. But it is clear from the contents of this book that the terms of reference for discussion about land and farming have changed remarkably fast. Notions of land abandonment and land surplus, common currency less than a decade ago, currently appear consigned to a footnote of agrarian history. After all, with the exception of the regulatory imposition of set-aside, farmers chose largely to ignore the theoretical projections of land surplus much as they chose to ignore the academic fantasy that they were now post-productivist (Evans et al, 2002). Elsewhere, we have examined aspects of the emerging discourse on sustainable agriculture in the early 2000s (Winter, 2006), through consideration of the 2002 Strategy for Sustainable Farming and Food (Defra, 2002). The strategy was implicitly built around three models for the future of farming and the rural economy: farmers as producers of food commodities in a global free market; farmers as multi-functional producers of public goods for which there is not an existing market, primarily through agri-environment schemes; and farmers as land-based entrepreneurs within a diversified rural economy. We are not suggesting that any of these models have been entirely superceded by the new terminology of security and services, but they have been significantly amended and in some respects undermined. The neo-liberal discourse around markets as the only conceptual foundation for food production has been undermined by both climate change, which scarcely featured in the 2002 Strategy, and the global recession. 'Soft' public goods, the multifunctional bi-products of farming, such as pleasing landscapes and wildlife, now take their place alongside the harder, more utilitarian notion of ecosystem services, encompassing carbon storage, water quality, etc. And rural development and the rural economy, so strong in the attempts in the 1990s to shift European policy away from agriculture, have diminished in rhetorical importance.

The new environmentalism of climate change is the root cause of these shifts and has put the traditional environmentalism of protecting biodiversity very much on the back foot. Critical to this is the utilitarian concept of the ecosystem services approach. As indicated in the contribution by Carruthers, the growth of utilitarian approaches to land use are in danger of eroding or displacing valuations of the natural world as of intrinsic value, with the danger of a corresponding diminution in traditional concerns with public access and environmental stewardship. We have not got the space here to consider all aspects of the new discourse. We will return at the end of the section to a brief account of our concerns with the notion of ecosystem services. But first we wish to identify the other key characteristics of the new discourse. Possibly the most important is that agricultural productivism is back. We can see this in the renewed self-confidence of the farming lobby in its espousal of the food security agenda and its new found alliances with energy interests. Conservationists have been on the defensive for the first time for many years and the agri-environment policy consensus that emerged so strongly in the 1980s and 1990s is in danger of fracturing. Even organic farmers appear to be more on the back foot, especially in the context of a challenging market place during the recession.

At the same time we should not forget that crises are prone to give rise to radicalism and exceptionalism. While there are elements of the farming lobby that seem keen for a return to 'a business as usual' model after the temporary blip of agri-environmental agriculture, the exceptions are the radicals who seek to shift the debate from climate change to peak oil and espouse a post-oil, post-industrial agriculture. From a return to horses to no-tillage agro-wood systems or micro-permaculture, the ideas are many. And the gap between the new radicals and the new productivists seeking to re-open the 'genetically modified' (GM) debate is a chasm.

But what of ecosystem services? It is strongly represented in this volume and it has a huge appeal that, in many ways, we find interesting. Its importance within a climate change-driven crisis is that it suggests very powerfully that food production, however significant the food security challenge, cannot be the only driver of land management decision-making. It offers new ways of conceptualizing certain important land uses, such as nature conservation sites, as part of a wider suite of functions. Thus conservation sites may be important also for storing carbon and managing water. Maybe in this pressured world the identification of functionality and utility really is the best way to approach the use and management of land. But there are problems. The notions of services is an economic one and is premised on value and price. However, as Heal (2000) has pointed out, ecosystem services don't easily fit into the framework of economic pricing, based on marginal adjustments to supply and demand. Services based on natural systems do not easily lend themselves to the marginal adjustments of markets. We should not forget in the current land debate that the origins of agricultural support mechanisms lay in part with the problems of market volatility caused by naturally induced variations in production levels.

Heal takes the valuation argument with regard to ecosystem services to its logical end point. Can we talk sensibly, he asks, 'of the value of preserving the climate system intact, or of the value of preserving biodiversity?':

> It will never make sense to ask about the value that we would lose if an entire and irreplaceable life support system were to be lost. The point is that if it is indeed a life-support system then its loss would lead to the end of all human life, and to put an economic value on that would

> seem foolish and inappropriate. (The numbers resulting from a recent attempt to do just this were described by one economist as 'a serious underestimate of infinity'.) … The conclusion that emerges from this analysis is that economics probably cannot really value the services of the earth's life-support systems in any way other than by using market prices, which value them in the sense of indicating the value of a small change in their availability. We should not be disappointed with this limited ability to value ecosystem services. If our concern is to conserve these services, then valuation is largely irrelevant. Let me emphasize this: Valuation is neither necessary nor sufficient for conservation. We conserve much that we do not value, and do not conserve much that we value. (Heal, 2000, pp28–29).

Heal stops short of asking whether our terminology should change as a result of these observations. We suspect that it might and Peter Carruthers offers pointers in this direction.

The evolving use of land

The projections of rapid population growth mean that, even in the absence of climate change and energy-security drivers, there will be significant pressures on the use of land to produce more food. It is not just a question of feeding a greater number of people, but also of feeding increasingly affluent (in relative terms) populations who will, not unreasonably, demand more livestock and dairy products in their diets. Some of this increased food production will be achieved through the expansion of agricultural lands, but much will have to be achieved through the further intensification of existing farmland, something environmentalists have long cautioned against. However, population growth is not the only driver. The 'perfect storm' of increasing demand for food, increased demand for energy and increased demand for water at the same time as adapting to and mitigating climate change, referred to recently by the UK Government's Chief Scientific Adviser, John Beddington (2009), means that in the near future changes in the way we think about and, significantly, changes in the way we use land are inevitable.

At this stage in the process, faced with what many regard as incontrovertible evidence of the need to adapt our use of rural land, the future can seem daunting. It is easy to be carried along on a wave of popular concern that our landscape will be overrun by wind turbines (although these are not permanent structures and can be removed when alternative energy sources become available), or that land will be used for fuel production instead of food production. In this context it is worth remembering that much agricultural land is currently not used directly to produce food for human consumption but grows feed for livestock, which ultimately, one way or another, find their way into the human diet. However, it is not long ago, just within living memory, that a significant proportion of the agricultural area of the UK was used for 'fuel' production. In the mid 1930s the approximately 700,000 horses that provided the engine of agriculture would have required significant areas for grazing and the production of supplementary feed. Thus, the notion that the use of agricultural land for the production of food for humans should be sacrosanct is relatively recent. At the same time, the idea that we could ever return to the widespread use of horses is equally troublesome due to the amount of land that would be required for feed. The point of this equine diversion is that change is daunting and challenging but that agricultural landscapes and our use of agricultural land have evolved over time. With careful planning and management of future change, it is possible to adapt the way we use land without completely sacrificing the 'old' concerns with landscape and nature conservation for the new environmentalism of climate change adaption and mitigation.

In considering our use of the land, the contributions by Karp et al and Morris et al point to considerable potential for land use change. In terms of energy crops, however, despite powerful drivers, a number of concerns have checked the domestic growth of the sector. Consequently, current analysis indicates that perennial energy crops, for instance, are unlikely to become the dominant enterprise on a farm (see Chapter 3) and are unlikely to transform the landscape other than at the very local level. Developments in the use of different feedstocks may see this change, however. On the other hand, farms have the potential to bring about a significant change in energy production through the widespread adoption of AD, although, as Banks et al point out in Chapter 5, further research is necessary in order to fully evaluate the potential of AD alongside the need to develop appropriate and effective support models.

As well as these 'new' uses of land, in their contribution Morris and colleagues argue that concerns with carbon sequestration, and a growing interest in water management alongside more long-standing concerns with agricultural production, landscape, recreation and amenity conservation, is already leading to a resurgence of interest in the neglected role and function of lowland floodplains. In this instance new environmental concerns could help reinforce and reinvigorate policy to support multifunctional land use. Indeed, there is evidence elsewhere that provides a powerful justification for continuing and expanding some long-standing areas of policy intervention. As Hopkins argues in Chapter 8, climate change requires an expansion of the network of protected areas in order to facilitate the transitions in wildlife populations that will occur over the coming years. Moreover, it is increasingly recognized that landscape and biodiversity conversation can be harnessed to support carbon sequestration and storage.

Thus, in some ways, the 'new environmentalism' of climate change adaptation and mitigation supports the 'old environmentalism' of biodiversity conservation and multifunctionality. However, the rapid re-emergence of food security on the political agenda means that, as Lowe and colleagues point out in Chapter 2, some commentators have argued for a new focus on productivism and a relaxation of environmental regulation. Increasing the global supply of nutritious and affordable food is essential if the growing population is to be sustained, but it should be achieved in a balanced manner that maintains the essential functions of ecosystems. New coalitions of interests are already forming around these issues and the public are being drawn into the debate through engagement with various forms of 'regenerative agriculture' (see Chapter 9), through deliberative processes of stakeholder engagement and more experimental approaches such as that outlined in the contribution by Firbank et al. The point is that if the new security agendas (food security, energy security, water security) and climate change challenge existing policy configurations, which they surely do, then it will be important to find new ways of encouraging stakeholder involvement in deciding how to use land and how to support that use in policy terms.

Finally, in Potter's rather bleak assessment of the evolving policy debate (see Chapter 12), climate change strengthens the public-good rationale for agricultural policy support at the same time as eclipsing some of the more traditional arguments for agricultural policy in favour of a more narrowly conceived, anthropocentric justification for policy intervention. Leaving aside for now arguments to justify policy intervention, it is unclear whether we have the knowledge base to design policies for carbon-sensitive farming, nor are we sufficiently well informed about how the largest group of resource managers on the planet (i.e. farmers) will react to emerging incentives and drivers.

What we need to know about land and land use: towards a new land use studies agenda

In the light of the speed of change outlined in the previous sections, it is hardly surprising to find that the research world relevant to the land debate is in some turmoil too. We must emphasize that what follows, by way of a finale to this book, contains some of our personal reflections. Although the purpose of the book has been largely to set out research-based 'state of the art' papers, we concede that we have not undertaken a systematic investigation of the strengths, weaknesses and capacity of UK land research and its fitness for tackling new priorities. We have not conducted any Delphi exercises or elicited expert opinions.

The study of land and land use has evolved through time, the main drivers of change in agricultural land use have themselves changed, but the central concern remains how to satisfy the multiple needs and desires associated with the use of land. During the last period when food security was a major domestic policy concern, Dudley Stamp observed that 'the problem of the moment and for the future is how to use our strictly limited land resources to satisfy the many needs of our people and fulfil their legitimate desires' (Stamp, 1955, p249). Stamp was a prolific writer about land, offering many ideas for how the above goal might be achieved. His approach was empirical, normative, prescriptive and applied but he did not compromise his underlying scholarship. Of course, he was a product of his time and his work pre-dates GIS, the emergence of environmental science and the explosion of a plurality of social science approaches from the 1960s. It is perhaps not surprising, therefore, that he is rarely quoted in today's land use debates, nor is he a frequent point of reference in his own discipline of geography. His blend of physical and human geography would be rather out of place in most geography departments today. Indeed the tradition of land use studies has largely disappeared from UK geography. As Sandy Mather et al (2006, p441) explain, 'land use, though a traditional focus of geographical concern, has attracted relatively little attention from human geographers in the UK in recent years'. This is by way of being an understatement. The 'cultural turn' in geography and the shift to political economy approaches that proceeded it have both left their mark. It is not so much that agriculture and rural issues have been neglected by cultural geographers and political economy. On the contrary rural geography is alive and well. It is that both approaches to human geography have left little room for the tradition of geographical inquiry that combines the physical and the human, one of geography's original strong selling points. Of course there are exceptions to this, but it is no exaggeration to say that land use, as a topic both spanning and unifying human and physical geography, is no longer a major topic for research or teaching in most geography departments.

Changes have taken place in agricultural research also and here they include institutional changes as well as disciplinary ones. Thirty years ago, UK agricultural research took place in three clearly demarcated sectors: ADAS, research institutes (RIs), and universities. ADAS, prior to privatization, was a national extension service that, in addition, undertook a significant portfolio of applied R&D, much of it taking place on a network of experimental husbandry farms, and all of it orientated towards the production and efficiency needs of the agricultural and horticultural industries. This capacity has now largely gone and in particular there has been a reduction in applied research linked directly to knowledge transfer activities (Leaver, 2009; Thirtle et al, 2004). ADAS still exists as a private company and does win some research contracts from government, but its network of farms has largely been disbanded. The RIs, funded primarily by research councils, were independent of both ADAS and the universities and were seen as a key strategic capacity supporting each of the main components of the land-based sector. Since then the RI world has undergone a series of changes, including reduction in research council core funding, closures, mergers and most recently absorption into universities. Thus the RI world is now fragmented in terms of governance arrangements. In Scotland the RIs, such as the Macaulay, have a stronger core funding relationship with central (Scottish) government than in England, but not the same relationship with research councils. South of the border, the Horticultural Research Institute is now fully incorporated into Warwick University; Rothamsted Research has not been taken into the university sector and remains a BBSRC-sponsored institute; the Grassland and Environmental Research Institute has been split in two, with part taken into Aberystwyth University and part (North Wyke Research) free-standing but part now merged with Rothamsted.

The universities have, in some respects, changed least. The big players in agricultural research 30 years ago – Reading, Newcastle, Nottingham – are still there, although their portfolios have diversified. Others, such as Oxford and Wye, have dropped out. But the continuing existence of some significant agricultural faculties belies an underlying trend towards a spreading of land-related research, where it does take place, across a much wider range of disciplines. This was initially prompted by the retreat from productivist agricultural research into environmental research or food research, which coincided with an expansion of environmental teaching. Some agricultural scientists and economists re-branded themselves under a range of headings such as 'environmental', 'natural resource management' or 'food'. With the (very) recent return of a productivist agenda, some research scientists are flexing their muscles again as they play to a food security agenda, but they are now as likely to be molecular biologists as agricultural scientists.

The key lesson to draw from this is that the context for expanding research on land cannot be assumed to be universally encouraging, at least in terms of the traditional centres for either geographical or agricultural research. On the other hand, capacity has emerged within other science disciplines and there is no doubt that much of the science relevant to our understanding of land will come from a wide range of disciplines, especially in the context of the new emphasis on ecosystem services and adaptation to climate change. That strong research on land use and climate change is taking place is not in doubt (see Benndorf et al, 2007; Berry et al, 2006; Rounsevell et al, 2006; Schulp et al, 2008). In recent years, amongst the most exciting of developments in land research has been work with GIS and related techniques and the management of large data sets. However, the concentration on large satellite or national socio-economic data sets produces spatial representations at a very coarse level of aggregation. For example, a spatial analysis of the whole of Europe combining socio-economic and land cover data, conducted by Dilly et al (see Helming et al, 2007), identifies just six zones (Devon and Cornwall lie entirely within one). A major source used in this work and in many other papers is the Coordination of Information on the Environment (CORINE) land cover map produced by the European Environment Agency, derived primarily from satellite images with some use of aerial photos and near-ground imaging (Bossard et al, 2000). Alongside the use of these coarse-grained data sets is an emphasis on modelling combining a range of large data sets (Helming et al, 2007).

There are dangers in these approaches. Description and/or modelling predictions do not necessarily help us towards a dynamic analysis of real-world changes. The inherent requirement to combine date sets often raises fundamental problems of incompatibility. They are remarkably ill-equipped to deal with the local drivers of change which we see as important in terms of both social and natural processes.

Where then do we see the future for land research? It is tempting to make a plea for more Relu-type research money but, welcome though that would be, there are some more fundamental requirements to be met. Despite Relu, our impression is that there is still some way to go to achieve the necessary level of cross-disciplinary integration. As social scientists, we are not convinced that the level of commitment to research integration from within the social science community is yet strong enough. Closer links between the universities and RIs and between different university departments is required, possibly through the creation of a 'virtual' land institute.

But institutional structures are only part of the story. At the outset of this concluding chapter we claimed a victory had been won in putting the issues of climate change and natural resource depletion on the global policy and science agendas. That we saw as the end of the beginning, but our view is that some large steps still need to be taken to position land as key to these debates, particularly, we would argue, in a UK context and, within the UK, especially in England. It seems self-evident to us that land is hugely important in terms of both mitigation and adaptation to climate change. But that is not always evident in the research priorities associated with climate change. If the rush to biofuels, induced by a tunnel-vision mentality of reducing the dependence on fossil fuels, teaches us anything, it is that land research needs to be at the heart of the climate change mitigation and adaptation agendas, both in terms of policy and science.

References

Barr, S. (2008) *Environment and Society: Sustainability, Policy and the Citizen*, Ashgate, Aldershot, UK

Beddington, J. (2009) Speech to GovNet Sustainable Development UK 2009 conference, http://www.govnet.co.uk/news/govnet/professor-sir-john-beddingtons-speech-at-sduk-09 accessed 28 May 2009

Benndorf, R., Federici, S., Forner, C., Pena, N., Rametsteiner, E., Sanz, M.J. and Somogyi, Z. (2007) 'Including land use, land-use change, and forestry in future climate change, agreements: Thinking outside the box', *Environmental Science and Policy*, vol 10, pp283–294

Berry, P.M., Rounsevell, M.D.A., Harrison, P.A. and Audsley, E. (2006) 'Assessing the vulnerability of agricultural land use and species to climate change and the role of policy in facilitating adaptation', *Environmental Science and Policy*, vol 9, pp189–204

Bossard, M., Feranec, J. and Otahel, J. (2000) *CORINE Land Cover Technical Guide*, European Environment Agency, Copenhagen

Buttel, F.H. (2000) 'Ecological modernization as social theory', *Geoforum*, vol 31, pp57–65

Defra (2002) *Strategy for Sustainable Farming and Food: Facing the future*, TSO, London

Evans, N., Morris, C. and Winter, M. (2002) 'Conceptualizing agriculture: A critique of post-productivism as the new orthodoxy', *Progress in Human Geography*, vol 26, pp313–332

Heal, G. (2000) 'Valuing ecosystem services', *Ecosystems*, vol 3, pp23–30

Helming, K., Pérez-Soba, M. and Tabbush, P. (eds) (2007) *Sustainability Impact Assessment of Land Use Change*, Springer, Berlin

Leaver, D. (2009) 'A new vision for UK agricultural research and development', *Journal of the Royal Agricultural Society of England*, vol 169, pp33–37

Lovelock, J. (2009) *The Vanishing Face of Gaia: A Final Warning*, Basic Books, New York

Mather, S., Hill, G. and Nijnik, M. (2006) 'Post-productivism and rural land use: Cul de sac or challenge for theorization?' *Journal of Rural Studies*, vol 22, no 4, pp441–455

Phillipson, J., Lowe, P. and Bullock, J. (2009) 'Navigating the social sciences: interdisciplinarity and ecology' *Journal of Applied Ecology*, vol 46, pp261–264

Rounsevell, M.D.A., Berry, P.M. and Harrison, P.A. (2006) 'Future environmental change impacts on rural land use and biodiversity: a synthesis of the ACCELERATES project', *Environmental Science and Policy*, vol 9, pp93–100

Schulp, C.J.A., Nabuurs, G.-J. and Verburg, P.H. (2008) 'Future carbon sequestration in Europe – effects of land use change', *Agriculture, Ecosystems and Environment*, vol 127, pp251–26

Stamp, L.D. (1955) *Man and the Land*, Collins, London

Thirtle, C., Lin, L., Holding, J., Jenkins, L. and Piesse, J. (2004) 'Explaining the decline in UK agricultural productivity growth', *Journal of Agricultural Economics*, vol 55, pp343–366

Winter, M. (2006) 'Rescaling rurality: Multilevel governance of the agro-food sector', *Political Geography*, vol 25, pp735–751

Index